A Primer of Ecological Genetics

A Primer of Ecological Genetics

Jeffrey K. Conner
Michigan State University

Daniel L. Hartl
Harvard University

Sinauer Associates, Inc. • *Publishers*
Sunderland, Massachusetts • U.S.A.

Cover photograph © Jacob Silverman Conner

A PRIMER OF ECOLOGICAL GENETICS

 For information address Sinauer Associates, Inc., 23 Plumtree Road, Sunderland, MA 01375 U.S.A.

Fax: 413-549-1118
Email: publish@sinauer.com
www.sinauer.com

Library of Congress Cataloging-in-Publication Data

Conner, Jeffrey K., 1957-
A primer of ecological genetics / Jeffrey K. Conner, Daniel L. Hartl.— 1st ed.
p. ; cm.
Includes bibliographical references and index.
ISBN 0-87893-202-X (pbk.)
1. Ecological genetics. 2. Population genetics. 3. Quantitative genetics.
[DNLM: 1. Genetics, Population. 2. Ecology. 3. Selection (Genetics) 4. Variation (Genetics) QH 456 C752p 2004] I. Hartl, Daniel L. II. Title.
QH456.C665 2004
576.5′8—dc22 2003022664

Printed in U.S.A.

6 5 4 3 2

Contents

6 Natural selection on phenotypes 189

7 Applied ecological genetics 231

Preface

This book covers basic concepts in population and quantitative genetics, including measuring selection on phenotypic characters, with a focus on methods applicable to field studies of ecologically important traits. It is designed for a broad audience of advanced undergraduates and graduate students, and is also aimed at providing professionals outside the field with an accessible introduction. The concepts included are critical for training students in ecology, evolution, conservation biology, agriculture, forestry, and wildlife management. The book should be useful as a textbook for courses on ecological and evolutionary genetics, conservation biology, and population biology, as well as a supplement to readings from the original literature in graduate-level courses in evolution.

The guiding principle of the book is to focus on clear explanations of the key concepts in the evolution of natural and managed populations. For example, we discuss the similarities and differences between inbreeding and random genetic drift, describe how dominance and epistasis represent interactions among alleles within and between loci, respectively, examine the most common misconceptions about heritability, and show the relationships among different methods of measuring selection. Mathematics and conceptual material are integrated and fully explained. The mathematics is used as a tool to improve understanding of the biological principles, not an end in itself. For most concepts, examples from the literature, mainly from studies of natural populations, are briefly described. These examples are not necessarily classic studies (although many are) but rather were chosen because they provide clear and simple illustrations of the concept under discussion. The solutions to these problems, worked out in full, can be found at the website www.sinauer.com/connerhartl.

We have also striven to synthesize and integrate the different chapters and sections to give students a clearer idea of the "big picture" of how different concepts relate to each other. Our main goal is to enable students to understand the concepts well enough so they can gain entry into the primary literature. For this reason, most chapters have sets of discussion questions keyed to a few original papers in ecological genetics, each of which represents an important contribution to the literature and is written clearly and simply enough to be accessible to a beginning student.

We welcome any comments or suggestions, particularly those that point out any errors or unclear concepts or descriptions. Please direct these remarks to connerj@msu.edu.

Acknowledgments

The authors thank Butch Brodie, Richard Frankham, and Mike Wade for reviewing one or more chapters, and especially Mike Whitlock for reviewing the entire manuscript. The book was also improved by comments from students in Conner's laboratory and in the graduate evolution course at Michigan State, especially Meghan Duffy, Frances Knapczyk, Angela Roles, Heather Sahli, James Sobel, Christy Stewart, and Rachel Williams. Nancy Haver drew the pen and ink drawings, Michele Ruschhaupt created the graphic art, and Roberta Lewis copyedited most of the book. The staff at Sinauer Associates did truly excellent work, including the production team led by Chris Small and Joan Gemme whose skill in design and layout are evident in this book. We particularly thank Sydney Carroll for overseeing the editing and production of the book and Andy Sinauer for guidance, encouragement, and extraordinary patience. Jeff Conner would like to thank his wife, Buffy Silverman, for trying to teach him to write well, and his children Jake and Emma for putting up with hearing about this book at too many dinners. Dan Hartl thanks Christine and Christopher for their enduring support, and all of the people in the Hartl laboratory for their continuing enthusiasm, creativity, and hard work.

Acronyms, Abbreviations, and Symbols

Note: Numbers in brackets refer to the chapter in which each symbol is introduced.

β	Linear selection gradient [6]
χ^2	Chi-square value [2]
γ	Non-linear (variance) selection gradient [6]
μ	Mutation rate at a locus per generation [3]
σ^2	Theoretical expectation of a variance [4]
a	Additive effect, which equals the genotypic value for the homozygotes [4]
A	Adenine, the purine base, or a nucleotide containing adenine [1]
AFLP	Amplified fragment length polymorphism [2]
ANOVA	Analysis of variance [4]
BLUP	Best linear unbiased prediction [4]
bp	Base pair, a unit of length in nucleic acids equal to 1 base pair [5]
Bt	The bacterium *Bacillus thuringiensis*, which produces insecticidal toxins [7]
c	Frequency of recombination between two gene loci, also denoted *r* [5]
C	Cytosine, the pyrimidine base, or a nucleotide containing cytosine [1]

CAPS	Cleaved amplified polymorphic site [2]
Cov	Covariance [5]
cM	Centimorgan, a unit of length in a genetic map equal to 1% recombination [5]
cpDNA	Chloroplast DNA
d	Genotypic value for the heterozygote [4]
d.f.	Degrees of freedom [2]
DNA	Deoxyribonucleic acid [1]
E	Environmental deviation [5]
e^2	Fraction of the phenotypic variance attributable to the environment; the complement of heritability [5]
ESU	Evolutionarily significant unit [7]
F_1	First offspring generation of a cross [5]
F_2	Second offspring generation usually resulting from self-fertilization of the F_1 generation [5]
F or F_{IS}	Inbreeding coefficient of individuals relative to the subpopulation [2, 3]
F_{ST}	Fixation index of a subpopulation relative to the total population [3]
F_{IT}	Metatpopulation increase in homozygosity due to inbreeding and population substructure [3]
5′ end	The end of a nucleic acid strand containing a free 5′ phosphate group on the sugar
G	Guanine, the purine base, or a nucleotide containing guanine [1]
G	Genotypic value [4]
G	Additive genetic variance/covariance matrix [6]
G × E	Genotype-by-environment interaction [5]
g-e	Genotype-by-environment interaction [5]
GE	Genetically engineered [7]
GM	Genetically modified [7]
GMO	Genetically modified organism [7]
G_{ST}	Fixation index (similar to F_{ST}) [5]
H	Heterozygosity, the frequency of heterozygous genotypes in a population [2]
H_0 or H_S	Heterozygosity expected at Hardy-Weinberg equilibrium ($2pq$) [2, 3]
H_O	Observed heterozygosity within subpopulations [3]
H_T	Expected metapopulation heterozygosity assuming random mating ($2p_0q_0 = 2\overline{p}\,\overline{q}$) [3]

h	Degree of dominance of a recessive allele [3]
h^2	Usual symbol for narrow-sense heritability [4]
h^2_B	Broad-sense heritability (V_G/V_P) [4]
h^2_N	Unambiguous symbol for narrow-sense heritability (V_A/V_P) [4]
HWE	Hardy-Weinberg equilibrium (genotypic frequencies p^2, $2pq$, q^2) [2]
IBD	Identical by descent [2]
indel	Insertion or deletion [3]
kb	Kilobase, a unit of length in nucleic acids equal to 10^3 bases or 10^3 base pairs [5]
LD	Linkage disequilibrium [5]
LE	Linkage equilibrium [5]
lod or LOD	The log odds score; logarithm of the likelihood ratio [5]
m	Migration rate [3]
MHC	Major histocompatibility complex [7]
ML	Maximum likelihood [4]
MS	Mean square (variance) [4]
mRNA	Messenger RNA [1]
mtDNA	Mitochondrial DNA [7]
n	Haploid chromosome number [2]
n or *N*	Sample size [2]
N_a	Actual population size [3]
N_e	Effective population size [3]
ns	In a statistical test, no significant difference from that expected with random sampling
p	Frequency of one allele, often the wildtype, dominant, or selectively favored allele [2]
P	Frequency of one homozygote [2]
P	Probability value in a statistical test (*P*-value) [2]
P	Phenotypic value [4]
PCR	Polymerase chain reaction [2]
q	Frequency of one allele, often the recessive, rare mutant, or deleterious allele [2]
Q	Frequency of one homozygote [2]
Q_{ST}	Quantitative trait differentiation [5]
QTL	Quantitative trait locus [5]
r_A	Additive genetic correlation [5]
r_E	Environmental correlation [5]

r_P	Phenotypic correlation [5]
r_{IJ}	Coefficient of relatedness [2]
r	Frequency of recombination between two gene loci; also denoted c [5]
R	Allele for resistance to a pesticide [7]
RAPD	Random amplified polymorphic DNA [2]
REML	Restricted maximum likelihood [4]
RFLP	Restriction fragment length polymorphism [2]
RNA	Ribonucleic acid [1]
s	Selection coefficient [3]
S	Allele for susceptibility to a pesticide [7]
SCAR	Sequence-characterized amplified region [2]
SNP	Single-nucleotide polymorphism [2]
SPAR	Single-primer amplified region [2]
SS	Sum of squares [4]
SSCP	Single-stranded conformational polymorphism [2]
SSR	Simple sequence repeat [2]
SSRP	Simple sequence repeat polymorphism [2]
STR	Simple tandem repeat [2]
STS	Sequence-tagged site [2]
T	Thymine, the pyrimidine base, or a nucleotide containing thymine [1]
3′ end	The end of a nucleic acid strand containing a free 3′ hydroxyl on the sugar [1]
U	The whole-genome mutation rate, especially for loci affecting fitness [7]
U	Uracil, the pyrimidine base, or a nucleotide containing uracil [1]
V_A	Additive genetic variance [4]
V_D	Dominance variance [4]
V_E	Environmental variance [4]
V_G	Genotypic variance [4]
V_I	Epistatic variance [4]
V_P	Phenotypic variance [4]
Var	Variance [4]
VNTR	Variable number of tandem repeats polymorphism [2]
w	Relative fitness [3]
z	Phenoypic trait value [6]
*	Statistical significance at the 5% probability level [5]
**	Statistical significance at the 1% probability level [5]

1

Introduction

What is Ecological Genetics?

Ecological genetics is at the interface of ecology, evolution, and genetics, and thus includes important elements from each of these fields. We can use two closely related definitions to help describe the scope of ecological genetics:

1. Ecological genetics is concerned with the *genetics of ecologically important traits, that is, those traits related to fitness* such as survival and reproduction. Ecology is the study of the distribution and abundance of organisms—in other words, how many individuals there are, where they live, and why. Distribution and abundance are determined by birth rates and death rates, which in turn are determined by interactions with the organism's biotic and abiotic environment. These interactions include predation, competition, and the ability to find mates, food, and shelter. Consider traits that would help an organism deal with each of these interactions. Cryptic coloration could help a beetle avoid being eaten, growing tall could help a plant compete with other plants for light, and a thick coat of fur might help a mouse survive winter cold. These are examples of *ecologically important traits*: those traits that are closely tied to fitness or, in other words, are important in determining an organism's adaptation to its natural environment, both biotic and abiotic.
2. Ecological genetics can also be defined as the *study of the process of phenotypic evolution occurring in present-day natural populations*. Phenotypic evolution can be defined as a change in the mean or variance of a trait

across generations due to changes in allele frequencies. The four processes that can cause evolution are **mutation, genetic drift, migration,** and **natural selection.** All of these processes are described in Chapter 3, and the last three in particular are closely related to ecology and therefore appear throughout the book. Ecological factors can cause population size to decline, and the resulting small population size causes genetic drift. Migration is clearly ecological, but how is natural selection related to ecology? Selection is caused by differences in fitness among organisms in a population, and these fitness differences are caused in part by interactions with the environment as previously mentioned.

Our two definitions are tied together by the concept of *adaptation*, which is the central theme of ecological genetics. An **adaptation** is a phenotypic trait that has evolved to help an organism deal with something in its environment. Like most ecologically important traits, the examples given above are adaptations. Natural selection is special among the four evolutionary processes because it is the only one that leads to adaptation. Mutation, genetic drift, and migration can either speed up or constrain the development of adaptations, but they cannot cause adaptation.

An overview of these ideas is shown in Figure 1.1, which summarizes much of what will be covered in this book. Beginning at the top, ecological factors, both biotic and abiotic, can cause fitness differences among organisms with different phenotypes within the population; this is natural selection. If mutation and recombination create genetic variation for these phenotypic traits, then the selection can act on this variation to change the

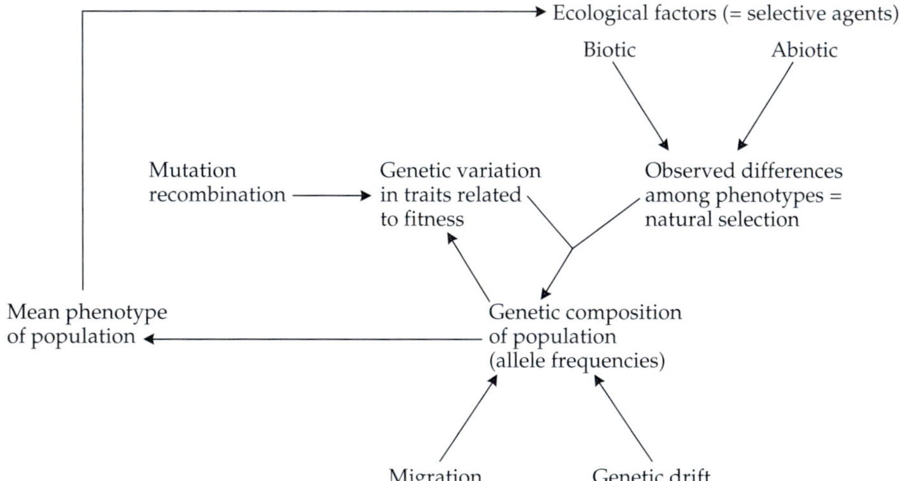

Figure 1.1 A schematic overview of key concepts in ecological genetics.

genetic composition of the population. The genetics of the population can also be affected by gene flow from other populations with different genetic composition, or by genetic drift if the population size is small. All these changes in genetic composition are likely to feed back and affect the genetic variation for the phenotypic traits, as well as change the average phenotype in the population across generations. These phenotypic changes can lead to an improvement in the ability of the population to survive and reproduce in its biotic and abiotic environment; that is, it can lead to adaptation.

As an example, a deer mouse is part of a beetle's biotic environment, and may cause beetles with increased defensive secretions to have higher fitness than those with less secretions, if the secretions deter deer mouse predation. If this phenotypic variation is caused at least in part by underlying genetic variation, then this will cause an increase in the frequency of alleles that increase defensive secretions. (An **allele** is a particular type of a given gene.) This increase in allele frequency across generations may be slowed by random genetic drift, or by gene flow from other beetle populations with low frequencies of the high-secretion alleles (perhaps because there are fewer mice coexisting with those other populations). Selection and drift may decrease genetic variation for secretion quantity, also slowing future evolution of secretions. If the average secretion quantity in the population increases in spite of these constraints, then this may reduce the impact of mouse predation in subsequent generations, increasing adaptation of the beetles to this environmental factor.

Overview of the Book

Chapters 2 and 3 cover the field of population genetics. **Population genetics** is the study of genetic variation within and among populations, focusing on the processes that affect genotypic and allele frequencies at one or a few gene loci. These processes include inbreeding, mutation, migration, drift, and selection; the genotypic and allele frequencies are revealed mainly through molecular markers. Population genetics for the most part does not focus on phenotypes, since the genes and alleles underlying most phenotypic traits are unknown, especially in natural populations. This is because most phenotypic traits are complex, being affected by several to many gene loci and by the environment.

Chapters 4 and 5 cover the field of **quantitative genetics,** which does focus on the phenotype, usually without knowing the genotypes underlying the traits. In the place of genotypic information, statistical abstractions such as variance, correlation, and heritability are used in quantitative genetics to help understand the genetics of complex phenotypes. QTL mapping (covered at the end of Chapter 5) is a marriage of molecular and statistical techniques for studying the genetics of complex phenotypic traits. QTL mapping is a first step in discovering the genes underlying phenotypic traits in natural populations, bringing together the fields of population and quantitative genetics. This convergence is very likely to lead to fundamental new insights in ecological genetics.

Chapter 6 is on techniques developed from quantitative genetics for studying natural selection on phenotypic traits (rather than on genotypes as in population genetics). These techniques have allowed biologists to measure the strength and direction of selection in natural populations, as well as help determine the ecological causes of the selection. Chapter 6 also synthesizes the quantitative genetic material in Chapters 4 through 6, and shows how short-term evolution can be predicted in natural populations using knowledge of genetic variance and the strength of selection.

Since ecological genetics is at the interface between ecology, evolution and genetics, it is a critical component of all three fields, as well as essential for the study of some of society's problems. In Chapter 7 we will discuss the importance of ecological genetic principles in conservation, the spread of invasive species, the evolution of pesticide, herbicide, and antibiotic resistance, and the environmental effects of genetically modified organisms used in agriculture.

The focus of the book will be on diploid sexual organisms. Most of the concepts covered also apply to asexual and haploid organisms, but there are important differences. Most of our examples will come from studies of plants and animals, because the ecological genetics of most microorganisms and fungi are not as well known.

Basic Genetic Terms

A **gene** is a stretch of **DNA** (deoxyribonucleic acid) coding for a polypeptide chain; one or more polypeptides make up a protein. The genetic information in DNA is coded in the sequence of four nucleotides, abbreviated according to the identity of the nitrogenous **base** that each contains: A (adenine), G (guanine), T (thymine), or C (cytosine). DNA molecules normally consist of two complementary helical strands held together by pairing between the bases: A in one strand is paired with T in the other strand, and G in one strand is paired with C in the other.

The process of creating proteins from the genetic code in DNA is called **gene expression.** The essentials of gene expression in the cells of eukaryotes are outlined in Figure 1.2. The first step is **transcription,** in which the sequence of nucleotides present in one DNA strand of a gene is faithfully copied into the nucleotides of an **RNA** (ribonucleic acid) molecule. As the RNA transcript is synthesized, each base in the DNA undergoes pairing with a base in an RNA nucleotide, which is then added to the growing RNA strand. The base-pairing rules are the same as those in DNA, except that in RNA nucleotides the base U (uracil) is found instead of T (thymine). The second step of gene expression is **RNA processing,** in which intervening sequences or **introns** are removed

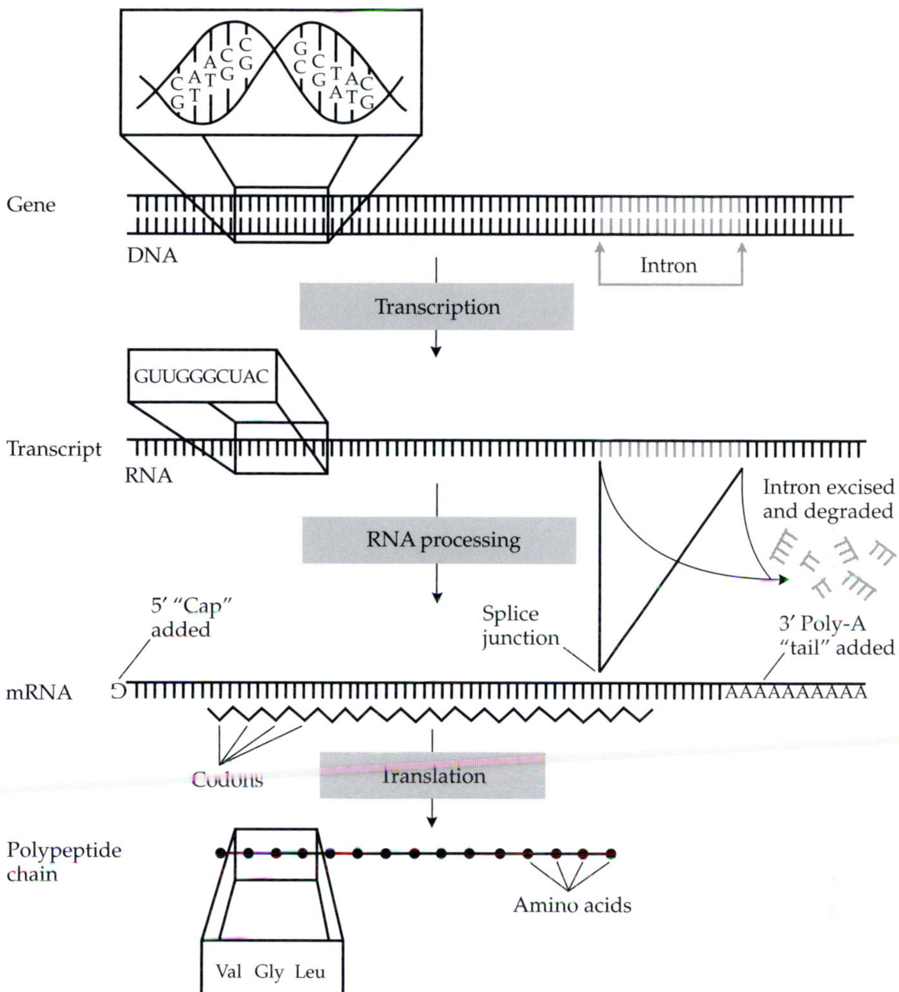

Figure 1.2 Principal processes in gene expression in eukaryotes.

from the RNA transcript by splicing and the ends of the transcript are modified. The regions between the introns that remain in the fully processed RNA are known as **exons;** these are the sequences that actually code for proteins. The fully processed RNA constitutes the **messenger RNA** (mRNA).

The messenger RNA undergoes **translation** on ribosomes in the cytoplasm to produce the polypeptide that is encoded in the sequence of nucleotides. In the translated part of the messenger RNA, each adjacent group of three nucleotides constitutes a coding group or **codon,** which specifies a corresponding amino acid subunit in the polypeptide chain. The standard **genetic code** showing which codons specify which amino acids is given in Table 1.1. After each three-letter codon are the three- and one-letter designations for the 20 amino acids. The three-letter and one-letter abbreviations are both established conventions. Note that in many cases changes in the third base in the codon do not change the amino acid that is specified; therefore, much variation at this position is not expressed (sometimes called the "silent" position). The codon AUG specifies methionine and also serves as the start codon for polypeptide synthesis. Any of three codons—UAA, UAG, or UGA—specify the end, or termination, of polypeptide synthesis, upon which the completed polypeptide chain is released from the ribosome. The start and stop codons are shaded in Table 1.1.

All the DNA in a cell is collectively called the **genome.** Genome size is typically expressed as the amount of DNA in a reproductive cell (sperm or egg), and it differs greatly among species. For example, the genome of *Arabidopsis thaliana*, a model plant for genetic studies, consists of about 120 million base pairs, whereas the genome of the lily *Fritillaria* is 1000 times as large, about 120 billion base pairs. The human genome is about 3 billion base

TABLE 1.1 *The standard genetic code*

First nucleotide in codon (5′ end)	Second nucleotide in codon				Third nucleotide in codon (3′ end)
	U	C	A	G	
U	UUU Phe/F	UCU Ser/S	UAU Tyr/Y	UGU Cys/C	U
	UUC Phe/F	UCC Ser/S	UAC Tyr/Y	UGC Cys/C	C
	UUA Leu/L	UCA Ser/S	UAA Stop	UGA Stop	A
	UUG Leu/L	UCG Ser/S	UAG Stop	UGG Trp/W	G
C	CUU Leu/L	CCU Pro/P	CAU His/H	CGU Arg/R	U
	CUC Leu/L	CCC Pro/P	CAC His/H	CGC Arg/R	C
	CUA Leu/L	CCA Pro/P	CAA Gln/Q	CGA Arg/R	A
	CUG Leu/L	CCG Pro/P	CAG Gln/Q	CGG Arg/R	G
A	AUU Ile/I	ACU Thr/T	AAU Asn/N	AGU Ser/S	U
	AUC Ile/I	ACC Thr/T	AAC Asn/N	AGC Ser/S	C
	AUA Ile/I	ACA Thr/T	AAA Lys/K	AGA Arg/R	A
	AUG Met/M	ACG Thr/T	AAG Lys/K	AGG Arg/R	G
G	GUU Val/V	GCU Ala/A	GAU Asp/D	GGU Gly/G	U
	GUC Val/V	GCC Ala/A	GAC Asp/D	GGC Gly/G	C
	GUA Val/V	GCA Ala/A	GAA Glu/E	GGA Gly/G	A
	GUG Val/V	GCG Ala/A	GAG Glu/E	GGG Gly/G	G

pairs. Genes are arranged in linear order along microscopic threadlike bodies called **chromosomes.** Each human **gamete** (sperm or egg) contains one complete set of 23 chromosomes; this is the **haploid** chromosome number, designated as n. Chromosome number can vary greatly: $n = 2$ in some scorpions and 127 in a species of hermit crab! A typical chromosome contains several thousand genes, in humans averaging approximately 1500 genes per chromosome. The position of a gene along a chromosome is called the **locus** of the gene. Sometimes the words gene and locus are used interchangeably, which can lead to confusion. **Recombination** between loci can occur during meiosis, which creates new combinations of alleles at these different loci. Recombination is rarer between loci that are close together on the chromosome; these loci are said to be genetically **linked.**

In most multicellular organisms, each individual cell contains two copies of each type of chromosome, one inherited from its mother through the egg and one inherited from its father through the sperm (so the **diploid** chromosome number, $2n$, is 46 in humans and 254 in hermit crabs). Note that these two copies of the chromosome are not the two complementary strands of DNA; each chromosome consists of a double-stranded DNA molecule. At any locus, therefore, every diploid individual contains two copies of the gene—one at each corresponding (homologous) position in the maternal and paternal chromosome. These two copies are the alleles of the gene in that individual. If the two alleles at a locus are indistinguishable according to any particular experimental criterion, then the individual is **homozygous** at the locus under consideration. If the two alleles at a locus are distinguishable by means of this criterion, then the individual is **heterozygous** at the locus.

The **genotype** of an individual is the diploid pair of alleles present at a given locus. Therefore, homozygous and heterozygous are the two major categories of genotypes. Typographically, genes are indicated in italics, and alleles are typically distinguished by uppercase or lowercase letters (A versus a), subscripts (A_1 versus A_2), superscripts (a^+ versus a^-), or sometimes just + and –. Using these symbols, the genotype of homozygous individuals would be portrayed by any of these formulas: AA, aa, A_1A_1, A_2A_2, a^+a^+, a^-a^-, +/+, or –/–. As in the last two examples, the slash is sometimes used to separate alleles present in homologous chromosomes to avoid ambiguity. The genotype of heterozygous individuals would be portrayed by any of the formulas Aa, A_1A_2, a^+a^-, or +/–.

The outward appearance of an organism for a given characteristic is its **phenotype.** Phenotypic traits can be defined at a number of hierarchical levels, each one dependent on a number of traits at lower levels. For example, the form of an enzyme encoded by a gene is a phenotype, as is a physiological function like metabolic rate that depends on a number of enzymes. A number of different physiological functions affect morphological traits like height, and physiology and morphology together can affect behavioral phenotypes such as courtship. Finally, all these lower level traits can affect life history traits like survival and reproduction, which determine the ultimate trait of individual fitness. The traits that are higher in this hierarchy are more complex and affected by more gene loci. The expression of most phenotypic traits, and especially the higher level ones, are also affected to varying degrees by the environment. This complexity means that the same genotype can produce different phenotypes, through the action of the environment. Conversely, the different genotypes can produce the same phenotypes, again due to the environment and also due to gene interactions. We will discuss complex phenotypic traits and fitness in more detail in Chapters 4 through 6.

2

Population genetics I: Genetic variation, random and nonrandom mating

What is Population Genetics?

In its broadest sense, **population genetics** is the study of naturally occurring genetic differences between organisms; these differences are called **genetic variation.** Genetic variation is important because it is the raw material for evolution. Genetic variation can occur at three hierarchical levels: within populations, between populations of the same species, and between different species. Therefore, to understand the purview of population genetics, we need to have a better understanding of populations and genetic variation.

What are populations and why are they important?

Populations are important because evolutionary processes occur primarily within populations; it is populations that evolve through changes in allele frequencies. In fact, most of the important concepts and processes in ecological genetics have meaning only at the level of the population, not the individual; these include genetic variation, allele and genotype frequencies, gene flow, drift, natural selection, heritability, and genetic correlation.

A species is rarely, if ever, a single interbreeding **panmictic** group, that is, one in which any member of the species can potentially mate with any other member of the opposite sex. Instead, species are divided into **populations.** A population can be defined as a local interbreeding (panmictic) group that has reduced gene flow with other groups of the same species. By reduced

gene flow we mean reduced movement of genes caused by migration and subsequent mating (see Chapter 3). The reduced gene flow between populations can be caused by **spatial structure,** which occurs when the individuals in a species are not evenly distributed in space (Figure 2.1).

Determining the boundaries of a population in nature can be quite simple when dealing with species that occur in discrete clumps in a patchy habitat, such as fish in small ponds or plants on small islands. This determination becomes more problematic with species that are not spatially structured but rather range continuously over a wide area, such as maple trees in a large forest or wide-ranging birds. Here the probability of interbreeding is a function of distance; individuals at opposite ends of the species distribution are certainly not interbreeding, but distinct population boundaries cannot be drawn. The key question from a genetic viewpoint is how much interbreeding and, thus, gene flow takes place at any given distance.

The terminology applied to populations can be confusing. Sometimes a group of populations connected by some level of gene flow is called a **metapopulation** (all the populations within the boundary in Figure 2.1), while other authors call the group a population and the subgroups within it **subpopulations, local populations,** or **demes.** In this book, whenever we are discussing spatially structured populations like those depicted in Figure 2.1, we will use *metapopulation* for the group and *subpopulation* for each unit to avoid ambiguity. Most mating occurs within subpopulations, but there is at least the potential for gene flow between the subpopulations in a single

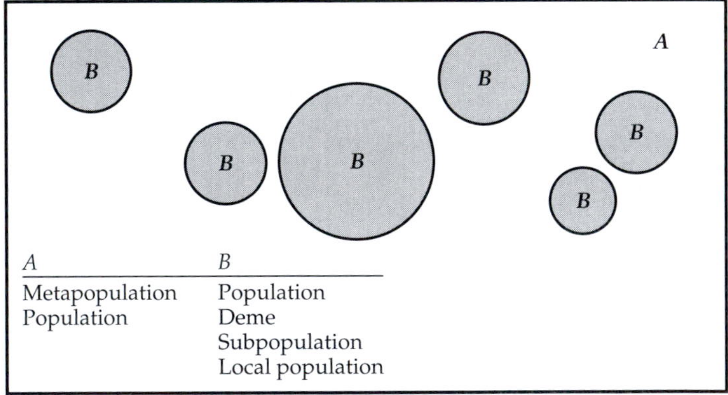

Figure 2.1 A simplified sketch of a spatially structured species. The shaded areas (denoted by *B*) are where the organisms live, whereas the area within which some gene flow occurs is denoted by *A*. *A* could potentially be as small as a cubic meter of soil for a microorganism, or perhaps as large as a continent for a migratory bird. The columns give names commonly used for *A* and *B*.

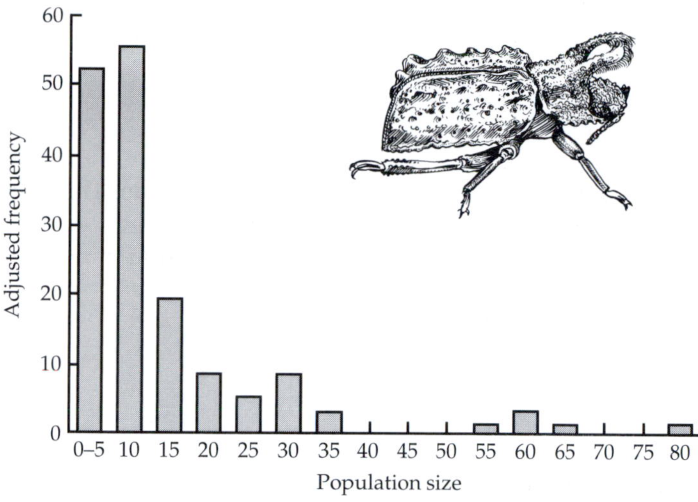

Figure 2.2 Frequency distribution showing the number of subpopulations at each size class in a metapopulation of fungus beetles, *Bolitotherus cornutus*. Most populations have very few beetles, but a few have many. (After Whitlock 1992.)

metapopulation. We will still use the word *population* in discussions not involving spatial structure.

A clear example of a structured population is given by fungus beetles in the forests of Virginia (Whitlock 1992). These beetles inhabit woody shelf fungi growing on dead trees, and only rarely move between trees. Therefore, the beetles on a group of fungi on a single tree represent a subpopulation, whereas all the beetles in a forested area are a metapopulation. There were 158 subpopulations in the metapopulation studied by Whitlock, with highly skewed subpopulation sizes (Figure 2.2); that is, two-thirds of the subpopulations had 10 or fewer beetles, but there were six subpopulations with more than 50 individuals. Subpopulation size was correlated with the age and size of fungi, with the larger subpopulations on the older and larger patches of fungi.

Genetic Variation

Genetic variation *within* a population occurs when there is more than one allele present in a population at a given locus; geneticists sometimes refer to this as a population that is **segregating** or **polymorphic** at that locus. Not all loci are variable—some are **fixed,** which means that all members of the population are homozygous for the same allele. Genetic variation is ubiquitous in natural populations, but not for all traits or loci. When variation occurs *between* populations of the same species it is called **genetic differentiation.** For example, two populations could be fixed for two different alleles at a given locus.

Population genetics has three interrelated goals:

1. To explain the **origin and maintenance** of genetic variation.
2. To explain the **patterns and organization** of genetic variation.
3. To understand the mechanisms that cause **changes in allele frequencies.**

The major processes studied by population geneticists are mutation, recombination, inbreeding, genetic drift, gene flow, and natural selection. Changes in allele frequencies within populations caused by natural selection can lead to adaptation. Genetic differentiation between populations can ultimately lead to the creation of new species. For these reasons, population genetics is central to the study of evolutionary change in nature.

Measuring Genetic Variation: Genetic Markers

The basic tools used to study genetic variation within and between populations are called **genetic markers.** Markers allow one to determine what alleles are present in populations, and are therefore extraordinarily useful for studying a wide variety of questions in ecology and evolution. Some of the main uses are (see Avise 1994; Mueller and Wolfenbarger 1999; Parker et al. 1998; Sunnucks 2000):

- Studying mating systems (discussed in this chapter). For example, how inbred is a population?
- Measuring gene flow and population structure (see Chapter 3). For example, how much migration occurs between subpopulations?
- Determining paternity to measure heritability (see Chapter 4) and male fitness (see Chapter 6). For example, are plants with larger flowers more successful at siring seeds?
- Producing genetic maps to find genes underlying complex traits (see Chapter 5). For example, how many loci affect body size?
- Conservation biology (see Chapter 7). For example, how genetically different are two populations of an endangered species?

In the descriptions of different types of markers discussed later in this chapter, the markers are discussed in the historical order in which they were developed. The first markers available revealed little of the underlying genetic variation; with each successive innovation more variation was revealed. This trend has culminated with gene sequencing, which reveals all the genetic variation, albeit only for a small portion of the genome for most organisms to date. The amount of marker variation necessary depends on the specific question being asked and the populations and organisms under study. Examples of the use of most of these types of markers can be found in later chapters.

Visible polymorphisms

The first type of marker used by geneticists was **visible (discrete) polymorphisms,** in which given phenotypic traits have only a few (usually two or three) distinct types or **morphs.** For these traits the majority of phenotypic variation is due to one or sometimes two gene loci; therefore, they are not strongly affected by the environment. The traits studied by Mendel in pea plants were visible polymorphisms—for example, white vs. purple flowers and round vs. wrinkled seeds. Visible polymorphisms were the only markers available to geneticists until the 1960s, so many of the classical studies in population and ecological genetics used these markers. Some visible polymorphisms are merely used as markers and are not traits of intrinsic interest, for example, round vs. wrinkled seeds in peas, red vs. white eyes in *Drosophila*. Others are directly of interest because they may be adaptations, such as flower color, heterostyly in plants (see Chapter 5), and melanic morphs in insects (see below).

When they are available, visible polymorphisms are still very useful for a variety of studies. However, only a very small fraction of phenotypic traits are controlled by one or two loci and little affected by the environment. Therefore, visible polymorphisms are not representative of the entire genome and do not reveal enough genetic variation for most research questions.

Molecular markers

Much of modern molecular genetics is based on some type of **electrophoresis,** in which macromolecules (proteins, RNA, DNA) are separated on a gel using an electric charge. Most empirical work in modern population genetics is based on molecular markers. All of the many molecular markers are based on the same basic principles. Samples containing proteins or DNA from a number of individuals are placed separately at one end of a gel (Figure 2.3). Then an electric field is applied to the gel, and different molecules in the sample move at different rates depending on differences in size, shape, molecular weight, and electric charge. Molecules that are identical or very similar end up clustered in the gel into **bands.** Finally, the gel is treated to **visualize** the bands. There are several materials that the gels can be made of, including starch, agarose, cellulose acetate, and polyacrylamide, but in all cases the purpose of the gel is to provide a matrix for the differential movement of the different molecules. There are also many types of visualization systems, including chemical stains, radioactivity, and

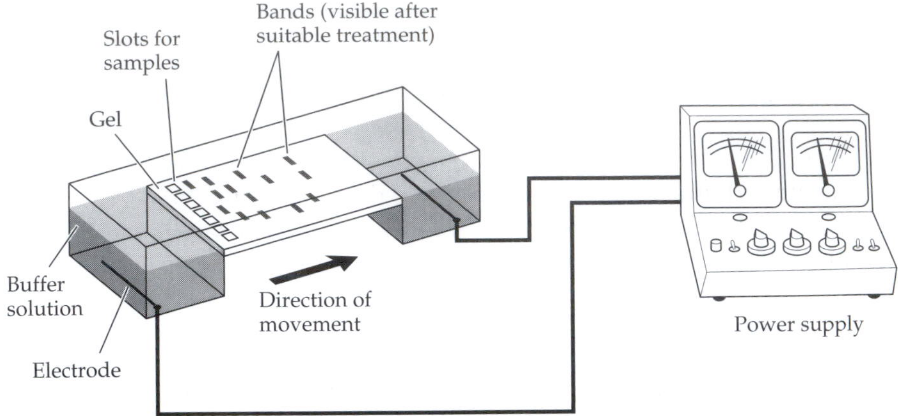

Figure 2.3 One type of laboratory apparatus for electrophoresis. Each sample slot contains material from one individual. The procedure is widely used to separate protein or DNA molecules. Proteins separate based mainly on differences in charge, while DNA fragments migrate in proportion to the logarithm of their size in base pairs (smaller fragments migrate faster).

fluorescence. For more information on many of these markers, see Hillis, Moritz, and Mable (1996).

Protein electrophoresis

The first molecular markers used in population genetics were proteins, usually soluble enzymes, also referred to as **allozymes** or **isozymes.** Lewontin and Hubby first used allozymes for population genetics in 1966, and the technique quickly revolutionized the field. For years, population geneticists had debated on theoretical grounds about how widespread genetic variation is in natural populations. Application of the allozyme technique to a variety of species revealed that variation is nearly ubiquitous in nature.

In protein electrophoresis, a part or all of the organism is typically ground up in a buffer (a solution that helps maintain a constant pH), and a sample of the buffer, now containing all of the soluble enzymes, is placed at the **origin** of the gel. After electrophoresis, a stain is used that takes advantage of the specificity of enzyme-substrate reactions to visualize only one enzyme. A variety of different stains can be used to visualize different enzymes on different gels. This solves a basic problem in all molecular marker techniques—how do you isolate one gene locus for study out of the myriad proteins and millions of bases of DNA in each cell? The allozyme staining solution typically contains the substrate and a dye that precipitates where the enzyme-catalyzed reaction takes place. When the gel is incubated in the staining solution, a colored band is produced wherever that one enzyme occurs in the gel (see Murphy et al. (1996) for technical details).

Some amino acids are positively, and some negatively, charged. When the electric field is applied to the gel, enzymes with different charges migrate at different rates. Because different allelic DNA sequences can code for slightly different amino acid sequences in the protein, this difference in mobility reveals underlying genetic variation. Note that we are talking about differences within the same enzyme, so different bands represent different forms of the same enzyme. These different forms are caused by differences in amino acid sequence, but these variants are all functional versions of the same enzyme. Figure 2.4 shows a typical banding pattern for a simple enzyme locus with two alleles present in this sample. This shows a key advantage of enzyme electrophoresis over many other markers: In most cases it produces **codominant** markers, which means that no allele is dominant over others, and therefore heterozygotes can be distinguished from both homozygotes. This is because heterozygotes appear as two distinct bands on the gel, while homozygotes produce only one band (see Figure 2.4).

Compared to visible polymorphisms, protein electrophoresis reveals far more of the underlying genetic variation, and this variation is also a more random sample of the genome, an important feature for the uses and questions listed above. Another advantage of allozymes is that they are almost always codominant. Their biggest disadvantage is that they reveal only a small subset of the actual variation in DNA sequences between individuals, for a variety of reasons:

- Since proteins are gene products, variation in proteins does not reveal variation in non-coding regions of the genome.
- For a large number of proteins, particularly structural proteins, specific stains are not available, so only the loci that code for a subset of the proteins in an organism can be examined.

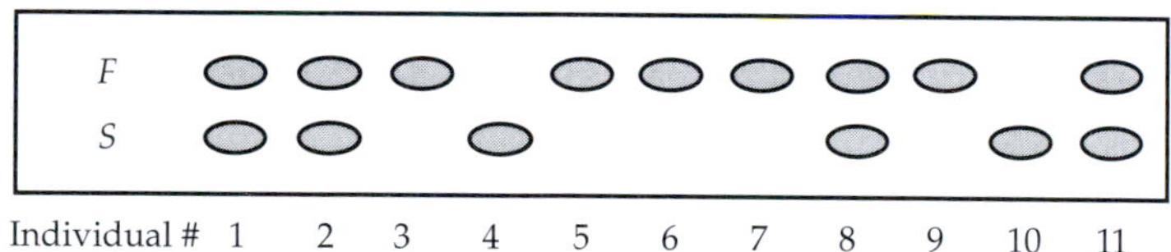

Figure 2.4 Banding pattern for a simple allozyme locus with two codominant alleles. There are 11 vertical lanes, each representing one individual. The two alleles are labeled *F* for faster migrating (the origin is at the bottom) and *S* for slower. Homozygotes appear as single bands, since the enzyme coded for by both alleles in an individual are identical and thus migrate the same distance. In heterozygotes, two bands appear, one for the enzyme coded for by each of the two different alleles present in heterozygous individuals.

- Even for proteins that can be stained, there can be many variations in amino acid sequence that do not cause differences in mobility on a gel, particularly if the overall charge of the proteins are the same.
- Many changes in the DNA sequence in a coding region do not cause changes in amino acid sequence of the protein. This is especially true of changes at the "silent," or third position, of the codon.

These are all consequences of the fact that protein electrophoresis examines the enzyme phenotype, rather than documenting the genotype directly. Therefore, the main limitation of protein electrophoresis is that it reveals only a small fraction of the genetic variation in a population. However, it is still a useful technique in organisms with a high degree of allozyme variation and for questions not requiring large amounts of genetic variation.

DNA markers

Molecular methods for examining variation in DNA sequences directly first became available in the 1970s, and new techniques continue to be developed at a dizzying pace. In essence, all these techniques are ways of converting genetic variation into pieces of DNA of different sizes, which can then be separated by gel electrophoresis. Proteins separate mainly based on differences in charge, whereas DNA separates based on differences in the size of the fragment. There are a wide array of techniques that are changing rapidly, so we will focus on the following techniques that have gained the widest use in ecological genetics.

RFLP. The first DNA technique to gain wide use in population genetics was DNA **restriction fragment length polymorphisms (RFLPs).** In this technique, a sample of DNA is mixed with restriction enzymes, which cut DNA at restriction sites with specific short DNA sequences (Figure 2.5A). The resulting DNA is then separated by electrophoresis and visualized with radioactively labeled or fluorescent DNA probes using a Southern blot (Figure 2.6). If the restriction site occurs rarely in the DNA sample, then the cut fragments will be large and the result will be bands that do not migrate very far during electrophoresis. If the restriction site occurs more frequently in the sample, then the DNA will be cut into smaller fragments that will migrate farther on the gel (Figure 2.5B). Thus, differences in banding patterns between individuals reveals genetic variation in the number and position of restriction sites that have the specific DNA sequence recognized by the restriction enzyme. RFLPs have the advantages that they are codominant and have been developed for many organisms, but they are considerably more labor-intensive and require more DNA than the more recent techniques based on PCR.

(A)

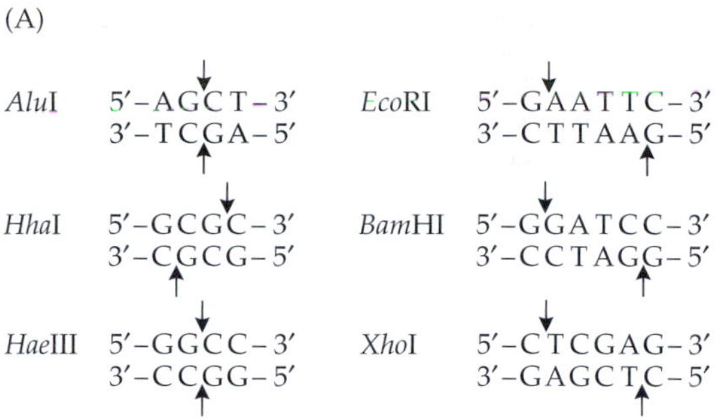

(B)

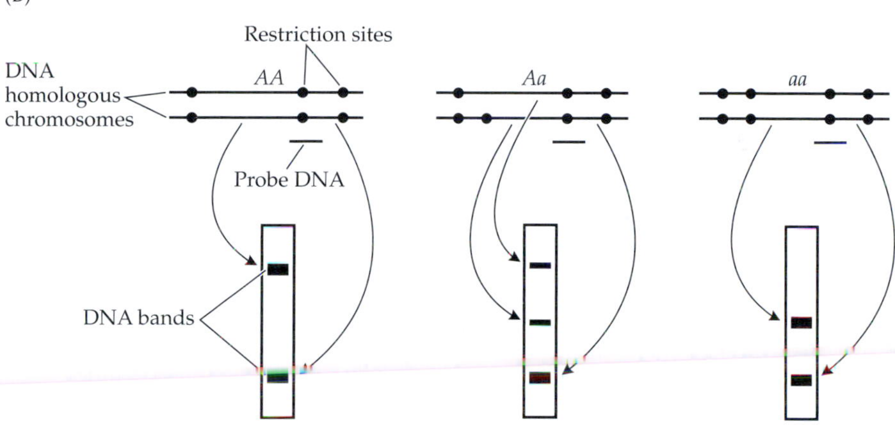

Figure 2.5 (A) The restriction sites for six of the more than 500 commercially available restriction enzymes. In each strand cleavage occurs at the position of the arrowhead. For example, the enzyme *Alu*I cuts DNA at sites containing the four-base sequence AGCT, and each strand is cleaved between the G and the C. By contrast, *Eco*RI cuts at the six-base sequence GAATTC, and each strand is cleaved between the G and the A. (B) RFLP analysis. Three genotypes are shown. At the top is a schematic of portions of a pair of homologous chromosomes with the location of restriction sites shown. The bottom shows the resulting banding patterns for each genotype on a Southern blot, with larger DNA fragments toward the top. One restriction site is missing in the *A* allele, resulting in a larger fragment than in the *a* allele. The smallest fragment is produced in both alleles. Arrows show which DNA fragment produces each band; thicker bands are those fragments present in both chromosomes (they sometimes do appear thicker on an actual gel). Only fragments that the probe DNA binds to are visualized as bands.

PCR. Almost all recent molecular markers are based on the **polymerase chain reaction (PCR),** which has revolutionized the entire field of genetics. PCR is a simple, automated method to produce many copies of a specific

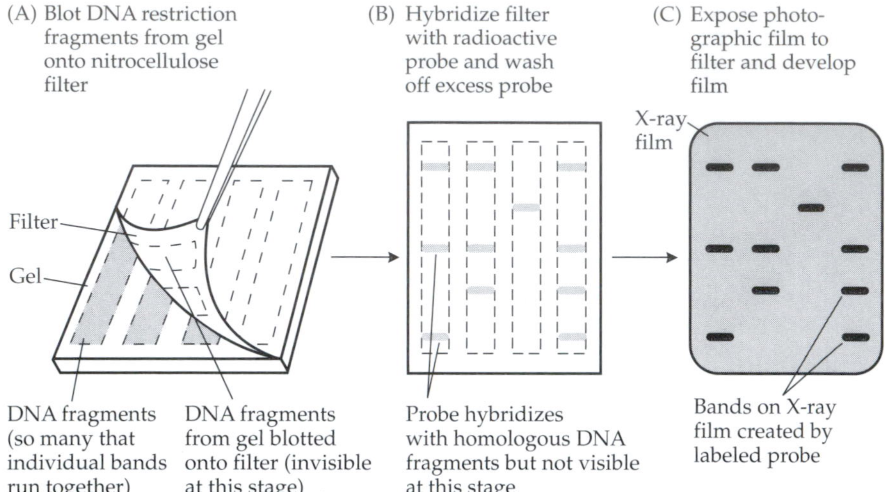

Figure 2.6 Southern blot procedure. (A) DNA fragments separated by electrophoresis are transferred and chemically attached to a filter. (B) The filter is mixed with radioactive or fluorescent probe DNA, which hybridizes with homologous DNA molecules on the filter. (C) After washing, the filter is exposed to photographic film, which develops dark bands caused by radioactive or light emissions from the probe.

fragment of DNA. To do this, short oligonucleotides (usually 20 to 30 nucleotides in length) that are complementary to the two ends of the fragment to be amplified are synthesized to be used as a primer for DNA synthesis. These are then mixed with a sample of the DNA to be studied along with DNA polymerase (to catalyze DNA synthesis) and large amounts of the four nucleotides (A, T, C, and G). This mixture is then put through repeated cycles of (1) denaturing the DNA sample into single strands to serve as a template for DNA synthesis, (2) binding (annealing) of the primers to these strands, and (3) synthesis of new copies of the DNA downstream of the primer (Figure 2.7). Since each new strand synthesized can serve as a template in the next round of synthesis, the number of copies of the fragment of interest doubles with each cycle, numerically overwhelming any other DNA in the sample.

In the PCR-based marker techniques described in the next section, it is only the target segment of DNA that is amplified. This means that the visualization step does not need to be specific as in allozymes and RFLP; any compound that stains DNA can be used, such as ethidium bromide, silver stain, or a variety of fluorescent dyes. This is a primary reason why the PCR-based markers are faster and simpler than most of the other techniques. Another advantage of PCR-based markers is that less sample DNA from each organism is needed

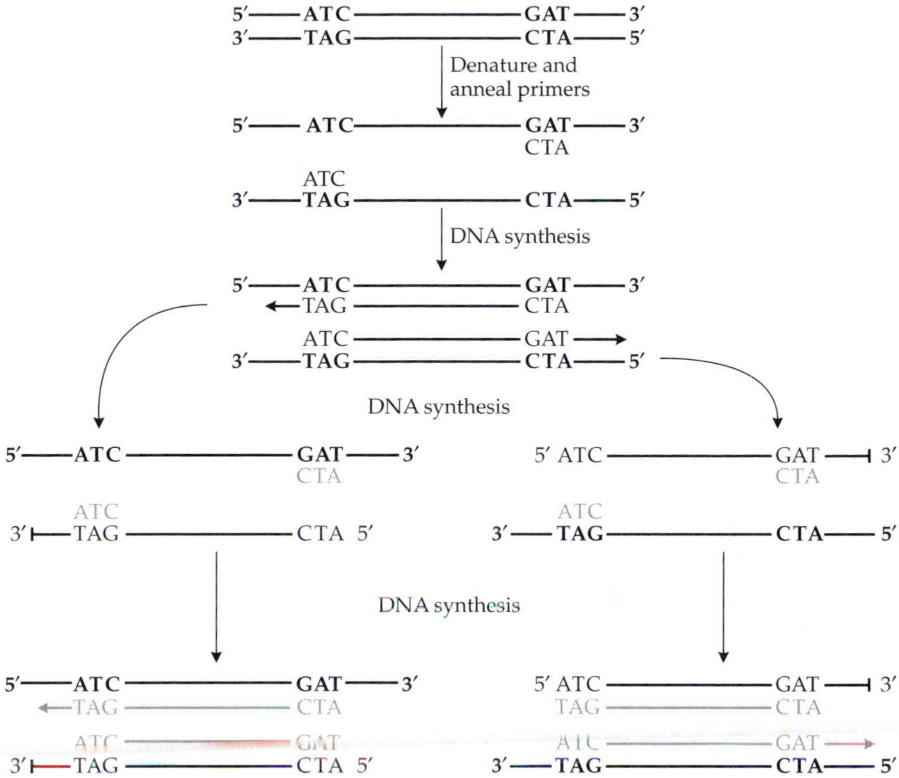

Figure 2.7 The polymerase chain reaction (PCR). PCR amplifies the fragment of DNA between the two primers exponentially, because the fragments produced in each cycle are used as templates in later cycles, leading to a doubling of the number of fragments in each cycle. The original sample DNA is shown in boldface type, the DNA synthesized in the first round in regular type, and the DNA from the second round in gray. After many cycles the vast majority of the fragments include only the primer sequences and the intervening region. For simplicity, only two cycles are shown, and primers are depicted as three bases long; in actual applications 20 or more cycles are common and primers are 10 or more bases long. Binding sites for the same primer sequence are shown close together on the two strands. Pairs of different-sequence primers for the two ends of the fragments are also used.

because it is amplified by the PCR reaction. There is a basic tradeoff between the markers discussed later in this chapter. Some, such as the VNTR and CAPS markers, have the great advantage of being codominant, but are much more difficult to develop for each new species because the PCR primer sequences must be determined. Others, such as RAPD and AFLP (to be discussed next), have the advantage of being based on random primer sequences, therefore, specific primer sequences do not need to be identified, but they have the drawback of being dominant and so heterozygotes cannot be distinguished.

RAPD. Perhaps the simplest and quickest method of generating markers using PCR is called **randomly amplified polymorphic DNA,** or **RAPD** (Williams et al. 1990). Here, randomly generated primers, 10 bases in length, are used for PCR with the entire genome of the study organism serving as a template. If a sequence complementary to the primer(s) occurs on both strands of DNA, and they are close enough together, then this fragment will be amplified and appear as a band on the gel when the products of the PCR reaction are electrophoresed (see Figure 2.7). If the binding sites on the two strands are too far apart, then synthesis of the fragment will not be completed when the synthesis phase of PCR ends, and the fragment will therefore not be amplified.

A panoply of different markers have come from the use of RAPD technology. If a fragment can be amplified using a single primer, then it is called a **SPAR (single primer amplified region).** When a particular amplified fragment is isolated and its nucleotide sequence determined, it becomes a **SCAR (sequence-characterized amplified region)**. A particular amplified fragment may also include a restriction-site polymorphism, in which case the polymorphism is called a **CAPS (cleaved amplified polymorphic site).** A CAPS is the analog of an RFLP, except that the genotype is identified by amplification and a nonspecific DNA stain, rather than in a Southern blot with a radioactive probe. This is why a CAPS is sometimes called a **PCR-RFLP.** CAPS alleles, like RFLP alleles, are codominant.

AFLP. This technique is a hybrid of RFLP and RAPD that combines many of the advantages of each. In **AFLP** (sometimes called **amplified fragment length polymorphism,** although it does not reveal length polymorphisms), the genome of the study organism is cut with restriction enzymes as in RFLP, and then some of these fragments are selectively amplified with PCR using random sequences as in RAPD (see Mueller and Wolfenbarger 1999 for details). Since AFLPs are based on PCR, they are faster and easier than RFLPs. They are more repeatable than RAPDs because the primers used for amplification are typically about twice as long, so errors in binding to the template DNA are less frequent.

Both RAPD and AFLP reveal genetic variation in a population by screening for the presence or absence of a specific random (or partially random) DNA sequence, rather than different lengths of fragments as in the other DNA markers. In RAPD, if there is a location in the genome with two sites on opposite strands that the random primers bind to, then this fragment is amplified and a band is formed on the gel. If another individual in the population has a different DNA sequence at one of these binding sites, then the primer will not bind there and no band will be formed. Similarly in AFLP, if a fragment produced by digestion with restriction enzymes has sequences at both ends matched by the primers, including the random portions, then that fragment will be amplified. In individuals with a different sequence at one or both ends, the fragment will not be amplified and no band will be produced. One consequence of this is that most RAPD and AFLP markers are inherited in a dominant fashion, meaning that heterozygotes and one of the homozygotes produce the same pattern. Heterozygotes produce a single band in exactly the same position as the band produced by one of the homozygotes (although there is twice as much DNA in the band from homozygotes). The other homozygote produces no band, which is often called a **null allele.** Therefore, only the band-absent (null) genotype can be scored definitively. Each band on a RAPD or AFLP gel is from a separate location on the genome, instead of having different bands representing different alleles at the same locus as in the other methods.

VNTRS (VARIABLE NUMBER OF TANDEM REPEATS). VNTRs are noncoding regions of the genome that consist of several to many copies of the same sequence. The repeated sequence can be anywhere from two to 64 nucleotides long. Longer sequences (from 10 to 64 nucleotides in the repeating unit) are called minisatellites, whereas shorter repeats (2 to 9 nucleotides) are called microsatellites, simple sequence repeats (SSR), simple sequence repeat polymorphisms (SSRP), or short tandem repeats (STR). There is a large amount of variation between individuals for the number of repeat units at each VNTR locus, especially for the shorter repeat units. For example, a common microsatellite repeat unit is AC, so one allele might have this unit repeated 10 times, ACACACACACACACACACAC, which is often written $(AC)_{10}$, and another allele might be $(AC)_{12}$. Note that these are double-stranded nuclear DNA, but for simplicity only the sequence of one of the strands is used to denote the marker.

For minisatellites, the DNA is cut with a restriction enzyme, and then a radioactively labeled DNA probe that has a complementary sequence to the repeated unit is used for visualization (similar to RFLP). For microsatellites, the regions on either side of the repetitive sequence (the **flanking regions**) are sequenced, and then complementary primers are constructed so that the microsatellite can be amplified by PCR and visualized with stain or fluorescence, similar to RAPDs. The different number of repeat units in different alleles causes each fragment to migrate a different distance on the gel, so genetic variability is revealed. Multilocus probes are often used for minisatellites, and microsatellite loci can have as many as 30 to 50 different alleles in a population, so these techniques can reveal a tremendous amount of variation. Indeed, these markers are often referred to as **DNA fingerprints,** because they are rarely exactly the same in two individuals. The main drawback is that these

TABLE 2.1 *Summary of genetic markers used most often in ecological genetics*

Marker	Codominant	Amount of variation revealed
Visible polymorphisms	Sometimes	Low
Allozymes	Yes	Low–moderate
RFLP	Yes	Moderate
RAPD	Rarely	High
AFLP	Rarely	High
VNTR	Yes	High

markers are difficult to develop for new species, because the VNTR regions must be found and the sequences of the flanking regions determined.

The ultimate level of resolution of genetic variation is **gene sequencing**. Sequencing is usually done by automated sequencing machines, and the entire genomes of a number of organisms (including humans) have been or are being sequenced. However, it is still too time consuming and expensive to use sequencing for examining genetic variation within most populations, since this usually requires sequencing a number of genes in a number of individuals. Another drawback for the application of sequencing to ecological genetics is that in most cases little is known about how specific sequences affect the phenotype. Sequencing has become the method of choice for reconstruction of phylogenies, however, since only one or a few individuals need to be sequenced for each taxon sampled.

Table 2.1 summarizes the different types of markers discussed above. Currently, microsatellites and AFLP are the most commonly used in ecological genetics, although many examples in later chapters are based on visible polymorphisms and allozymes, because there is a larger literature on these and they provide simpler examples. Microsatellites are preferred if a sufficient number of primer sequences are known or can be generated for the organism under study; they are codominant and can have many alleles per locus in a population. AFLP markers are based on random sequences so, unlike microsatellites, they take very little time to develop for each species, a key advantage for studies of nonmodel organisms. Another advantage for AFLP is that many loci can be scored on each gel. The disadvantages of AFLP are that they are dominant and have only two alleles per locus.

Measuring Genetic Variation: Simple Summary Statistics

The genetic variation uncovered by genetic markers can be quantified in a number of ways. The most fundamental measures in population genetics are

allele and **genotypic frequencies.** The word *frequency* in population genetics is used to mean proportion, so the **frequency** of a given allele or genotype is simply its proportion out of all the alleles or genotypes present at the locus in the population. The term *gene frequency* is often used interchangeably with *allele frequency*; this is unfortunate, because it is inaccurate and leads to confusion. Genes do not differ in frequency, alleles do.

If we have only two alleles at a locus, designated A and a, then the usual symbols for their frequencies are p and q, respectively. Since frequencies (proportions) must sum to 1, then

$$\begin{aligned} p + q &= 1 \\ q &= 1 - p \end{aligned} \tag{2.1}$$

These equations are very simple, but they are extremely important to keep in mind when thinking about allele frequencies. Genotypic frequencies are usually denoted by uppercase letters, and they also must add to 1. In the case of a locus with just two alleles, P is the frequency of AA, H is the frequency of Aa, and Q is the frequency of aa. These must also sum to 1; that is, $P + H + Q = 1$. To calculate allele frequencies from genetic marker data, divide the number of each allele by the total number of alleles in the sample:

$$p = \frac{2n_{AA} + n_{Aa}}{2n} \tag{2.2}$$

where n = the number of individuals in the sample and n_{AA} and n_{Aa} are the numbers of AA and Aa genotypes in the sample, respectively. In Equation 2.2 the 2 in the numerator is present because there are two A alleles in each homozygous individual, and the 2 in the denominator is present because there are two alleles at each locus in each diploid individual.

We can rearrange Equation 2.2 to obtain the allele frequency in terms of the genotype frequencies:

$$\begin{aligned} p &= \frac{2n_{AA}}{2n} + \frac{n_{Aa}}{2n} \\ p &= \frac{n_{AA}}{n} + \frac{1}{2}\frac{n_{Aa}}{n} \\ p &= f(AA) + \tfrac{1}{2} f(Aa) \\ &= P + \frac{H}{2} \end{aligned} \tag{2.3}$$

Here $f(AA)$ stands for the frequency (proportion) of homozygotes.

EXERCISE: Write the equations for q in terms of both number and frequency of genotypes.

ANSWER:

$$q = \frac{2n_{aa} + n_{Aa}}{2n} \qquad 2.4$$

$$q = f(aa) + \tfrac{1}{2} f(Aa)$$

$$= Q + \frac{H}{2} \qquad 2.5$$

As an example, we can calculate allele frequencies in a natural population of water fleas (*Daphnia obtusa*; Spitze 1993). In the Ojibway Pond population, the following genotypes were scored at the allozyme locus *PGM*: 57 individuals were *MM*, 53 *MS*, and 18 *SS* (*M* and *S* stand for medium and slow). Using Equations 2.2 and 2.4, and p for the frequency of *M*, we find:

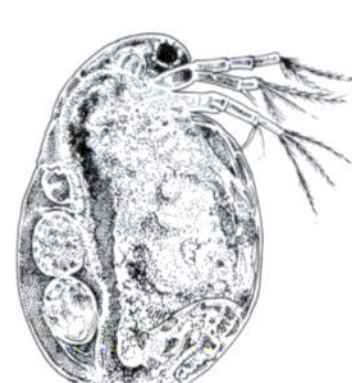

$$p = \frac{(2 \times 57) + 53}{2 \times 128} = 0.652$$

$$q = \frac{(2 \times 18) + 53}{2 \times 128} = 0.348$$

Note that p and q sum to 1, and that 128 is the total number of individuals in the population (57+53+18).

EXERCISE: Calculate the allele and genotype frequencies in Figure 2.4.

ANSWER: There are five *FF* homozygotes, two *SS* homozygotes, and four *FS* heterozygotes. Using Equations 2.2 and 2.4, $p = 0.636$ and $q = 0.364$. Always check that $p + q = 1$, which it does here. Genotypic frequencies $P = 5/11 = 0.454$, $H = 4/11 = 0.364$, $Q = 2/11 = 0.182$, which also sum to 1.

Organization of Genetic Variation within Populations

In this section we discuss the different ways that mating systems organize alleles into diploid genotypes (Figure 2.8). Under different mating systems (e.g., random mating vs. inbreeding) different genotypic frequencies are generated from the same allele frequencies.

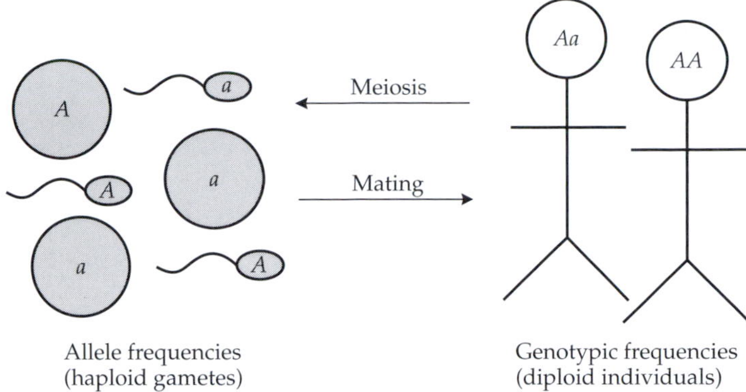

Figure 2.8 A schematic showing how mating combines alleles (in the haploid pool of gametes) into genotypes (in the diploid individuals). The genotypic frequencies produced by a given set of allele frequencies depends on the mating system. The cycle repeats itself when the diploid individuals produce haploid gametes through meiosis.

Random mating

Our starting point will be the simplest case, **random mating.** When mating is random, the chance that an individual mates with another individual of a given genotype is equal to the frequency of that genotype. A more precise definition comes from probability: Under random mating, the frequency of matings between two genotypes is equal to the product of their individual frequencies. When mating is random and a number of other assumptions are met, then the genotypic frequencies for a given set of allele frequencies can be calculated with the **Hardy-Weinberg model.** Godfrey Hardy and Wilhelm Weinberg each independently published the model in 1908, laying the foundation for the field of population genetics. An introduction to models is presented in Box 2.1.

Assumptions of the Hardy-Weinberg model

The Hardy-Weinberg model has eight assumptions. They are listed here along with some brief explanations.

1. The organism is **diploid, sexual,** and has **discrete generations.** Discrete generations refer to a life history like that of an annual plant, in which the parental generation has died by the time the offspring generation reproduces.
2. Allele frequencies are the same in both sexes.
3. **Mendelian segregation** occurs, which means that heterozygotes produce equal numbers of gametes containing each allele. For example, an

BOX 2.1 *What Is a Model?*

A **model** is a theoretical abstraction of the real world that has two principal uses:

- To reduce complexity, allowing us to see important underlying patterns; that is, to see the forest in spite of the trees.
- To make specific predictions to test with experiments or observations. Thus, models can guide **empirical** studies by suggesting which data are most important to gather.

However, it is important to remember that models cannot provide direct information about what is actually occurring in the real world. The predictions made by models need to be supported or refuted by empirical data.

There are three main types of models—verbal, simulation, and analytical. **Verbal models** tend to take the form of "if this condition holds, then logically this should happen." Although some people have a low opinion of verbal models because they lack mathematical rigor, these models can be extraordinarily powerful. Darwin's theory of natural selection was a verbal model, yet it revolutionized biology. Darwin's verbal model stated that if there is variation in a trait that is related to fitness, and this variation is heritable, then the trait will change across generations; that is, it will evolve.

In **simulation models** the system to be modeled is simulated in a computer. For example, the mating system diagrammed in Figure 2.8 could be simulated in the computer, starting with frequencies of different alleles in the gametes. The computer could then be instructed to combine gametes randomly for random mating and calculate the resulting diploid genotypic frequencies.

Analytical models define the entire system with equations that can then be solved for different values of the input variables to make predictions about the behavior of the system. Analytical models are the most difficult to construct, but can also be the most powerful. For example, the Hardy-Weinberg model is an analytical model of allele and genotypic frequencies under random mating.

All models make starting **assumptions**, either explicitly or implicitly, to simplify the system. These assumptions are partly to make the model mathematically or computationally tractable, especially in the case of analytical models, but also to make the models easier to understand—remember that simplification of the real world is one of the primary goals of models. It is important to consider the assumptions of any model carefully, because they determine how well the model fits the real world. This does not mean that a model is worthless for understanding an organism that violates some of the model's assumptions, because very often the predictions of a model will be **robust** to these violations. To be *robust* means that the predictions do not change much when some of the assumptions are violated.

Aa individual produces equal numbers of *A* and *a* gametes. There are a few genes that violate this assumption; this condition is known as **meiotic drive** or **segregation distortion.** When meiotic drive occurs, one allele in heterozygous individuals is overrepresented in the gametes.

Examples include the *T* allele in mice (*Mus*) and the *segregation distorter* locus in *Drosophila*.

4. **Random mating** occurs, meaning that mating is random with respect to the genotypes under consideration (it may be nonrandom with respect to genotypes at other loci). Later in this chapter we will discuss two types of nonrandom mating, assortative mating and inbreeding.
5. No **mutation**
6. No **migration**
7. No **random genetic drift (large population size)**
8. No **natural selection**

The Hardy-Weinberg model is robust to many violations, especially of the last four assumptions. After considering allele and genotype frequencies under all these assumptions, we will **relax** (remove) one or more of the last five assumptions and examine the consequences of each of these in turn. In fact, much of population genetics involves study of the consequences of relaxing these last five assumptions, to better understand how mating systems, mutation, migration, drift, and natural selection affect the genetics and evolution of populations.

Deriving the Hardy-Weinberg equations

Recall our previous comment that we will often want to calculate genotypic frequencies from allele frequencies and vice versa. Given the assumptions above, this conversion back and forth is simple, and is the essence of the **Hardy-Weinberg principle:** Genotypic frequencies in the offspring can be calculated directly from allele frequencies in the parents after only one generation of random mating. There are two principal ways of deriving the Hardy-Weinberg equations, and each provides insight into how genotypic frequencies arise under random mating.

The first derivation starts with allele frequencies in the gametes produced by parents. Given the assumptions above, the frequencies of alleles in the pool of gametes produced by the parents are the same as the allele frequencies in the parental population. With random mating, these gametes unite randomly during fertilization, so that the genotypic frequencies of the offspring are the products of the gametic allele frequencies. Figure 2.9 shows cross-multiplication squares drawn with the sides of the squares proportional to the allele frequencies. Because the areas of the squares are the products of the lengths of their sides, the areas are proportional to the offspring genotypic frequencies. Note that there are two ways to make a heterozygote—an *A* sperm and an *a* egg or an *a* sperm and an *A* egg—so that the genotype frequencies of those two boxes are added together.

Therefore, under Hardy-Weinberg assumptions the frequency of *AA* homozygotes (P', where the prime denotes the offspring generation) is p^2,

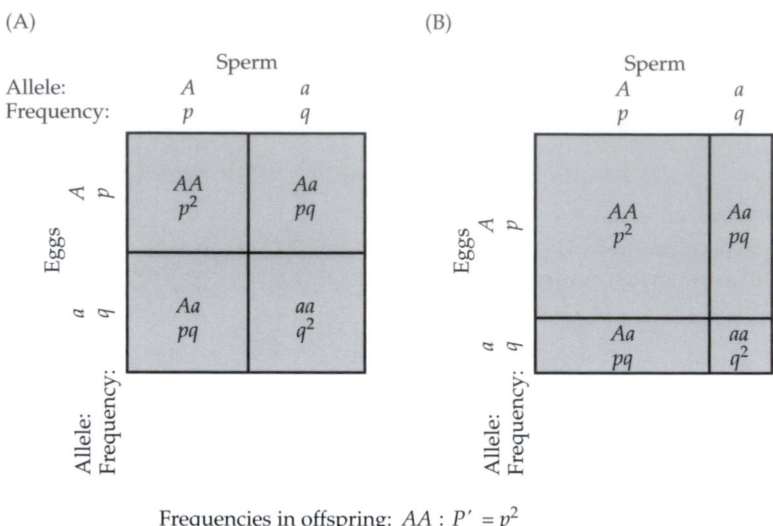

Figure 2.9 Cross-multiplication squares showing graphically how genotypic frequencies (inside boxes) are products of allele frequencies in gametes under Hardy-Weinberg assumptions. In the square on the left, the allele frequencies are equal (i. e., $p = q = 0.5$), while in the right-hand square the *A* allele is three times as common as *a* ($p = 0.75$; $q = 0.25$).

the frequency of *aa* (Q') is q^2, and the frequency of heterozygotes (H') is $2pq$. Figure 2.9A shows the case where the two alleles are at equal frequency ($p = q = 0.5$), so that the frequencies of the homozygotes are each 0.25 and the frequency of the heterozygote is 0.5.

Now we can determine the allele frequencies in the offspring using equation 2.3:

$$
\begin{aligned}
p' &= P' + \frac{H'}{2} \\
&= p^2 + \frac{(2pq)}{2} && \text{substitute from Hardy–Weinberg equation} \\
&= p^2 + pq && \text{cancel} \\
&= p(p+q) && \text{factor} \\
&= p && \text{remember } p+q=1
\end{aligned}
$$

EXERCISE: Derive q' in terms of q.

These derivations show that under Hardy-Weinberg assumptions, allele and genotypic frequencies remain the same across generations. This means that Mendelian inheritance, by itself, does not change allele frequencies. This is why geneticists often refer to the **Hardy-Weinberg equilibrium (HWE).** If the genotypic frequencies are changed without changing the allele frequencies, then the genotypic frequencies will return to the Hardy-Weinberg values (p^2, $2pq$, q^2) after one generation of random mating. If some other force changes the allele frequencies (see Chapter 3), however, a new HWE occurs with genotypic frequencies corresponding to the new allele frequencies (i.e., new p and q produce new p^2, $2pq$, q^2), again after one generation of random mating.

An example of this is given in Figure 2.9B. Here $p = 0.75$ and $q = 0.25$. As long as the assumptions of Hardy-Weinberg are met, the genotypic frequencies are still p^2, $2pq$, and q^2, which with the new allele frequencies are approximately 0.562, 0.375, and 0.062. This illustrates a key point: Hardy-Weinberg equilibria can occur for any allele frequency. It is a common misconception that HWE means equal allele frequencies, $p = q = 0.5$, but even if one allele is fixed the population is in HWE (p, p^2, and P all equal 1; q, $2pq$, H, q^2, and Q all equal 0).

Figure 2.10 shows the frequencies of all three genotypes under HWE at a locus with two alleles for all possible allele frequencies. This graph helps us understand a wide variety of fundamental concepts in population genetics;

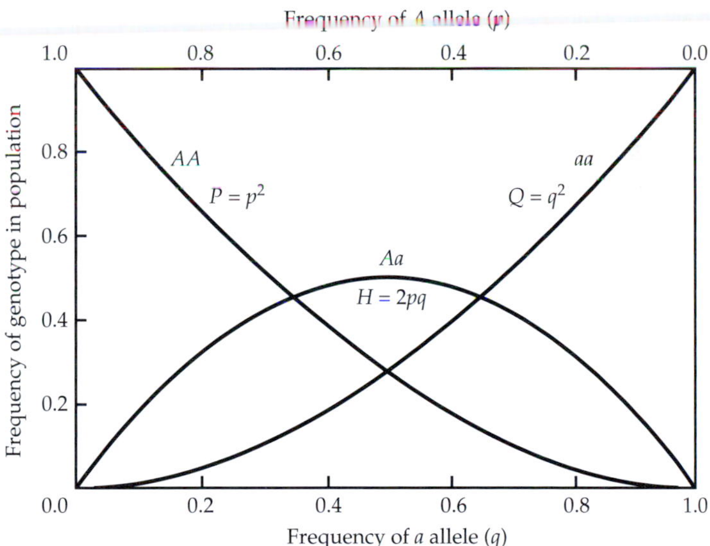

Figure 2.10 Genotypic frequencies across all allele frequencies for two alleles under Hardy-Weinberg equilibrium. Note that when an allele is rare, the frequency of homozygotes for that allele is much lower than the frequency of heterozygotes.

we will refer to it often. There are three important points to note. First, these are the genotypic frequencies only at HWE; factors like inbreeding that cause deviations from HWE will change these genotypic frequencies. Second, at intermediate allele frequencies, heterozygotes are more common than either homozygote. Third, when the frequency of an allele is less than 0.1, virtually all of these alleles are contained in heterozygotes and there are very few homozygotes for the allele. As we will see in more detail in Chapter 3, this is how deleterious recessive alleles (for example, lethal disease alleles in humans) are maintained in the population. Because selection maintains the harmful recessive alleles at a low frequency, almost all of the harmful alleles are contained in heterozygous individuals, in which the recessive disease is not expressed.

Another way to derive the Hardy-Weinberg equations is to start with the matings between the diploid parents rather than with the gametes they produce. With three genotypes there are six possible types of matings. For example, the mating $AA \times AA$ occurs only when an *AA* male mates with an *AA* female, and this occurs a proportion $P \times P$ (or P^2) of the time. Similarly, an $AA \times Aa$ mating occurs when an *AA* female mates with an *Aa* male (proportion $P \times H$), or when an *Aa* female mates with an *AA* male (proportion $H \times P$)—so the overall proportion of $AA \times Aa$ matings is $PH + HP = 2PH$. The frequencies of these and the other types of matings are given in the second column of Table 2.2.

The offspring genotypes produced by the matings are given in the last three columns of Table 2.2. The offspring frequencies follow from the assumption of Mendelian segregation, so that an *Aa* heterozygote produces

TABLE 2.2 *Demonstration of the Hardy-Weinberg principle*

Mating	Frequency of mating	Offspring genotype frequencies		
		AA	*Aa*	*aa*
$AA \times AA$	P^2	1	0	0
$AA \times Aa$	$2PH$	$^1/_2$	$^1/_2$	0
$AA \times aa$	$2PQ$	0	1	0
$Aa \times Aa$	H^2	$^1/_4$	$^1/_2$	$^1/_4$
$Aa \times aa$	$2HQ$	0	$^1/_2$	$^1/_2$
$aa \times aa$	Q^2	0	0	1
	Totals (next generation)	P'	H'	Q'

where: $P' = P^2 + 2PH/2 + H^2/4 = (P + H/2)^2 = p^2$

$H' = 2PH/2 + 2PQ + H^2/2 + 2HQ/2 = 2(P + H/2)(Q + H/2) = 2pq$

$Q' = H^2/4 + 2HQ/2 + Q^2 = (Q + H/2)^2 = q^2$

an equal number of *A*-bearing and *a*-bearing gametes. Homozygous *AA* genotypes produce only *A*-bearing gametes, and homozygous *aa* genotypes produce only *a*-bearing gametes. Therefore, a mating of *AA* with *aa* produces all *Aa* offspring, a mating of *AA* with *Aa* produces $\frac{1}{2}$ *AA* and $\frac{1}{2}$ *Aa* offspring, a mating of *Aa* with *Aa* produces $\frac{1}{4}$ *AA*, $\frac{1}{2}$ *Aa*, and $\frac{1}{4}$ *aa* offspring, and so forth.

The genotype frequencies of *AA, Aa,* and *aa* after one generation of random mating are denoted in Table 2.2 as P', H', and Q' respectively. The new genotype frequencies are calculated as the sum of the products shown at the bottom of the table. For each genotype, the frequency of each mating producing the genotype is multiplied by the fraction of that genotype produced by that mating. The new genotype frequencies P', H', and Q' simplify to p^2, $2pq$, and q^2, the Hardy-Weinberg frequencies.

Box 2.2 introduces some of the basics of statistical analysis, which we will use throughout the rest of the book.

Uses of the Hardy-Weinberg Model: Tests for Departure from HWE

It is often useful to test whether genotypic frequencies in a real population are in HWE, because if they differ significantly, this suggests that one or more assumptions of HWE have been violated and therefore a potentially important evolutionary or ecological process is causing the deviation. The converse is not true, however, because HWE is robust to violations of assumptions and the chi square test is not particularly powerful. Therefore, finding no significant deviation from HWE does not mean that none of the assumptions of HWE have been violated.

As an example, we will use the chi-square test to determine whether Spitze's (1993) PGM allozyme genotype frequencies from *Daphnia*, discussed on page 24, deviate significantly from HWE. The first step is to determine the numbers of each genotype *expected* under HWE given the allele frequencies that we calculated above. We will then use the chi-square test to see how closely the *observed* numbers of the three genotypes fit this expectation. With allele frequencies $p = 0.652$ and $q = 0.348$, the expected genotypic frequencies are $p^2 = 0.425$, $2pq = 0.454$, and $q^2 = 0.121$ (which sum to 1). Multiplying each of these by the sample size (128 individuals) gives the expected numbers as 54.4, 58.1, and 15.5. This conversion is necessary because the chi-square test must be based on the observed *numbers*, not frequencies. (Using genotypic frequencies instead of numbers in the chi-square test is a common error.) The comparison is thus between the observed (*obs*) and expected (*exp*) numbers:

	MM	*MS*	*SS*	Totals
obs	57	53	18	128
exp	54.4	58.1	15.5	128

BOX 2.2 *A Brief Introduction to Statistics*

Ecological geneticists use statistics to find patterns among the sometimes bewildering variability of the natural world. Thus, the goals of statistical analysis are similar to the goals of modeling, but while modeling is based on theoretical abstractions, statistics is a way of understanding empirical data collected in the real world. These data are usually collected from a random sample of a population, and statistical tests are used to determine if patterns in the sample are due to random chance or some real biological phenomenon. The larger the sample size (often denoted by n or N), the more confident we can be that the patterns are real.

We will discuss three basic kinds of statistical analyses in this book, which are designed to answer three different types of questions. In part, these types of questions depend on the kind of data that are collected. Data can be either **discrete** (also called **qualitative** or **categorical**) or **continuous** (**quantitative**). Discrete data can be easily grouped into distinct groups or classes. All the genetic markers we discussed above, both visible and molecular, produce distinct genotypes that are good examples of discrete data. Most phenotypic traits, however, are continuously distributed; that is, there are not a few distinct types. Examples include virtually any measure of size of most traits of any organism, such as weight, height, length, or width. These traits are called quantitative because they need to be measured; they will be discussed further in Chapters 4–6. The three types of questions and corresponding statistical analyses are:

1. Is there a difference between two or more discrete groups in some continuously variable character? For example, are there differences between male beetles in the mass of the offspring they father? Are there differences in flower shape between several populations of plants when grown in a common environment? Statistical tests used for these kinds of questions include *t*-tests and analysis of variance (ANOVA); we will discuss these techniques in Chapter 4.
2. Is there a relationship between two continuous variables? Examples of these kinds of questions are: Is the tarsus length of a bird related to how many offspring it produces? Are the leg lengths of parent and offspring frogs correlated? These types of questions are often addressed using the statistical tools of correlation and regression (see Chapters 4 and 5).
3. Does a pattern in a population deviate from that expected from a given model? This is often called a goodness-of-fit test. An example of this is a test of whether genotypic frequencies in a real population deviate from Hardy-Weinberg expectations. A common way to do this is with a chi-square test, described in the main text below.

All statistical methods produce a test statistic. Typically, the larger the test statistic, the more confidence we have that the difference or relationship we are testing was not caused merely by a chance occurrence in our sample, but rather reflects a real pattern in the population. This level of confidence is quantified by the *P*-value, which is the probability that this difference or relationship could

Box 2.2 continued

have been produced by chance. Therefore, the smaller the *P*-value is, the more confident we are that we have a real difference or relationship.

At what *P*-value do we conclude that a pattern is real? By convention, tests with *P*-values less than 0.05 are judged statistically significant, meaning we have confidence that they are real and not just due to chance. However, a *P*-value of 0.05 means we have a 5% chance of being wrong when we conclude there is a real pattern. Also, there is nothing magic about 0.05, and in reality we have almost the same confidence in a result with a *P*-value of 0.06 as one with a *P*-value of 0.04. The bottom line is that the lower the *P*-value is, the more confidence we have that the result reflects a real difference or relationship. The larger the sample size, the lower the *P*-value will be on average for the same magnitude of pattern, because we have higher confidence when we have sampled more of the population.

(Calculating the totals for the observed and expected numbers is a useful crosscheck.) Thus, the fit to HWE in the observed is close but not exact. The *P*-value from the chi-square test will tell us how likely it is that this amount of deviation from HWE could be due to chance alone. In comparisons of this type, the value of the chi-square test statistic is calculated as:

$$\chi^2 = \sum \frac{(obs - exp)^2}{exp} \qquad 2.6$$

where the summation sign Σ means summation over all classes of data, in this case all three genotypes. The resulting value of

$$\chi^2 = \frac{(57-54.4)^2}{54.4} + \frac{(53-58.1)^2}{58.1} + \frac{(18-15.5)^2}{15.5} = 0.98$$

is the test statistic.

Associated with any χ^2 value is a second number called the **degrees of freedom (*d.f.*).** In general, the number of degrees of freedom associated with a χ^2 equals the number of classes of data (in this case, 3 genotypes) minus 1 (because the observed and expected totals must be equal), minus the number of parameters estimated from the data (in this case, 1, because the parameter p was estimated from the data). Thus, the number of degrees of freedom for our chi-square value is 3 – 1 – 1 = 1. (Note: A degree of freedom is not deducted for estimating q because of the relation $q = 1 - p$; that is, once p has been estimated, the estimate of q is automatically fixed, so we deduct just the one degree of freedom corresponding to p.)

The actual assessment of goodness of fit is determined from Figure 2.11. To use the chart, find the value of χ^2 along the horizontal axis; then move

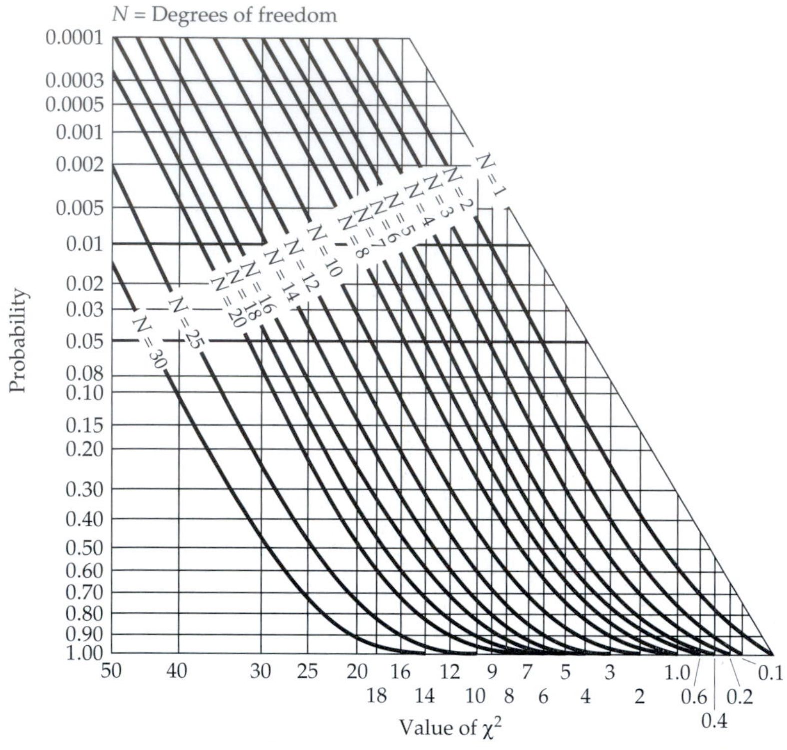

Figure 2.11 Graph of χ^2. To use the graph, find the value of χ^2 along the x (horizontal) axis, then move straight up until you intersect the diagonal line corresponding to the appropriate number of degrees of freedom. The P-value is read from the y (vertical) axis at that intersection point. (Courtesy of James F. Crow.)

vertically from this value until the proper degrees-of-freedom line is intersected; then move horizontally from this point of intersection to the vertical axis and read the corresponding probability value. In the present case, with $\chi^2 = 0.98$ and one degree of freedom, the corresponding probability value is about $P = 0.36$. This P-value represents the probability that chance alone could produce a deviation between the observed and expected values at least as great as the deviation actually obtained. Thus, if the probability is large, it means that chance alone could account for the deviation, and it strengthens our confidence in the validity of the model used to obtain the expectations—in this case HWE. On the other hand, if the probability associated with the χ^2 is small, it means that chance alone is not likely to lead to a deviation as large as actually obtained, and it undermines our confidence in the validity of the model. Because the probability in the *Daphnia* example is 0.36, which means there is a 36% probability that the deviations from HWE could have been due to chance, we have no reason to reject the hypothesis that the genotype frequencies are in HWE for this gene locus.

EXERCISE: At the *PGI* locus in the Ojibway Pond population of *Daphnia*, Spitze found two alleles, *S* and *S–*, and the number of individuals with each genotype was 42 *SS*, 48 *SS–*, and 38 *S–S–*. Test these for deviation from HWE.

ANSWER: Allele frequencies (from Equations 2.2 and 2.4) are p = 0.516 and q = 0.484 (which sum to 1). Expected genotypic frequencies are p^2 = 0.266, $2pq$ = 0.500, and q^2 = 0.234 (which sum to 1). Multiplying by the sample size gives the expected numbers: 34.0, 64.0, and 30.0. The χ^2 test statistic calculated from the observed and expected numbers using Equation 2.6 is 8.0. With one *d.f.*, the *P*-value for this chi-square is 0.005, which gives us confidence that there is a real deviation from HWE. There is a deficit of heterozygotes, and an equal excess of homozygotes; inbreeding, as well as other factors, can cause this pattern. (See the section on Nonrandom Mating later in this chapter.)

Recessive alleles hidden in heterozygotes

HWE sometimes helps solve the dilemma that arises when studying dominant markers, such as many visible polymorphisms, RAPDs, and AFLPs. The dilemma is that dominant homozygous and heterozygous genotypes cannot be distinguished, and so the genotypic and allele frequencies cannot be estimated directly. The solution is that, if one is willing to assume HWE, then the allele frequencies can be estimated anyway. The trick is to use the observed frequency of *recessive homozygotes* to estimate the q^2 term of the HWE. That is, if Q is the frequency of homozygous recessive genotypes in a sample, then q is estimated as the square root of Q because the expected value of q^2 equals Q. Since HWE is robust to violations of many of the assumptions, this estimate can work in real populations even though some of the assumptions are likely to be violated. However, many loci in natural populations are not in HWE, and there is no way to know this with a dominant marker.

An example of using HWE to estimate allele frequencies with a dominant marker is in **industrial melanism,** which may be the best-known example of evolution in action and is a classic example in ecological genetics as well (Majerus 1998). The term *industrial melanism* refers to an observed increase in the frequency of dark-pigmented (melanic) morphs of a number of insect species, resulting from coal-burning during the industrial revolution that blackened insect resting places with soot. In most cases, the melanic color pattern is due to a single dominant allele. In one study of a heavily polluted area near Birmingham, England, Kettlewell (1956) observed a frequency of 87% melanic peppered moths (*Biston betularia*). Therefore, the frequency of recessive homozygotes is 0.13, and assuming HWE, the frequency q of the recessive allele is estimated as the square root of 0.13, or 0.36.

EXERCISE: Assuming HWE, calculate the frequency p of the dominant allele and the frequencies of the three genotypes.

ANSWER: $p = 1 - q = 0.64$, and the frequencies of the genotypes are $P = 0.41$, $H = 0.46$, and $Q = 0.13$ (p^2, $2pq$, q^2). It does not make sense to try to test the fit of these genotypic frequencies to Hardy-Weinberg frequencies, because they were calculated using the Hardy-Weinberg formula, so the fit must be perfect.

Nonrandom Mating

For the rest of this Chapter and the beginning of the next, we will be relaxing one assumption of Hardy-Weinberg at a time, while keeping all the other assumptions in place. This will allow us to examine the genetic results of each process in isolation. In Chapter 3 we will relax more than one assumption simultaneously, to see how the different genetic processes interact with each other. In the next two sections we discuss two types of nonrandom mating, assortative mating and inbreeding. Except for negative assortative mating (discussed in the next section), nonrandom mating does not change allele frequencies, but rather changes how the alleles are distributed into diploid genotypes.

Assortative mating

Assortative mating means nonrandom mating based on some phenotypic trait, and can be either positive or negative. In **positive assortative mating** (often simply called assortative mating) mates are phenotypically more similar to each other than expected by chance. Positive assortative mating is very common. You may have noticed positive assortative mating for height in humans—tall people often date and marry other tall people, and shorter people do the same. Humans also date and marry assortatively for a number of measures of socioeconomic status, including the number of rooms in the couples' parents' houses! Positive assortative mating for size occurs in a wide variety of animals, especially arthropods (Crespi 1989). Plants mate assortatively for flowering time, because early flowering plants are often no longer flowering when late flowering plants are in bloom (Waser 1993).

The genetic effect of positive assortative mating is to increase **homozygosity** (the frequency of homozygous genotypes) and decrease **heterozygosity.** Think of a locus without dominance in a moth, where *AA* moths tend to be larger than *Aa* moths, which in turn are larger than *aa* moths. With positive assortative mating, *AA* moths will tend to mate with other *AA*, and *aa* with *aa*; all these matings will produce only homozygous offspring. Even matings between *Aa* moths will produce half homozygous offspring. There will be very few of the *AA* × *aa* matings that produce all heterozygotes, so the frequency of heterozygotes decreases.

The genetic effects of assortative mating are only at those loci that affect the phenotypic trait by which the organisms are mating assortatively (and at loci linked to these). Thus, positive assortative mating for flowering time will increase homozygosity and decrease heterozygosity only at gene loci that affect flowering time. Loci that do not affect the mating trait are not affected, because mating is random with respect to genotypes at those loci. This fact distinguishes assortative mating from inbreeding, which affects all loci equally (discussed in the next section). Like inbreeding, positive assortative mating changes genotypic frequencies without changing allele frequencies.

Negative assortative mating, also called disassortative mating, occurs when mates are phenotypically less similar to each other than expected by chance. A very common and familiar example of disassortative mating is the two sexes—males tend to mate with females and vice versa. Otherwise, disassortative mating is far less common than positive assortative mating, especially in animals. An important and well-studied example of disassortative mating in plants is self-incompatibility. A number of large plant families have specific self-incompatibility loci. Pollen that shares alleles at this locus in common with the plant it lands on cannot successfully fertilize that plant (Barrett 1988). This prevents self-fertilization and sometimes prevents matings between close relatives, since close relatives often share identical alleles.

Not surprisingly, the genetic effect of negative assortative mating is the opposite of the effect of positive assortative mating—it increases the frequency of heterozygotes and decreases homozygosity. This is caused by *AA* and *aa* homozygotes mating with each other more often than expected by chance because they are phenotypically dissimilar. Note that, as with positive assortative mating, the genetic effect is only at loci that affect the phenotypic trait used in mating as well as loci linked to these. Unlike all the other factors we discuss in this chapter, negative assortative mating can alter allele frequencies through a process called negative frequency-dependent selection, which we will discuss further in Chapter 3.

Inbreeding

There are many different ways in which the word *inbreeding* is used and defined (see Chapter 3; Keller and Waller 2002; Templeton and Read 1994).

In this book we define **inbreeding** as mating between individuals in a population that are more closely related than expected by random chance. *The importance of inbreeding is due to its genetic effect, which is to increase homozygosity at all loci.* The reason that inbreeding affects all loci equally is that related individuals are genetically similar by common ancestry, and therefore they are more likely than unrelated individuals to share alleles throughout the genome. This contrasts with assortative mating, in which mating is between individuals that are *phenotypically* similar, and thus tend to share alleles only at loci affecting the phenotypic traits used in mating.

To understand how inbreeding increases homozygosity, it is simplest to examine the most extreme case of inbreeding, **self-fertilization.** Self-fertilization (or **selfing**) can occur only in hermaphrodites. Selfing is common in plants and in internal parasites such as tapeworms. In selfing, male and female gametes are produced normally through meiosis; the difference is that pollen or sperm fertilizes ovules or eggs on the same individual. Because there is recombination, selfed progeny are not genetically identical to the parent. If we start with a population in HWE with equal allele frequencies at a given locus ($p = q = 0.5$), then the genotypic frequencies are 0.25 for the two homozygotes and 0.5 heterozygotes (Figure 2.12). With self-fertilization, homozygotes can only produce homozygous offspring, whereas heterozygotes produce 25% of each homozygote and 50% heterozygotes. (Self-fertilization in a heterozygote is the same as a mating between two heterozygotes.) The arrows in Figure 2.12 show why the fre-

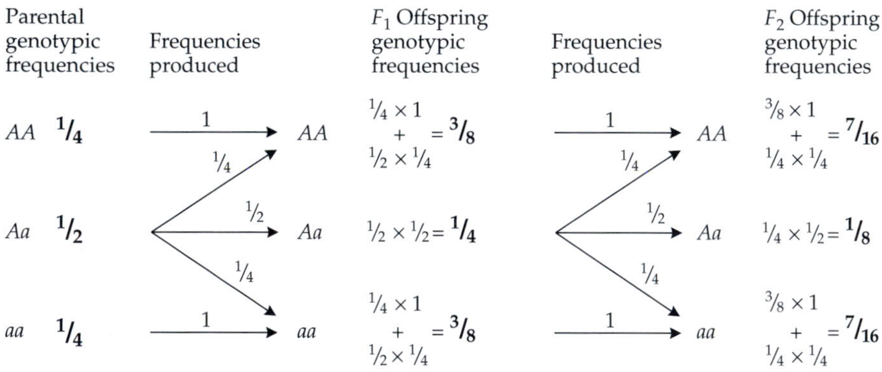

Figure 2.12 Genetic effects of inbreeding illustrated with self-fertilization. To calculate the offspring genotypic frequencies, multiply the parental frequency by the frequency of each offspring genotype produced by that parental genotype (above each arrow). Both homozygous and heterozygous parents produce homozygous offspring, so the frequencies need to be summed to get the total frequencies of homozygous offspring.

quency of homozygotes increases with self-fertilization: Whereas all offspring of homozygous parents are homozygous, only half of the offspring of heterozygotes are heterozygous, with the other half divided equally between the two homozygotes. A similar process occurs, but at a slower rate, with less severe forms of inbreeding (e.g., sibling or first-cousin mating); that is, the frequency of homozygotes increases at the expense of the frequency of heterozygotes.

Inbreeding by itself *does not change allele frequencies*. One way to explain this is to think about alleles as being shuffled into homozygotes without changing the frequencies of these alleles. To check this principle, use Equations 2.3 and 2.5 to calculate the allele frequencies in the offspring generation. The increase in homozygosity and lack of change in allele frequencies with inbreeding occur *regardless of what the allele frequencies are*—these results do not depend on the allele frequencies being 0.5 as in our example. This fact is illustrated in Figure 2.13, which shows the genotypic frequencies at all allele frequencies after one generation of selfing, as well as those under HWE from Figure 2.10.

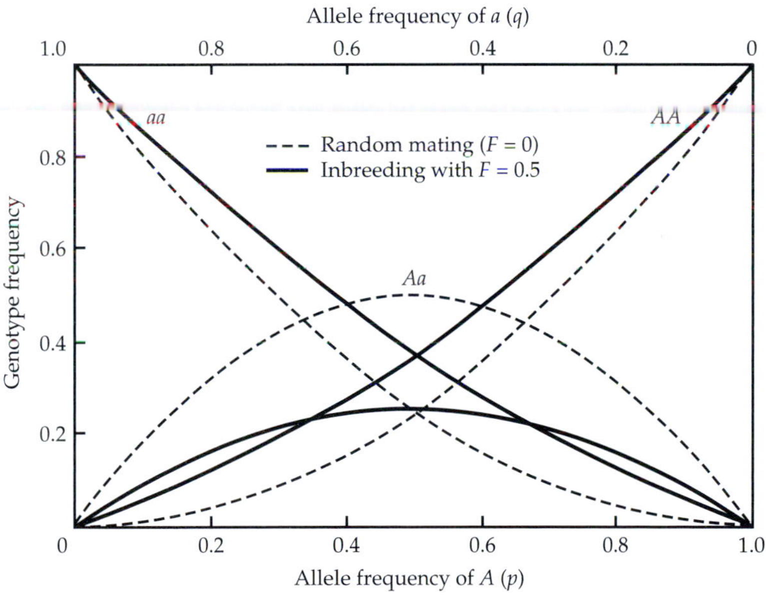

Figure 2.13 Changes in genotypic frequencies across all allele frequencies due to inbreeding. The dashed lines are genotypic frequencies under HWE from Figure 2.10, and the solid lines are those after one generation of selfing ($F = 0.5$). Note that with inbreeding the curves for homozygotes move up and the curve for heterozygotes moves down.

EXERCISE: Work through the effects of one generation of selfing starting from a Hardy-Weinberg population in which $p = 0.75$ and $q = 0.25$. Calculate genotypic and allele frequencies in the offspring.

ANSWER: Genotypic frequencies in the parental generation are 0.562, 0.375, and 0.062. After one generation of selfing, the heterozygote frequency is halved, with one quarter of the heterozygotes being converted to each homozygote. Therefore, the new genotypic frequencies are 0.656, 0.188, and 0.156, and the allele frequencies are unchanged (by Equations 2.3 and 2.5).

We can use the predictable increase in homozygosity and decrease in heterozygosity resulting from inbreeding to measure the effects of inbreeding. In the following equation, the **inbreeding coefficient, *F*,** measures the amount of inbreeding by comparing the frequency of heterozygotes in the population to the frequency expected under random mating:

$$F = \frac{H_0 - H}{H_0} \qquad \textbf{2.7}$$

As before, H equals the frequency of heterozygotes in the population, which can be measured using a variety of genetic markers, and H_0 is the expected frequency under HWE, that is, $2pq$. So the inbreeding coefficient equals the

EXERCISE: What is F after one generation of self-fertilization with $p = q = 0.5$ and with $p = 0.75$ and $q = 0.25$ (our two examples above)?

ANSWER:

$$F = \frac{0.5 - 0.25}{0.5} = 0.5$$

for $p = q = 0.5$, and

$$F = \frac{0.375 - 0.188}{0.375} = 0.5$$

for $p = 0.75$ and $q = 0.25$;

It is the same for any starting allele frequency, since $H = H_0/2$ after one generation of selfing regardless of allele frequency.

fractional reduction in heterozygosity due to inbreeding, relative to a random mating population with the same allele frequencies.

F can range between –1 and 1. When the population is in HWE, H_0 equals *H* and $F = 0$; that is, no inbreeding. If inbreeding continues, the population can become entirely homozygous, that is, $H = 0$ so $F = 1$. This happens very rapidly with selfing (Figure 2.14), and less rapidly with less severe forms of inbreeding, but note that with half-sibling mating (mating between two individuals that share only one of their parents in common) the inbreeding coefficient is more than 0.8 after only 20 generations. Figure 2.14 shows that there is no one value of *F* for a given mating system, because *F* also depends on the number of generations over which inbreeding has taken place. When *F* is negative, it means that the population is more outbred than expected by chance, that is, active avoidance of mating with relatives.

EXERCISE: Repeat the selfing exercise on the bottom of page 40 for two more generations, and calculate *F*, genotypic, and allele frequencies for both offspring generations.

Figure 2.14 shows that the inbreeding coefficient increases more rapidly with more closely related mates. The degree of relatedness is quantified as the coefficient of relatedness (r_{IJ}), which is the proportion of genes shared between two individuals *I* and *J* due to common descent (often referred to as *identical by descent* or IBD; Maynard Smith 1998). For example, r_{IJ} for a parent and each of their offspring is 0.5 in a sexual species, because half of each

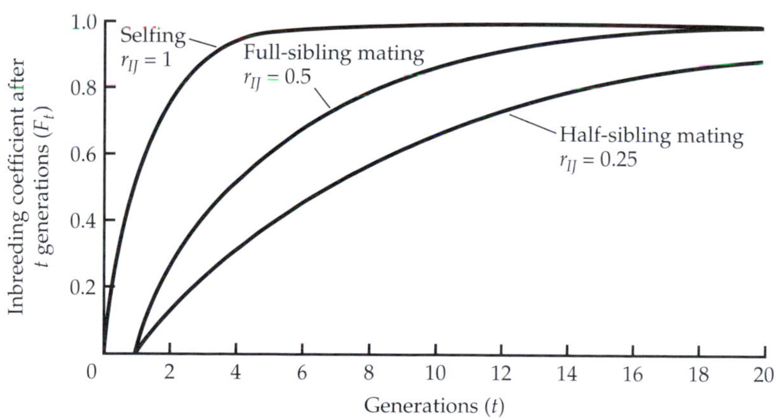

Figure 2.14 Theoretical increase in the inbreeding coefficient *F* for regular systems of mating: selfing, full-sibling mating, and half-sibling mating.

TABLE 2.3 *Values of the coefficient of relatedness (r_{IJ}) for different relationships*

Relationship	r_{IJ}
Self and monozygotic (identical) twins	1.0
Parents and offspring	0.5
Grandparents and offspring	0.25
Full siblings	0.5
Half siblings	0.25
Aunt or uncle to niece or nephew	0.25
First cousins	0.125

individual's genes come from each parent. The value of r_{IJ} for full-siblings is also 0.5, but only on average, because each sibling received half of their genes from each of the same pair of parents, but the haploid set of genes that is put in each parental gamete is a random sample of half the parental genome due to recombination. More distant coefficients of relatedness can be calculated as products. For example, because parents and offspring share half their genes, the r_{IJ} for grandparents and offspring is the product of the two coefficients ($0.5 \times 0.5 = 0.25$). Table 2.3 gives values of the coefficient of relatedness for a variety of different relationships.

Dole and Ritland (1993) estimated F using seven allozyme loci in natural populations of two species of monkey flower. *Mimulus guttatus* is primarily **outcrossing** (outbreeding), while *M. platycalyx* is primarily selfing (Figure 2.15). As expected, F was much higher in the latter, averaging 0.54 compared to an average F of 0.17 for *M. guttatus.*

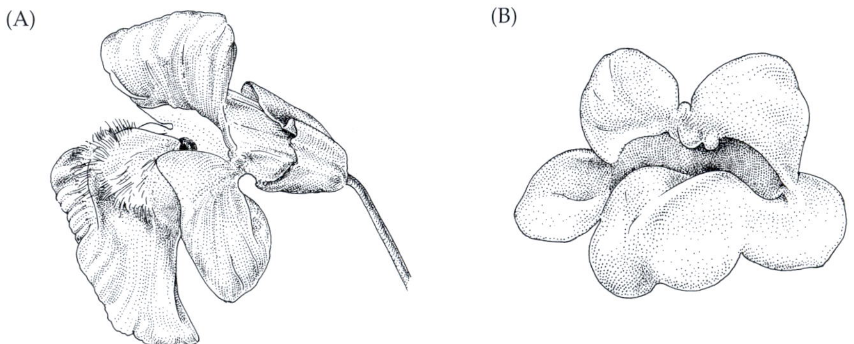

Figure 2.15 Flowers of (A) *Mimulus guttatus* and (B) *M. platycalyx.*

Inbreeding can also be estimated and studied directly using pedigrees. This approach is very useful if detailed pedigrees are known for the study organism, but this information is rarely available in natural populations, especially since paternity can often be uncertain. Pedigree analysis is described in detail in the population genetics books listed in the Suggested Readings at the end of this chapter.

The most important effect of the genome-wide increase in homozygosity caused by inbreeding is called **inbreeding depression,** which refers to a decrease in fitness that often accompanies inbreeding. Inbreeding and inbreeding depression are important in a number of contexts, and therefore will be discussed further in later chapters.

PROBLEMS

Whenever a statistical test is needed, report the test statistic (e.g., value of chi-square), the *P*-value and explain what the statistical result means biologically. The following electrophoresis gel diagrams are used in Problems 2.3 and 2.5. Above each lane is the number of individuals in the population sample with the banding pattern illustrated.

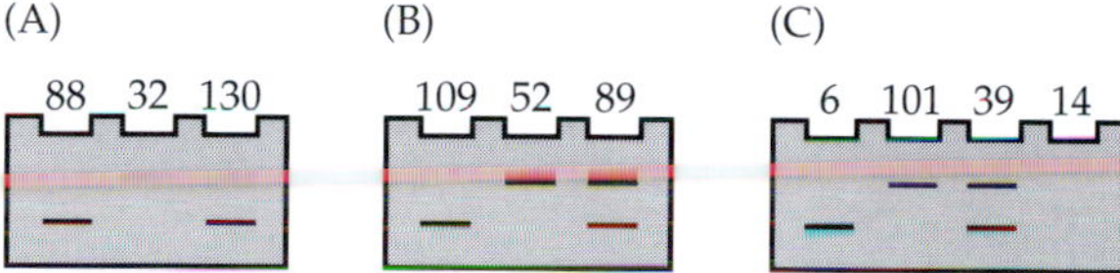

2.1 Spitze (1993) reported the following numbers of genotypes at the *PGI* locus in the *Daphnia* population in Nothing Pond:

SS	11
SS-	55
S-S-	61

a. What are the observed genotype and allele frequencies?

b. Given the observed allele frequencies, what are the genotypic frequencies expected under Hardy-Weinberg? Using a chi-square test, how well do the observed genotypic frequencies agree with the Hardy-Weinberg expectations?

2.2 Calculate *F* for the *Daphnia PGI* locus in the Ojibway Pond (page 35; note that there are data for *PGM* in the text as well) and the Nothing Pond (Problem 2.1 above) populations. What is a possible biological interpretation of these data? Explain your reasoning.

2.3 Gels A and B show banding patterns from RFLP markers in two related plant species.

a. Estimate allele and genotypic frequencies and test for HWE in each species.

b. Estimate F for each species.

c. What is a possible biological interpretation of the data from parts a and b? Explain your reasoning.

2.4 RAPD and AFLP markers have only two genotypes—presence or absence of a band. When a band is present, the genotype can be homozygous or heterozygous. Therefore, only the band-absent (null) genotype can be scored definitively. In a study of a selfing plant (*Medicago truncatula*), Bonnin et al. (1996) reported band-presence at a frequency of 0.59 at their RAPD locus B6-600 in a population from Aude, France.

a. Assuming HWE, what are the frequencies of the two alleles and the three genotypes?

b. Because this is a selfing plant, assuming HWE is not valid. Is your estimate of the frequency of the band-present allele an over- or underestimate? Explain your reasoning.

2.5 Gel C shows the banding patterns from two AFLP markers (the upper and lower sets of bands).

a. Estimate the frequency q of the null allele of each of the two AFLP markers assuming HWE.

b. Estimate the percentage of *band-present* individuals (not the overall frequencies) that are heterozygous at each of the two markers. What biological principle does the difference between these two percentages illustrate?

SUGGESTED READINGS

Avise, J. C. 1994. *Molecular Markers, Natural History, and Evolution*. Chapman and Hall, New York. A review of many ways that molecular markers can be used to study natural populations.

Hartl, D. L. 2000. *A Primer of Population Genetics*, 3rd Edition. Sinauer Associates, Sunderland, MA. A more mathematical introduction to the field.

Hartl, D. L., and A. G. Clark. 1997. *Principles of Population Genetics*, 3rd Edition. Sinauer Associates, Sunderland, MA. A detailed and comprehensive treatment of population genetics.

CHAPTER REFERENCES

Barrett, S. C. H. 1988. The evolution, maintenance, and loss of self-incompatibility systems. Pp. 98–124 *in* J. L. Doust, and L. L. Doust, eds. *Plant Reproductive Ecology*. Oxford University Press, New York.

Bonnin, I., T. Huguet, M. Gherardi, J.-M. Prosperi, and I. Olivieri. 1996. High level of polymorphism and spatial structure in a selfing plant species, *Medicago truncatula* (Leguminosae), shown using RAPD markers. Am. J. of Botany 83:843–855.

Crespi, B. J. 1989. Causes of assortative mating in arthropods. Anim. Behav. 38:980–1000.

Dole, J., and K. Ritland. 1993. Inbreeding depression in two *Mimulus* taxa measured by multigenerational changes in the inbreeding coefficient. Evolution 47:361–373.

Hardy, G. 1908. Mendelian proportions in a mixed population. Science 28: 49–50.

Hillis, D. M., C. Moritz, and B. K. Mable. 1996. *Molecular Systematics*, 2nd Edition. Sinauer Associates, Sunderland, MA.

Keller, L. F., and D. M. Waller. 2002. Inbreeding effects in wild populations. Trends in Ecology and Evolution 17:230–241.

Kettlewell, H. B. D. 1956. Further selection experiments on industrial melanism in the *Lepidoptera*. Heredity 10:287–301.

Lewontin, R. C., and J. L. Hubby. 1966. A molecular approach to the study of genic heterozygosity in natural populations. II. Amount of variation and degree of heterozygosity in natural populations of *Drosophila pseudoobscura*. Genetics 54: 595–609.

Majerus, M. E. N. 1998. *Melanism: Evolution in Action*. Oxford University Press, New York.

Maynard Smith, J. 1998. *Evolutionary Genetics*. Oxford University Press, New York.

Mueller, U. G., and L. L. Wolfenbarger. 1999. AFLP genotyping and fingerprinting. Trends in Ecology and Evolution 14:389–394.

Murphy, R. W., J. W. Sites, Jr., D. G. Buth, and C. H. Haufler. 1996. Proteins: Isozyme electrophoresis. Pp. 45–126 *in* D. M. Hillis, C. Moritz, and B. K. Mable, eds. *Molecular Systematics*, 2nd Edition. Sinauer Associates, Sunderland, MA.

Parker, P. G., A. A. Snow, M. D. Schug, G. C. Booton, and P. A. Fuerst. 1998. What molecules can tell us about populations: Choosing and using a molecular marker. Ecology 79:361–382.

Spitze, K. 1993. Population structure in *Daphnia obtusa*—quantitative genetic and allozymic variation. Genetics 135:367–374.

Sunnucks, P. 2000. Efficient genetic markers for population biology. Trends in Ecology and Evolution 15:199–203.

Templeton, A. R., and B. Read. 1994. Inbreeding: One word, several meanings, much confusion. Pp. 91–105 *in* V. Loeschske, J. Tomiuk, and S. K. Jain, eds. *Conservation Genetics*. Birkhauser Verlag, Basel, Switzerland.

Waser, N. M. 1993. Population structure, optimal outbreeding, and assortative mating in angiosperms. Pp. 173–199 *in* N. Thornhill, ed. *The Natural History of Inbreeding and Outbreeding*. University of Chicago Press, Chicago.

Weinberg, W. 1908. On the demonstration of heredity in man. Naturkunde in Wurttemberg Stuttgart. 64: 368–382. [Original in German. Translated in S. H. Boyer IV (ed.), 1963, "Papers on Human Genetics," Prentice Hall, Englewood Cliffs, NJ.]

Whitlock, M. C. 1992. Nonequilibrium population structure in forked fungus beetles: Extinction, colonization, and the genetic variance among populations. Am. Nat. 139:952–970.

Williams, J. G. K., A. R. Kubelik, K. J. Livak, J. A. Rafalski, and S. V. Tingey. 1990. DNA polymorphisms amplified by arbitrary primers are useful as genetic markers. Nucleic Acids Research 18:6531–6535.

3

Population genetics II: Changes in allele frequency

In this chapter we will examine the four major factors that change allele frequencies in more detail—mutation, gene flow (migration), genetic drift, and natural selection. As in the section on nonrandom mating in Chapter 2, we will begin by relaxing just one Hardy-Weinberg assumption at a time to see the effects of each of the four factors in isolation. At the end of the chapter, we will relax two or three at a time to see how the factors interact with one another.

Mutation

Mutation is important because it is the ultimate source of all genetic variation. The simplest type of mutation is a **base substitution,** when one nucleotide base is substituted for another in a DNA sequence (e.g., A substituted for G). Many base substitutions have little or no phenotypic effect. For example, a base substitution in the third position of a codon often does not change the amino acid specified by that codon (see Table 1.1).

Insertions and **deletions** occur when a portion of DNA is added or removed from a sequence. These may have large phenotypic effects, particularly when the number of bases inserted or deleted is not divisible by three; this can result in a **frame shift** in which all downstream codons in the gene locus are read incorrectly. An important cause of insertions and deletions are moveable fragments of DNA called **transposable elements** or **transposons.** A final class of mutations are **chromosomal rearrangements,** including

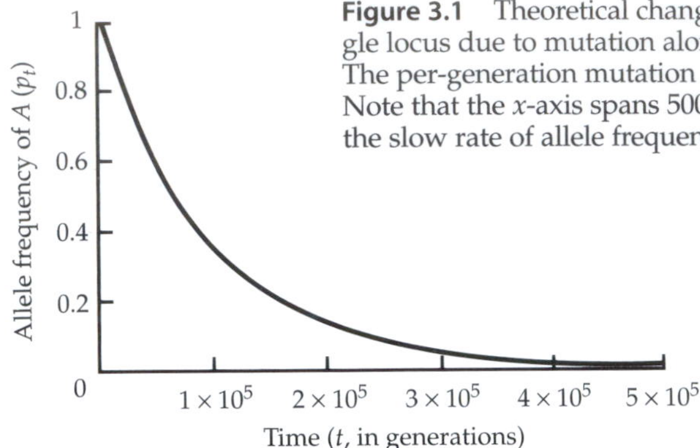

Figure 3.1 Theoretical change in allele frequency at a single locus due to mutation alone, based on Equation 3.1. The per-generation mutation rate, μ, is 10^{-5} in this example. Note that the *x*-axis spans 500,000 generations, illustrating the slow rate of allele frequency change by mutation.

inversions and **translocations**. These may or may not have major phenotypic effects, depending on whether genes or gene complexes are disrupted by the rearrangement.

Unlike inbreeding, mutation can change allele frequencies, but these changes are usually very slow. The single-locus mutation rate per generation is denoted by the symbol μ, which for most whole gene loci is 10^{-4} to 10^{-6}; that is, one in ten thousand to one in a million. This represents the probability that a mutation will occur somewhere in a given gene locus in any generation, so clearly individual mutations are rare events. The theoretical change in allele frequency at a locus due to mutation alone can be calculated as:

$$p_t = p_0(1 - \mu)^t \quad \textbf{3.1}$$

where p_t is the frequency of the **wildtype** allele (the most common allele in nature) at generation t, p_0 is the initial frequency, and $1 - \mu$ is the proportion of wildtype alleles that do not mutate each generation. Figure 3.1 is a graph of this equation for $\mu = 10^{-5}$. Note the scale on the *x*-axis of Figure 3.1; this shows clearly that the effect of mutation on allele frequency is exceedingly slow. Mutation rates are not constant across loci, however; there are "hot spots" of higher mutation rates in the genomes of many, perhaps all, organisms, and mutation rates vary across species (Graur and Li 2000).

Migration

Migration is defined as the movement of individuals between populations, whereas **gene flow** is the movement of genes between populations. Since individuals carry genes, these terms are used interchangeably by most population geneticists. However, if an immigrant does not successfully reproduce in its new habitat, then migration has occurred without the new genes being established in the population.

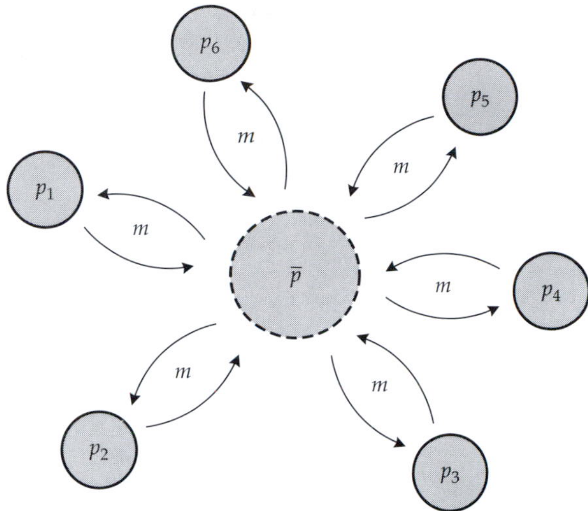

Figure 3.2 Schematic of the island model of gene flow. Each solid circle represents a subpopulation with allele frequency p_i, which contributes a proportion m of its individuals to the migrant pool (dashed circle). The allele frequency of the migrant pool is the average p_i over all subpopulations, $\bar{p}$.

The simplest model of gene flow is the **island model** (Figure 3.2). In this model there are a number of subpopulations, each with its own allele frequencies p_i and q_i (i = 1 for subpopulation 1, 2 for subpopulation 2, etc.), which when averaged give the allele frequencies of the overall metapopulation $\bar{p}$ and $\bar{q}$. (By convention, a symbol with a bar over it denotes the mean or average of that quantity.) In the simplest version of the island model, a proportion, m, of individuals leave each of their subpopulations and form a migrant pool. The migrants are chosen randomly with respect to their genotypes, so that the allele frequencies of the migrant pool are the average allele frequencies in the metapopulation, $\bar{p}$ and $\bar{q}$. The migrants then immigrate into the subpopulations, again randomly with respect to their genotypes, so immigrants into each subpopulation also have the metapopulation average allele frequencies. The new allele frequency at generation t in a subpopulation is then given by:

$$p_t = p_{t-1}(1-m) + \bar{p}m \tag{3.2}$$

The first term, $p_{t-1}(1 - m)$, represents the alleles from the individuals that did not migrate, because p_{t-1} is the allele frequency in the previous generation before migration and $1 - m$ is the proportion of individuals that did not migrate. The second term, $\bar{p}m$, represents the new alleles from the immigrants.

Some simple algebra gives the equation for the rate of change in allele frequency in a subpopulation:

$$\Delta p = p_t - p_{t-1} = m(\bar{p} - p_{t-1}) \qquad \textbf{3.3}$$

The delta symbol (Δ) denotes amount of change in a variable, and in population genetics it is often used to measure change across a generation, as it is here. This equation shows that the rate of allele frequency change depends on m, the migration rate, and $\bar{p} - p_{t-1}$, the degree of difference in allele frequency between the immigrants and the subpopulation of interest in the previous generation. Note that if the allele frequency in the subpopulation before migration is identical to the allele frequency of the migrant pool (i.e., $\bar{p} = p_{t-1}$), then $\Delta p = 0$ and migration causes no change in allele frequency in that subpopulation.

EXERCISE: Derive Equation 3.3 from Equation 3.2.

ANSWER:

$$p_t = p_{t-1}(1-m) + \bar{p}m$$

$$p_t = p_{t-1} - p_{t-1}m + \bar{p}m$$

$$p_t - p_{t-1} = \bar{p}m - p_{t-1}m$$

$$\Delta p = p_t - p_{t-1} = m(\bar{p} - p_{t-1})$$

Under this simple model, the following equation can be used to determine allele frequencies in the subpopulation of interest after any number of generations (t) in the future:

$$p_t = \bar{p} + (p_0 - \bar{p})(1-m)^t \qquad \textbf{3.4}$$

As an example, we can use this equation to model what happens when migration occurs among subpopulations that each start out at a different allele frequency (Figure 3.3). The subpopulations converge rapidly to $\bar{p}$, the average allele frequency in the metapopulation. The equilibrium allele frequency is determined by the initial conditions, that is, $\bar{p}$. Remember that we are assuming no mutation, genetic drift, or selection, so that $\bar{p}$ does not change; migration changes allele frequencies in individual subpopulations but not in the metapopulation as a whole. Again note the scale on the x-axis, which shows that migration can change allele frequencies in subpopulations very quickly, much more quickly than mutation does (compare with Figure 3.1).

Recall that in Chapter 2 we learned that a prime concern in population genetics and evolution is genetic variation among populations or **population differentiation,** which can be defined as differences in allele frequencies

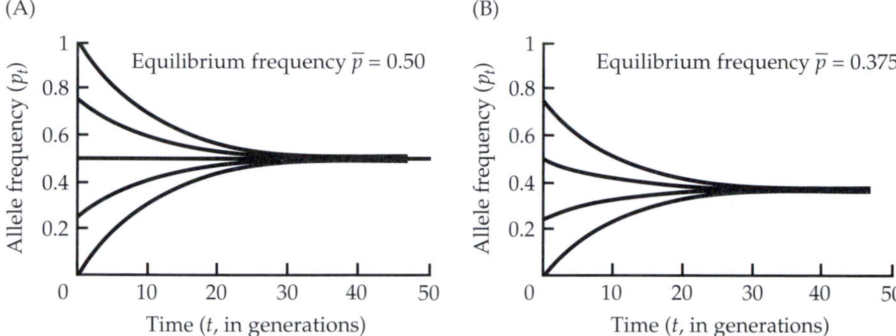

Figure 3.3 Theoretical changes in allele frequency over time with migration using Equation 3.4. The lines in each graph represent changes in allele frequencies in subpopulations that each started at a different initial allele frequency. The migration rate, m, is 0.1. (A) Five subpopulations; the average frequency of both alleles ($\bar{p}$ and $\bar{q}$) is 0.5. (B) The results when the subpopulation starting at $p = 1$ is removed. The four remaining subpopulations converge to the new average allele frequencies ($\bar{p} = 0.375$ and $\bar{q} = 0.625$).

among populations. Figure 3.3 shows clearly that *migration is a potent force in reducing the level of differentiation*. After less than 40 generations with a migration rate (m) of 0.1, the subpopulations that were highly differentiated at time zero converge to the same allele frequency ($\bar{p}$). Note that the subpopulations will converge to any $\bar{p}$; in the examples in Figure 3.3 they happen to be 0.5 and 0.375, but if the initial subpopulations had a different average allele frequency, then they would converge on that $\bar{p}$.

We can extend our understanding of the effects of migration on genetic variation by comparing Figure 3.3 with Figure 2.10. Note from Figure 2.10 that within-population variation (as measured by heterozygosity, H) is at a maximum at intermediate allele frequencies ($p = q = 0.5$). In the two populations in Figure 3.3A that start out fixed for alternate alleles ($p = 0$ and $p = 1$) there is zero within-population variation and maximum among-population variation. After 50 generations both populations are at $p = 0.5$, so there is maximum within-population variation and no among-population variation (differentiation). Therefore, migration converts among-population variation into within-population variation, decreasing the former and increasing the latter. In fact, as we will see below, the most common method of measuring gene flow is by looking at its genetic effects, that is, by measuring differentiation. For example, Caccone (1985) measured gene flow among subpopulations of 11 species of cave-dwelling arthropods. She found that troglobites (obligate cave-dwellers) had lower levels of gene flow, and thus higher levels of differentiation at allozyme markers, than species that do not always inhabit caves.

Genetic Drift

At the beginning of Chapter 2 we introduced the idea of spatial structuring of species into metapopulations, and metapopulations into smaller subpopulations. To better understand the genetic consequences of this structuring, we will use a simplified model (Figure 3.4). In this model we start with a large panmictic base population with allele frequencies p_0 and q_0 at a two-allele locus. From this base population we randomly create a number of subpopulations of equal size. Allele frequencies in the subpopulations are p_i and q_i, where the subscript i again denotes the subpopulation. The average allele frequencies across all subpopulations are $\overline{p}$ and $\overline{q}$, which equal p_0 and q_0 because the subpopulations were a random sample of the base population. Note the similarities to the island model of migration in Figure 3.2. However, in our discussion of genetic drift we will initially assume no migration among subpopulations, as well as random mating within subpopulations (no inbreeding), no mutation or selection, and the same numbers of individuals in each subpopulation of each generation (no population growth).

Random genetic drift

The term **random genetic drift** (or just **genetic drift**) refers to fluctuations in allele frequency which occur by chance, particularly in small subpopulations, as a result of sampling error. The fundamental reason for genetic drift is illustrated in Figure 3.4. A population of N diploid zygotes arises from a sample of $2N$ gametes chosen from an essentially infinite pool of gametes. For example, if the subpopulation size $N = 10$, then only 20 gametes out of a very large number (all the sperm and eggs produced by the 10 individuals over their lifetimes) are used to create the next generation. With such a small proportion of gametes that are successful, the allele frequencies in these gametes will often deviate from the allele frequencies in the parents by chance alone. This is the same situation as flipping a coin 10 times; if you do this repeatedly you will often get large deviations from five heads and five tails. However, if you flip the coin 100 times repeatedly, the deviations from 50% heads will be much smaller.

The ultimate result of genetic drift is random fixation of alleles and genetic differentiation among subpopulations. An experimental example is shown in Figure 3.5. Forty-eight laboratory subpopulations of the flour beetle *Tribolium castaneum* were set up, 12 each at $N = 10$, 20, 50, and 100. The beetles used to found the populations were heterozygous at a visible polymorphism for body color in which the wild type homozygotes (b^+b^+) are reddish, the mutant homozygotes (bb) are black, and the heterozygotes (bb^+) are an intermediate brown color. After only 20 generations, seven of the 12 subpopulations of 10 had become fixed at the marker locus, whereas three, one, and none of the subpopulations of 20, 50, and 100, respectively, had fixed.

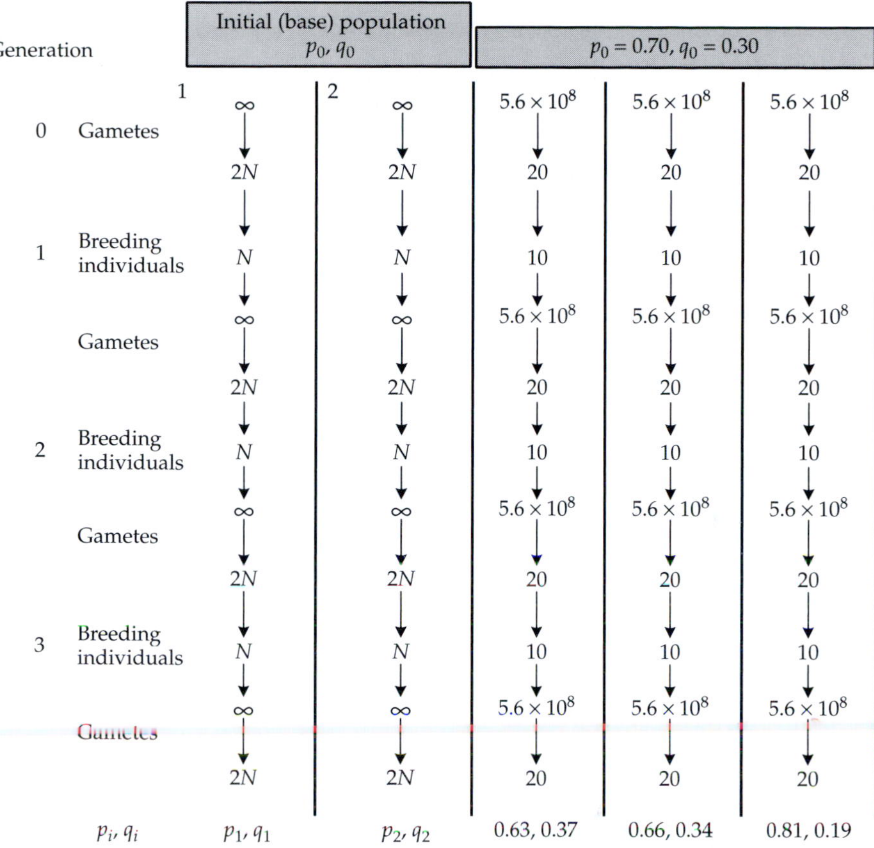

Figure 3.4 Model for analyzing the effects of random genetic drift on allele and genotype frequencies. Each of the subpopulations (vertical columns) is founded from the same initial large population (top). The left two columns show the general model using symbols, and the right three columns give a hypothetical example with numbers and frequencies. Each subpopulation consisting of *N* individuals produces a very large number of gametes, of which 2*N* are chosen at random to form the next generation's subpopulations; random genetic drift results from sampling error during this process. Mutation, migration among the subpopulations, and selection are assumed absent, and *N* remains at the same small number throughout. The gamete numbers in the example are the numbers of gametes expected to be produced by $N = 10$ individuals of wild radish, which on average produce about 700 flowers and 80,000 pollen grains (plus 8 ovules) per flower. After the three generations, each of these hypothetical radish subpopulations has drifted to a different allele frequency shown at the bottom. (Data from Conner et al. 1995; 1996.)

The fixation is due to the large amplitude of allele frequency fluctuation across generations in the small subpopulations. Once the allele frequency goes to 0 or 1, then the subpopulation is stuck there until mutation or migra-

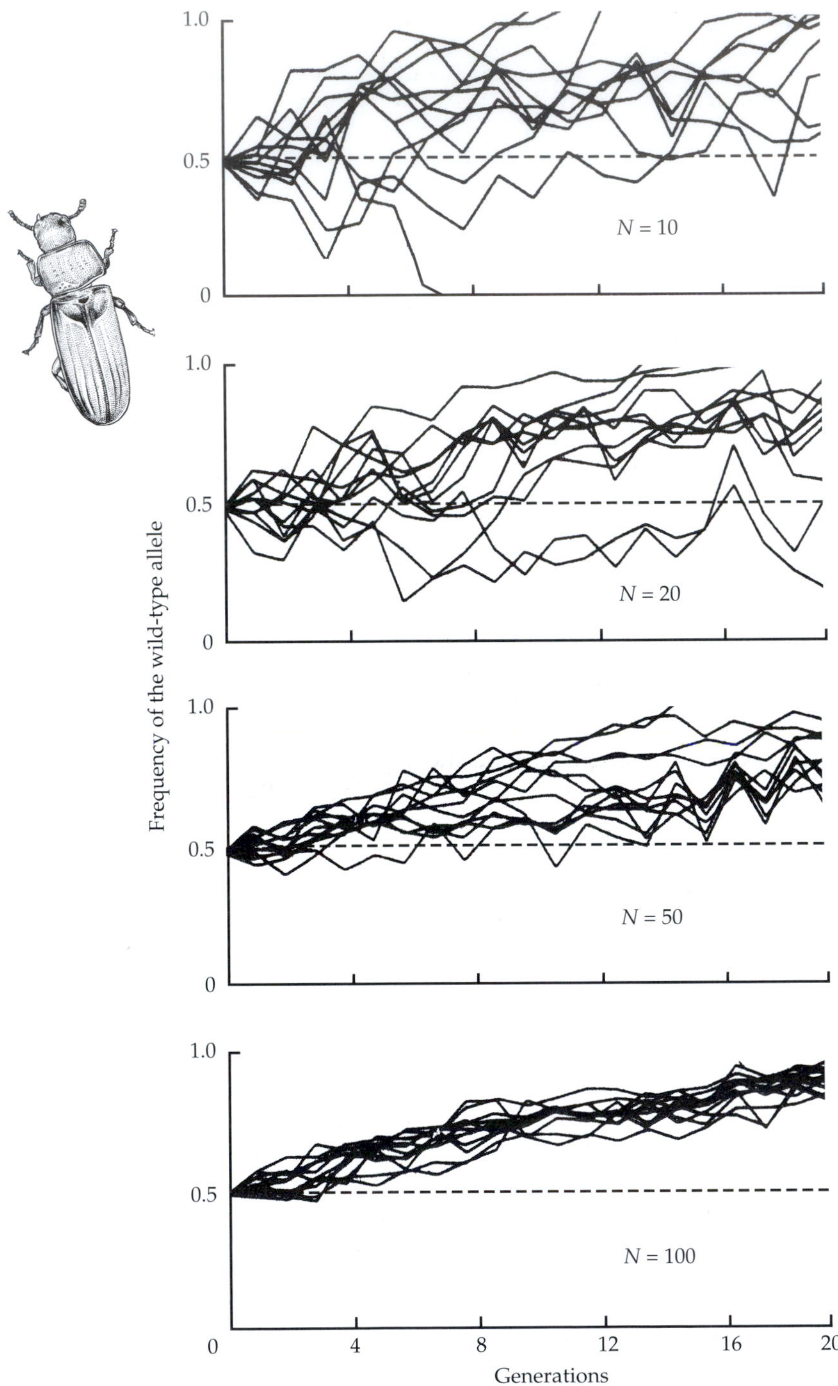
1.0
0.5
0
N = 10
1.0
0.5
0
N = 20
1.0
0.5
0
N = 50
1.0
0.5
0
N = 100
0
4
8
12
16
20
Generations
Frequency of the wild-type allele

◀ **Figure 3.5** Experimental demonstration of the dependence of genetic drift on population size in flour beetles (*Tribolium castaneum*). Note that the magnitude of fluctuation in allele frequency, as well as the frequency of fixation, decreases with increasing population size (N). (From Rich, Bell, and Wilson 1979.)

tion introduces a new allele into the subpopulation. The difference in amplitude of the fluctuations between the four sets of subpopulations that differ only in size clearly shows that the magnitude of random genetic drift depends solely on N, and that drift can have rapid and major effects on allele frequencies when subpopulation sizes are very small.

You may be wondering why the frequency of the wildtype increases steadily over time; this is especially evident in the subpopulations of 100 beetles. This increase cannot be due to genetic drift, because genetic drift is random and this change is directional and repeatable across all the subpopulations. In fact, this change is due to natural selection acting against the mutant black phenotype, so that this example shows the effects of both genetic drift and selection simultaneously. This selection is why, with $N = 10$, of the seven subpopulations that became fixed, six fixed for the wildtype allele. (Note that one did fix for the deleterious black allele.) If there were no selection, the average allele frequency ($\overline{p}$) would remain at 0.5, and approximately equal numbers of subpopulations would have fixed for each allele. In general, under the effects of drift alone, if p is the frequency of the A allele, $\overline{p}$ is also the expected proportion of subpopulations that become fixed for A out of all subpopulations that are fixed. If we started with an average allele frequency of 0.75, three-fourths of the subpopulations that fix would become fixed for A in the absence of selection. Similarly, as $\overline{p}$ increases due to selection in Figure 3.5, the probability of fixing the wildtype allele also increases.

Population differentiation

Random genetic drift tends to create subpopulation differentiation. We can see this in the top panel of Figure 3.5, but it is even clearer in a different example (Figure 3.6). Buri (1956) set up 107 subpopulations of *Drosophila melanogaster*, each with N = 16 bw^{75}/bw heterozygous flies (bw = *brown eyes* mutation). Thus, $\overline{p}$, $\overline{q}$, p_0, q_0 all equal 0.5. These subpopulations were maintained at a constant size of N =16 for 19 generations by randomly choosing eight males and eight females for the next generation in each subpopulation. The plot in Figure 3.6 shows a **frequency distribution** for each generation, that is, the number of subpopulations having 0, 1, 2,...32 bw^{75} alleles. At generation 0 (not shown in Figure 3.6), there is no differentiation because all the subpopulations have exactly the same allele frequency at the bw locus. As

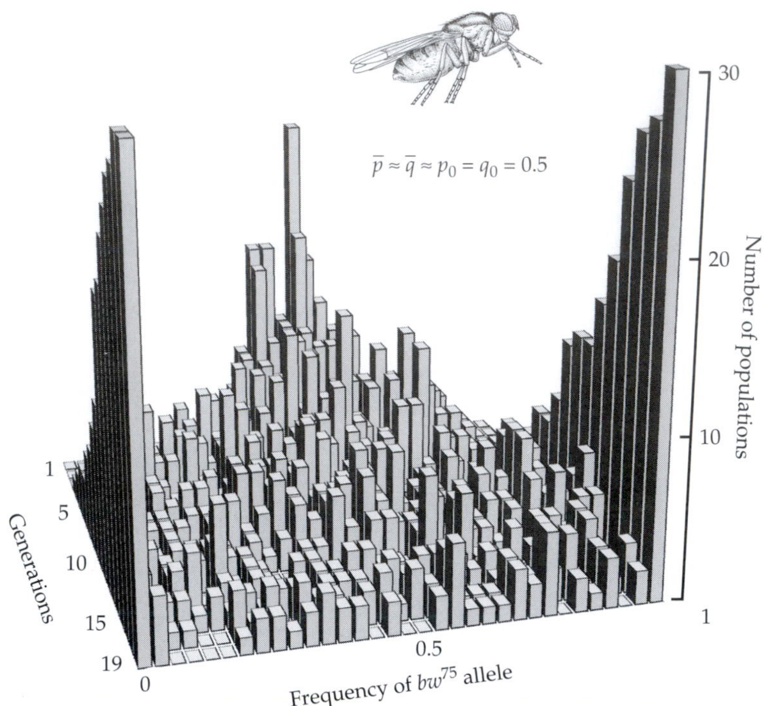

Figure 3.6 Population differentiation due to genetic drift in *Drosophila melanogaster*. Shown are frequency distributions of allele frequencies in each generation for 107 subpopulations maintained at $N = 16$ throughout. Allele frequencies averaged over all subpopulations remained very close to the initial allele frequencies of 0.5, as equal numbers of subpopulations fixed for each allele. (Data from Buri 1956.)

time passes (moving forward on the generation axis), the allele frequencies spread out, and by generation 6 some of the populations have fixed. The number of fixed populations increases rapidly, and roughly equal proportions fix for each allele, suggesting a lack of selection in this case. Therefore, in only 19 generations of genetic drift, the subpopulations went from no differentiation to a very high degree of differentiation, that is, many subpopulations fixed for one allele or the other.

Comparing Figure 3.6 to Figure 3.3, we see that genetic drift and migration have opposite effects—genetic drift increases among-population variation (differentiation) and decreases within-population variation, whereas migration decreases among-population variation and tends to increase within-population variation. In the experiment shown in Figure 3.6, what is the effect of genetic drift on genotypic frequencies in the entire metapopulation? There is a decrease in heterozygosity because the fixed subpopulations are entirely homozygous for one of the alleles and because segregating populations with one allele common are made up mostly of homozygotes for that allele (see Figure 2.10). What about the genotypic frequencies within each subpopulation? They are likely close to HWE because *Drosophila* tends to avoid inbreeding. Note that the many subpopulations that are fixed for one allele or the other are still in HWE; $p = p^2 = 1$ and $q = q^2 = 1$ in these subpopulations. Therefore, the excess homozygosity in the metapopulation due to drift differs from the effects of inbreeding. Inbreeding causes an excess of homozygotes *within* each subpopulation, and so the genotypic frequencies within subpopulations deviate from HWE. Whether or not HWE is present within subpopulations is thus a critical distinction between genetic drift and inbreeding.

The average allele frequency across the entire metapopulation, $\bar{p}$, does not change if there are enough subpopulations and no selection, because the allele frequencies will drift up or down at random in each subpopulation, leaving the mean unchanged. When viewed from a metapopulation perspective, then, genetic drift increases homozygosity and decreases heterozygosity without changing allele frequencies. This is exactly the pattern produced by inbreeding within subpopulations, and therefore most authors refer to this as an *inbreeding effect*. However, this "inbreeding" is not due to a higher frequency of mating between related individuals than expected by chance within the subpopulations. There is nonrandom mating at the level of the metapopulation: Mating is restricted to pairs of individuals within subpopulations, so that individuals in different subpopulations do not mate. The patterns of allele and genotype frequencies with genetic drift are summarized in Table 3.1.

F-statistics

One of the founders of population genetics, Sewall Wright, used this similarity between genetic drift and inbreeding to create *F*-statistics, which provide an integrated view of genetic variation at three hierarchical levels of population structure: within subpopulations, among subpopulations, and

TABLE 3.1 *Summary of allele and genotypic frequencies with genetic drift, assuming random mating within subpopulations*

	Allele frequencies	Genotypic frequencies
Within subpopulations	$p_i q_i$ (vary from zero to one)	p^2, $2pq$, q^2 (HWE) (If *A* is fixed, $p^2 = 1$, $2pq = q^2 = 0$, still HWE)
Overall metapopulation	$\bar{p}$, $\bar{q}$ (= p_0, q_0 if no selection) = 0.5 in Figure 3.6 example	Excess homozygosity due to fixation by genetic drift

the total variation in the metapopulation. All of these measure loss of heterozygosity relative to HWE, using three different measures of H:

- H_I, defined as heterozygosity *observed* within subpopulations; this is the H we calculated in Chapter 2 as the proportion of heterozygotes. Ideally, more than one marker locus is used, so H_I and F-statistics can be calculated for each locus or averaged over all loci.
- H_S, defined as the heterozygosity *expected* with random mating in a subpopulation, which from Chapter 2 we know to be $2p_iq_i$. (We referred to this as H_0 in Chapter 2.)
- H_T, the expected heterozygosity if the entire metapopulation were undivided and undergoing random mating (i.e., panmictic). H_T equals $2p_0q_0 = 2\overline{p}\overline{q}$.

Note that all these heterozygosities are expressed as frequencies, not numbers. Using the frequencies is crucial for the proper calculation of F-statistics, which is opposite to the use of numbers for chi-square tests of HWE (Chapter 2). We can define the F-statistics for the three hierarchical levels, starting with individuals (I) within subpopulations (S):

$$F_{IS} = \frac{\overline{H}_S - \overline{H}_I}{\overline{H}_S} \qquad \textbf{3.5}$$

From this equation it is clear that F_{IS} measures the proportional reduction in heterozygotes within subpopulations due to inbreeding, and thus it is the same as the inbreeding coefficient, F, from Chapter 2. This can be calculated for individual subpopulations, but to compare variation at different levels of population structure, we need to use H_S and H_I averaged over all subpopulations. The bar over each H denotes the average. To measure differentiation among populations we use F for subpopulations (S) relative to the total metapopulation (T):

$$F_{ST} = \frac{H_T - \overline{H}_S}{H_T} \qquad \textbf{3.6}$$

This equation shows that F_{ST} measures the proportional reduction in the heterozygosity of the metapopulation due to differentiation among subpopulations, relative to the expectation with no population subdivision. Note that again H_S is averaged over all subpopulations; F_{ST} is meaningless for one subpopulation because it measures differentiation among multiple subpopulations. There is only one H_T because it refers to the entire metapopulation.

There is another equivalent definition of F_{ST}:

$$F_{ST} = \frac{V}{\overline{p}\overline{q}} \qquad \textbf{3.7}$$

where V is the **variance** in allele frequencies among the subpopulations. We will discuss variance in more detail in the next chapter; at this point it suffices to say that it is what it sounds like—a measure of variation. Therefore, if all subpopulations have exactly the same allele frequencies, then the variance and F_{ST} both equal zero. As differentiation increases, so does the variance, so F_{ST} increases. There are a number of additional ways of calculating F_{ST} and an analogous measure G_{ST}; all are closely related measures of subpopulation differentiation.

F_{ST} is often called the **fixation index** because it increases as more subpopulations become fixed (or close to fixation) for one allele. This view of F_{ST} also helps explain how it measures subpopulation differentiation. As subpopulations drift toward fixation, that is, as they deviate from $p = q = 0.5$, F_{ST} increases because H_S decreases. Why does H_S decline? Recall that H_S for any subpopulation i equals $2p_iq_i$, which from Figure 2.10 reaches its maximum when $p = 0.5$. What happens to H_T as the subpopulations drift? It does not change because the allele frequencies in the overall metapopulation do not change. F_{ST} cannot be negative, at least theoretically, because the average H_S should never be larger than H_T (Equation 3.6), and variances cannot be negative (Equation 3.7).

For most organisms, F_{ST} is typically 0.15 or less, but values up to 0.7 have been recorded. Wright referred to values of F_{ST} from 0.05 to 0.15 as moderate differentiation, from 0.15 to 0.25 as great differentiation, and greater than 0.25 as very great differentiation. For example, F_{ST} measured with allozymes is low in one species of willow leaf beetle (*Chrysomela aeneicollis*), and declines with the scale at which populations are sampled (Rank 1992). Among different river drainages, F_{ST} is 0.04, but among localities within river drainages it is only 0.02, demonstrating increased gene flow at shorter distances. In contrast, F_{ST} was 0.38 for the *Pgm-2* allozyme locus in 43 subpopulations of *Phlox cuspidata* in Texas (Box 3.1; Levin 1978); the lower gene flow in this species is due to its high degree of self-fertilization, which reduces gene flow through pollen movement.

Chrysomela aeneicollis

The final hierarchical F-statistic measures the reduction in heterozygosity in an individual (I) relative to the total metapopulation (T) due to both nonrandom mating within subpopulations and genetic drift among subpopulations:

$$F_{IT} = \frac{H_T - \overline{H}_I}{H_T} \qquad \textbf{3.8}$$

F_{IT} is not usually as informative as F_{IS} or F_{ST} because it does not separate the effects of inbreeding and genetic drift. The three F-statistics are related to each other by:

$$(1 - F_{IS})(1 - F_{ST}) = (1 - F_{IT}) \qquad \textbf{3.9}$$

BOX 3.1 *Calculation of F-statistics*

Levin (1978) scored allele frequencies at the *Pgm-2* locus in 43 Texas subpopulations of *Phlox cuspidata*. Forty of these subpopulations were fixed for the *b* allele (listed together in the first row of the table below). In the other three subpopulations the frequencies of *b* were 0.49, 0.83, and 0.91, with observed heterozygote frequencies of 0.17, 0.06, and 0.06, respectively:

Subpopulation	p_i	H_I
1–40	1	0
41	0.49	0.17
42	0.83	0.06
43	0.91	0.06

Phlox cuspidata

From these data we can calculate the three hierarchical *F*-statistics and their components as follows:

$\overline{H}_I$ = the observed proportion (frequency, not numbers) of heterozygotes within subpopulations, averaged over all subpopulations:

$$\overline{H}_I = \frac{(40 \times 0) + 0.17 + 0.06 + 0.06}{43} = 0.0067$$

$\overline{H}_S$ = the expected proportion of heterozygotes within subpopulations, assuming random mating (= $2p_iq_i$), averaged over all n subpopulations:

$$\overline{H}_S = \frac{\sum_{i=1}^{n} 2p_iq_i}{n} =$$

$$\frac{(40 \times 0) + 2(0.49 \times 0.51) + 2(0.83 \times 0.17) + 2(0.91 \times 0.09)}{43} = 0.0220$$

H_T = the expected proportion of heterozygotes over the entire metapopulation (=$2\overline{p}\,\overline{q}$). H_T is not itself an average because there is only one metapopulation, but since there is a unique value for p and q for each subpopulation, the average allele frequencies across all subpopulations are used:

$$\overline{p} = \frac{(40 \times 1) + 0.49 + 0.83 + 0.91}{43} = 0.9821$$

$$1 - \overline{p} = \overline{q} = 0.0179$$

$$H_T = 2\overline{p}\overline{q} = 2 \times 0.9821 \times 0.0179 = 0.0352$$

Box 3.1 continued

We can now calculate the three *F*-statistics using the definitions given in the text:

$$F_{IS} = \frac{\overline{H}_S - \overline{H}_I}{\overline{H}_S} = \frac{0.0220 - 0.0067}{0.0220} = 0.70$$

$$F_{ST} = \frac{H_T - \overline{H}_S}{H_T} = \frac{0.0352 - 0.0220}{0.0352} = 0.38$$

$$F_{IT} = \frac{H_T - \overline{H}_I}{H_T} = \frac{0.0352 - 0.0067}{0.0352} = 0.81$$

Check these using equation 3.9:

$$(1 - F_{IS})(1 - F_{ST}) = (1 - F_{IT})$$

$$(1 - 0.7)(1 - 0.38) = (1 - 0.81)$$

$$(0.3)(0.62) = 0.19$$

$$0.19 = 0.19$$

F_{ST} or F_{IT} are metapopulation level measures of population structure, quantifying the degree of subpopulation differentiation within the total population, and the overall amount of reduction in heterozygosity. Therefore, it doesn't make sense to calculate separate F_{ST} or F_{IT} for each subpopulation. Each subpopulation can have its own value for F_{IS}, however, because this is just the inbreeding coefficient we calculated before. To compare the reduction in heterozygosity across the three hierarchical levels, we use the average F_{IS} calculated previously.

In this example F_{IS} is very large, demonstrating a high level of inbreeding in these self-fertilizing plants. This high F_{IS} comes entirely from the three unfixed subpopulations. The forty fixed subpopulations do not contribute to this estimate of inbreeding, because there is no genetic variation and thus no heterozygotes to be reduced in frequency. Mathematically, these 40 subpopulations are represented by zeros in both H_I and H_S, so they do not affect F_{IS}. F_{ST} is also quite large, which is also likely due to the high level of self-fertilization. Much of gene flow in plants is through movement of pollen by wind or animal pollinators, so when most pollen stays on the same plant, as it does in highly selfing species, it greatly reduces gene flow from pollen movement among subpopulations. Note that this high differentiation is mainly due to subpopulation #41, which is the only one with a lower frequency of the *b* allele.

Effective population size (N_e)

Recall that the magnitude of genetic drift, and thus the amount of subpopulation differentiation measured by F_{ST}, depends on population size. But the magnitude of genetic drift does not depend on a simple count of the number of individuals in the subpopulation, except with unrealistic assumptions:

1. Equal numbers of each sex (equal **sex ratio**).
2. No sexual or natural selection (i.e., each individual has an equal probability of successfully contributing gametes to the next generation).
3. Subpopulation size remains the same in each generation.

To determine the magnitude of genetic drift in real-world cases when the above assumptions are violated, we need the concept of **effective population size, N_e.** Effective population size is the size of an idealized population (all three assumptions above met) that would experience the same magnitude of genetic drift as the actual population of interest. When assumption 1 above is violated, effective population size is reduced relative to the actual number of individuals because every offspring must have one mother and one father. So when there are fewer males than females, which is the usual case in biased sex ratios, the smaller number of males still contributes half of the genes in the next generation, increasing genetic drift. To estimate effective population size with unequal sex ratio the following equation can be used:

$$N_e = \frac{4N_m N_f}{N_m + N_f} \tag{3.10}$$

N_m and N_f are the number of males and females in the subpopulation, respectively. The denominator is the actual number of individuals in the population, N_a. For example, if $N_m = N_f = 10$, then $N_e = 400/20 = 20 = N_a$. If we decrease N_m to 5, then $N_a = 15$, but N_e is only $200/15 = 13.3$. A classic example of skewed sex ratios in nature comes from fig wasps. Fig wasp females lay their eggs inside figs and commonly have strongly female-biased sex ratios. In one set of 12 figs that each had eggs from five different females, the average number of wasps (N_a) in each fig was 240, but only 23% of these were male (Frank 1985).

EXERCISE: Calculate the average N_e for these fig wasp subpopulations.

ANSWER: $N_e = \dfrac{4 \times (240 \times 0.23) \times (240 \times 0.77)}{240} = 170$

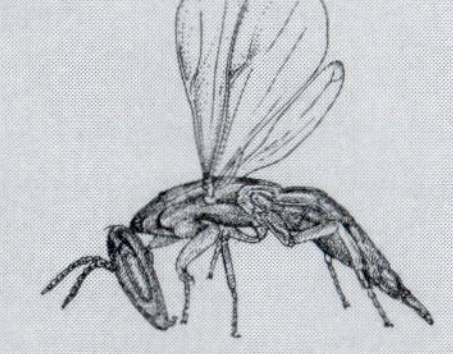

Thus, N_e is 70 individuals fewer than N_a, a reduction of almost a third.

A related case is one in which assumption 2 is violated—for example, when a few dominant males carry out most of the matings. Here effective population size is also decreased because the males that do not mate are effectively eliminated from the population. Any variance in reproductive success among individuals greater than random expectations, a commonplace occurrence in natural populations, reduces effective population size because some individuals are overrepresented genetically and some are underrepresented. In this case, N_e can be estimated with the following equation:

$$N_e \approx \frac{8N_a}{V_m + V_f + 4} \qquad \textbf{3.11}$$

V_m and V_f are the variances in the number of offspring produced by males and females. Because these variances are in the denominator, when males or females differ greatly in reproductive success, these variances are high and effective population size becomes smaller. Therefore, effective population size is smaller with stronger sexual selection, that is, when there is lots of variance in mating success. For example, Clutton-Brock, Albon and Guinness (1988) reported over four times greater variance in lifetime reproductive success for males (V_m = 41.9, N_m = 33) than for females (V_f = 9.1, N_f = 35) in red deer (*Cervus elaphus*).

Cervus elaphus

Plugging these numbers into Equation 3.11 gives:

$$N_e \approx \frac{8 \times (33 + 35)}{41.9 + 9.1 + 4} = 9.9$$

Thus, effective population size is about 1/7 actual population size due to variance in reproductive success. Variance in reproductive success is often the most important factor that reduces N_e relative to N_a.

Violations of assumption 3, that is, when fluctuations in population size occur, also decrease N_e. An extreme decrease in population size is called a **bottleneck.** To calculate N_e with fluctuations, the harmonic mean population size is used:

$$\frac{1}{N_e} = \frac{1}{t}\left(\frac{1}{N_1} + \frac{1}{N_2} + \cdots + \frac{1}{N_t}\right) \qquad \textbf{3.12}$$

where N_t refers to the population size in generation t; note that N_e from Equations 3.10 or 3.11 can be used for each generation. Harmonic means are strongly affected by small numbers such as those that occur in a bottleneck. For example, if population sizes in a population over three generations are

1000, 10, and 1000, then $N_e = 29$. The arithmetic mean is 670. This points out a big problem in determining the effects of fluctuations on N_e —we need to know the history of population sizes, but this is extremely difficult data to obtain. A rare example of this kind of data comes from the classic studies of E. B. Ford and his colleagues on the moth *Panaxia dominula* in England. For a simplified example, we can examine the population numbers for four years of the study, 1954 to 1957, during which there was a bottleneck:

Year	1954	1955	1956	1957
N	11,000	2000	11,000	16,000

EXERCISE: Calculate N_e from these data, and compare to the average N_a.

ANSWER: $$\frac{1}{N_e} = \frac{1}{4}\left(\frac{1}{11{,}000} + \frac{1}{2000} + \frac{1}{11{,}000} + \frac{1}{16{,}000}\right)$$

$$\frac{1}{N_e} = 0.000186$$

$$N_e = 5374$$

The arithmetic average N_a is 10,000, so the N_e is only a little more than half that value.

The estimate of N_e for *Panaxia* above is certainly an overestimate because we used N_a for each year. A more realistic estimate of N_e could be made if estimates of N_e for each year were known (from Equation 3.10 or 3.11, for example) and plugged into Equation 3.12. Unfortunately, data that detailed for a number of years is very difficult to obtain, although a few long-term studies like the red deer study mentioned earlier may have the necessary data.

An important type of bottleneck is a **founder event,** defined as the creation of a new population by a small number of colonists. Very often the allele frequencies in this small group are different by chance from the population from which they originated, and the small number of initial colonists can cause additional genetic drift. An example of the effects of founder events is given later in the population genetics synthesis section.

We have stated that the magnitude of genetic drift depends on N_e, but so far, we have not quantified this. There are two relevant equations:

$$\Delta F_{ST} = \frac{1}{2N_e} \qquad \textbf{3.13}$$

This gives the increase in F_{ST} due to genetic drift over one generation. Note that N_e is in the denominator; this makes sense, because smaller effective population size leads to greater genetic drift and thus a more rapid increase in F_{ST}. N_e here refers to the effective size of the subpopulations. To project this t generations into the future, use the following:

$$F_t = 1-(1-\Delta F_{ST})^t \qquad \textbf{3.14}$$

where F_t is F_{ST} at generation t. This equation assumes no change in N_e over the period modeled.

Inbreeding with random mating in small populations

The term *inbreeding* can cause a tremendous amount of confusion, in large part because it is used in more than one way (Templeton and Read 1994). As we noted above, the different reductions in heterozygosity caused by *true inbreeding* (mating among relatives more often than expected by chance) and genetic drift are both referred to as inbreeding, and the symbol F (or sometimes also f) is used for both meanings, often without Wright's very useful subscripts.

Recall that small population size by itself does not increase F_{IS} if mating within the population is random (i.e., proportional to genotypic frequencies). The "inbreeding" due to small population size is actually a consequence of genetic drift, not mating with relatives more often than expected by chance (see Box 3.2), although population subdivision does create nonrandom mating at the level of the metapopulation. This nonrandom mating occurs because individuals in different subpopulations rarely, if ever, mate. Recall that true inbreeding and genetic drift differ in their effects on genotypic frequencies within subpopulations. Inbreeding increases the frequencies of both homozygotes, and therefore the subpopulation deviates from HWE. On the other hand, genetic drift increases the frequency of only one of the homozygotes as one allele randomly heads to fixation, and the subpopulation stays in HWE. Therefore, small subpopulation size does not increase F_{IS}, but it does increase F_{ST} in the metapopulation. This is because with genetic drift and random mating actual heterozygosity within subpopulations always equals expected heterozygosity ($H_I = H_S$), but both of these decline, so $F_{IS} = 0$ and F_{ST} increases (see Equations 3.5 and 3.6).

Therefore, although effects of genetic drift and inbreeding on genotypic frequencies are distinguishable, their effects on the phenotype can be similar. Because changes in allele frequencies due to genetic drift have nothing to do with adaptation, deleterious alleles can drift to high frequencies and thus be expressed in homozygotes even at HWE. For example, in the *Tribolium* beetle example in Figure 3.5, one subpopulation fixed for the deleterious

BOX 3.2 *Example of Inbreeding Depression Due to Genetic Drift*

Templeton and Read (1994) describe a captive breeding program for endangered Speke's gazelles. The captive population was founded by four wild animals, one male and three females, who were presumably unrelated individuals. This represents an extreme founder event in which N_e is further reduced by the sex ratio bias ($N_e = 3$ by Equation 3.10). All these animals were likely to be carrying deleterious recessive alleles at a number of loci, and these alleles would have been rare in the wild population due to selection. For any of the deleterious recessives that the male is carrying, the frequency in this founding or parental population suddenly jumps to $1/8$ because there are only eight alleles at each locus in the four individuals. This is already a much higher frequency than most deleterious recessives would be maintained at in a large natural population. In the offspring (F_1) generation, however, the frequency of the deleterious recessives will jump to $1/4$. This is because half of all the alleles in the F_1 generation will be from the male (he is the father of all offspring), and half of these will be the deleterious recessives (at the loci at which the male carried deleterious recessives). If the F_1 individuals mate randomly, $1/16$ (q^2) of the F_2 will be homozygous for the male's deleterious recessive, even though the population is in Hardy-Weinberg equilibrium due to the random mating. Therefore, there is inbreeding in one sense, because in a small population like this, relatives are forced to mate with each other even with random mating. But it is not inbreeding in the way we defined it, that is, relatives mating with each other more often than by chance, which would lead to an excess of homozygotes compared to HWE and a positive F_{IS}.

black allele. This means that, with genetic drift alone, reduction in fitness due to the expression of deleterious recessives occurs without nonrandom mating within the subpopulations, and this is still called inbreeding depression even though the subpopulations are in HWE (see Box 3.2 for an example). It is more accurate to say that genetic drift can increase the frequency of deleterious alleles.

Natural Selection: Selection on Genotypes

The basic population genetic models of natural selection examine the effects on allele frequencies of selection on genotypes at one or a few loci. These models are very difficult to use in empirical studies of natural populations, because selection acts directly on the phenotype, and most phenotypic traits are affected by more than one locus and by the environment. Still, since selection ultimately acts on genotypes (through the phenotype) and changes allele frequencies, these models are very important for understanding evolution by natural selection. These models also are the basis for studies of

DNA sequence evolution. Up to this point in the book, we have been concerned with allele and genotypic frequencies almost entirely, and have ignored the phenotype (the principal exception being assortative mating). To discuss selection, we need to consider the phenotype, in particular, the highest trait in the phenotypic hierarchy, fitness.

Fitness

Fitness is the key concept in understanding natural selection, because selection is caused by differences in fitness among individuals with different phenotypes or genotypes. Unfortunately, fitness is very difficult to define. Our hypothetical example in Table 3.2 uses lifetime reproductive success (offspring production); we will discuss fitness in more detail in Chapter 6. The second column in Table 3.2 shows the average number of offspring produced for the three genotypes. This is the measure of **absolute fitnesses.** More useful is the measure of **relative fitnesses, w,** which is defined relative to the most favored genotype. Therefore, the relative fitness is simply the absolute fitness divided by the highest absolute fitness of all the genotypes (60 in Table 3.2). Relative fitness is used because selection operates through *differences* among individuals in the population, so that the absolute value of survival or any other fitness measure is irrelevant. What is relevant is an individual's or genotype's fitness relative to other individuals or genotypes in the population. Finally, the selection coefficient, s, is $1 - w$, and therefore measures the strength of selection against a given genotype. The selection coefficient of 0.2 in Table 3.2 is very strong, and thus would lead to a rapid decrease in frequency of the *a* allele.

To understand the genetics of phenotypic traits (including fitness), we need to consider modes of **gene action,** that is, how genotypes affect the phenotype. For now we will consider two modes of gene action, **additivity** and **dominance.** The simplest form of dominance is **complete dominance,** in which the heterozygote has the same phenotype as one of the homozygotes (as in Table 3.2). In this example, the *A* allele completely masks the fitness effects of the *a* allele in the heterozygote. Complete dominance is just one

TABLE 3.2 *Hypothetical data illustrating the calculation of relative fitness and the selection coefficient from absolute fitness data*

Genotype	Absolute fitness (no. of offspring)	Relative fitness (w)	Selection coefficient ($s = 1 - w$)
AA	60	1	0
Aa	60	1	0
aa	48	0.8	0.2

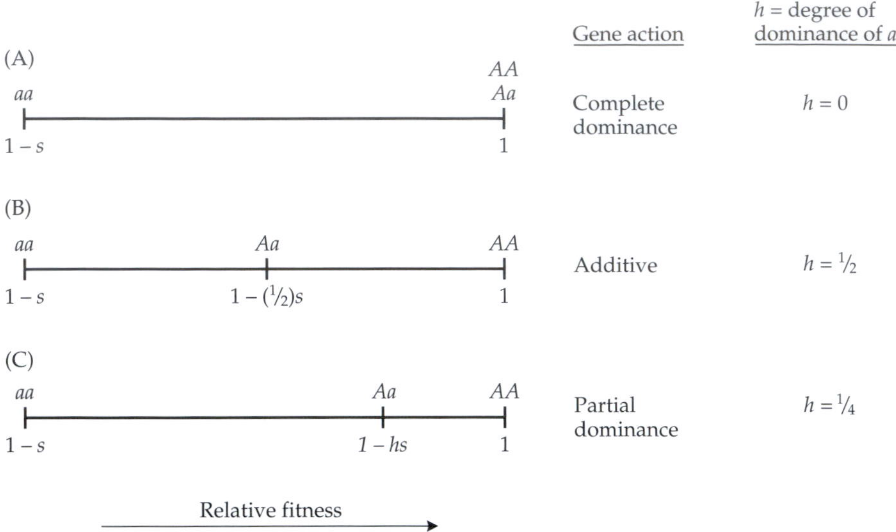

Figure 3.7 Different degrees of dominance for fitness. In each case, the fitness of each of the three genotypes at one locus with two alleles is depicted by its horizontal placement and is also given below each line. (After Falconer and Mackay 1996.)

extreme, however; the degree of dominance of one allele relative to another at the same locus can vary continuously. The effect of different degrees of dominance on the relative fitness of the three genotypes at a locus with two alleles is shown in Figure 3.7. The dominance we are discussing here is dominance for fitness itself, not some other phenotypic trait. As above, s is the selection coefficient measuring the strength of selection against the *aa* genotype. The degree of dominance for fitness of the *a* allele is represented by h; note that this is 1 minus the dominance of the *A* allele, a point that causes considerable confusion. The scale is relative fitness, which increases as you move to the right.

As in Table 3.2, the relative fitness of *AA* individuals is 1 and that of *aa* individuals is $1 - s$. The fitness of heterozygotes is given by $1 - hs$. In the example in Table 3.2, the *a* allele was completely recessive for fitness, so $h = 0$; that is, there is no dominance of the *a* allele. Therefore, heterozygote fitness $1 - hs = 1$; that is, there is complete dominance of the *A* allele—the heterozygote has the same fitness as the dominant homozygote (Figure 3.7A).

Figure 3.7B shows another important case, that of **additivity,** in which neither allele shows any dominance over the other. In this case, $h = 1/2$, and the heterozygote fitness is exactly intermediate between the fitnesses of the two homozygotes. This is sometimes called **genic selection.** There can also be **partial dominance** as shown in Figure 3.7C; in other words, h can take any value between 0 and 1 (it is ¼ in this example). We will discuss another case, overdominance, below.

How does selection at one locus change allele frequencies at that locus?

Figure 3.8 shows theoretical changes in allele frequency over time with moderately weak selection ($s = 0.05$) and different degrees of dominance with respect to fitness. The y-axis is the frequency of the favored allele; since the allele starts at a very low frequency, this figure models the increase in frequency of a new favorable mutant. Note that if the favored allele is dominant or the locus is additive, the favored allele increases rapidly to fixation or near fixation. In the case of a favored recessive, the increase is extremely slow until a threshold frequency is reached, and then it goes rapidly to fixation. To understand these patterns we need to refer back to Figure 2.10, which shows genotypic frequencies with different allele frequencies under HWE. Because selection acts on phenotypes, we need to consider the diploid genotypes that influence those phenotypes.

Why does the allele frequency increase earliest when the favored allele is dominant? Recall from Figure 2.10 that when an allele is rare, as a new mutation is, it is almost entirely in the heterozygous state. When the favored mutation is dominant, each individual heterozygous for that allele completely expresses the favored phenotype, which has the highest fitness (see Figure 3.7). When the favored allele is additive, heterozygote fitness is intermediate between the two homozygote fitnesses (Figure 3.7 again), so they increase more slowly. However, note that with additive gene action the

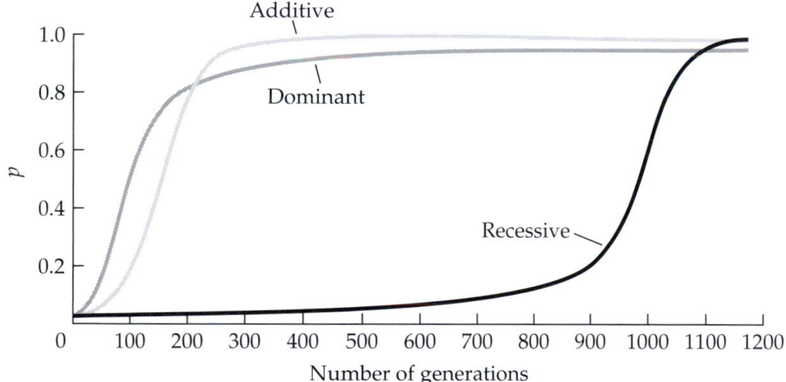

Figure 3.8 Expected change in frequency over time of a favored allele ($s = 0.05$) with dominant, additive, or recessive effects on fitness.

to fixation in about 300 generations, whereas the dominant favored allele has still not fixed after 1200 generations. The reason for this is that when the favored allele moves to high frequency, the now rare deleterious recessive allele is almost entirely in heterozygotes, where its unfavorable effects are not expressed in the phenotype and therefore are invisible to selection. This is the case we referred to in Chapter 2, in which rare deleterious recessive alleles are maintained in so-called carriers, that is, heterozygous individuals that do not express the effects of the allele.

The reverse of the dominant pattern is shown in the bottom curve in Figure 3.8, where the favored allele is recessive. The extremely long period in which the allele remains rare is due to the fact that it is primarily in heterozygotes, so its *beneficial* effects are not expressed and selection cannot act to increase its frequency. At about 900 generations the allele reaches a frequency of 0.2, after which it increases rapidly. By referring to Figure 2.10, you can see that 0.2 is the minimum frequency for an appreciable number of homozygotes to occur; this is a prerequisite for a favored recessive to be increased by selection. Note that after this point the favored recessive goes rapidly to fixation, because the deleterious dominant is always expressed so selection can remove it.

Students are often confused by the advantageous recessive example because they share a common misconception that dominant alleles must be advantageous and recessive must be deleterious. This is often, but not always, true in natural populations for two reasons. First, new deleterious dominant mutations are rapidly eliminated by selection because they have lower fitness in the heterozygous state. Second, a new advantageous recessive mutation is often lost through drift in the long period before it reaches a high enough frequency to be found in appreciable numbers of homozygotes. The curves in Figure 3.8 are smooth because they are based on simple equations that assume no genetic drift. But imagine if they showed the random fluctuations illustrated in Figure 3.5—in the vast majority of cases the line for the advantageous recessive would intersect zero, that is, the allele would go extinct before it increased in frequency. This is called **Haldane's sieve,** after the British population geneticist J. B. S. Haldane who first demonstrated it theoretically. Of course, the loss through drift is less likely as effective population size increases. Remember that Figure 3.8 is based on populations in HWE; if there were true inbreeding, the frequency of homozygotes would increase, and the advantageous recessive would have a better chance of being increased in frequency by selection before being removed by genetic drift. Therefore, we expect advantageous recessive alleles to occur somewhat more often in species that are inbred and have high effective population sizes, but even under these conditions, advantageous recessives will still be less common than advantageous dominants.

Figure 3.8 shows allele frequency **dynamics,** that is, changes over time. An **equilibrium** value of p is a value for which $\Delta p = 0$, which means that the allele frequency remains unchanged generation after generation. There are, however, several types of equilibria depending on what happens when the allele frequency is **perturbed,** that is, moved away from the equilibrium value. Consider first the case when the initial allele frequency is near (but not equal to) the equilibrium. If the allele frequency moves progressively farther away from the equilibrium in subsequent generations, the equilibrium is said to be **unstable.** If the allele frequency moves progressively closer to the equilibrium in subsequent generations, the equilibrium is said to be **locally stable.** A locally stable equilibrium might also be **globally stable,** which means that, whatever the initial allele frequency, it always moves progressively closer to that equilibrium. HWE is an example of a **semistable** or **neutrally stable equilibrium,** in which every allele frequency represents an equilibrium because it remains unchanged until perturbed.

Are there equilibria in Figure 3.8? Yes, both $p = 0$ and $p = 1$ are equilibria in the absence of mutation or migration, because only one allele is present. Are these stable or unstable equilibria? The equilibrium at $p = 0$ is unstable, because if a mutation introduces the favored allele, selection moves the population away from $p = 0$. The equilibrium at $p = 1$ is globally stable, because selection always moves the population toward that point.

Overdominance or heterozygote advantage

Overdominance or **heterozygote advantage** occurs when the heterozygous genotype has a higher fitness than both homozygous genotypes. A hypothetical example of this pattern of relative fitnesses is depicted in Figure 3.9A. The dynamics of overdominance are shown in Figure 3.9B. Here there are three equilibria, but only one is stable. As always, $p = 0$ and $p = 1$ are equilibria, but in this case they are unstable. With any perturbation (introduction of the other allele through mutation or migration), the allele frequency in the population moves rapidly to an intermediate value. The reasons why this happens are shown more clearly in Figure 3.9C, in which the population allele frequency is plotted on the x-axis (instead of the y-axis as in Figure 3.9B), and the y-axis represents population mean fitness. The mean fitness of a population is defined as:

$$\bar{w} = Pw_{AA} + Hw_{Aa} + Qw_{aa} \qquad \textbf{3.15}$$

Recall that P, H, and Q are the frequencies of the three genotypes; w_{AA} is the fitness of AA, w_{Aa} is the fitness of Aa, and w_{aa} is the fitness of aa. This is a **weighted mean,** in which the fitness value for each genotype is weighted by the frequency of that genotype. For example, if P is 0.8, then w_{AA} will tend to

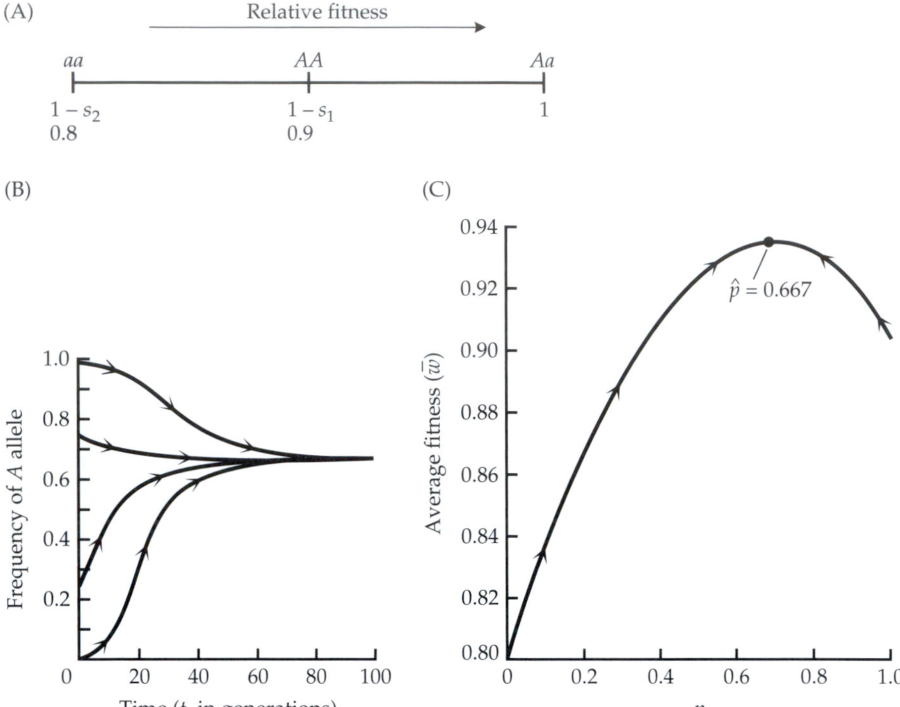

Figure 3.9 A hypothetical example of overdominance (heterozygote advantage). (A) Relative fitnesses of three genotypes depicted as in Figure 3.7; note that the heterozygote has highest fitness. The selection coefficients are $s_1 = 0.1$ and $s_2 = 0.2$. (B) Allele frequency changes over time (dynamics) for the relative fitnesses shown in part A. Note that allele frequencies converge to one globally stable equilibrium regardless of initial frequency. (C) Population mean fitness versus allele frequency, again for the same relative fitnesses. Mean fitness is maximized at the equilibrium, which is why it is globally stable.

dominate the mean fitness of the population. This makes sense because most individuals in the population have this genotype and thus they have this fitness.

At $p = 0$, the mean fitness equals w_{aa}, which is 0.8 in this hypothetical example, and at $p = 1$, $\bar{w} = w_{aa} = 0.9$. The arrows in Figure 3.9 show the direction that allele frequencies evolve due to selection; Figure 3.9C illustrates the very important point that, with constant genotypic fitness values, *selection increases population mean fitness*. (This is not true for frequency-dependent selection discussed in the next section.) The intermediate allele frequency is a globally stable equilibrium because it represents the only **adaptive peak,** that is, a population mean fitness maximum. At an adaptive peak, the allele frequency is said to be at an **optimum.**

The optimum allele frequency depends on the relative fitnesses of the three genotypes. In the example in Figure 3.9, the *A* allele is more common at equilibrium than the *a* allele because the *AA* genotype is more fit than *aa*. What would the equilibrium allele frequency be if relative fitnesses for both homozygotes were equal, say, $w = 0.8$? It would be $p = 0.5$. Regardless of the exact equilibrium, the most important feature of heterozygote advantage is that both alleles are maintained at a stable equilibrium. In the case when one homozygote is favored (e.g., Figure 3.8), a plot like Figure 3.9C would be monotonically (constantly) increasing, so that only one allele is maintained at equilibrium, regardless of the degree of dominance for fitness of that allele.

This is one of the reasons why overdominance has been discussed a great deal by population geneticists—it can maintain genetic variation in natural populations. However, to date there are only a few well-established examples of overdominance in nature. The classic example is sickle cell anemia in humans, in which heterozygotes for the trait (+/*s*) are less susceptible to malaria than nonsickle homozygotes (+/+). Homozygotes for the sickle allele (*s*/*s*) have reduced fitness due to severe anemia. Another example is resistance by rats to warfarin, a poison used in pest control. Heterozygotes (*SR*) are more resistant to warfarin than homozygous susceptible genotypes (*SS*), whereas homozygotes for the resistance allele (*RR*) have reduced fitness because they need large amounts of vitamin K.

A common feature of both of these examples is that the selection against each homozygote comes from different factors. In sickle cell anemia, selection against +/+ individuals is from malaria and selection against *s*/*s* from anemia. In warfarin resistance, *SS* are killed by warfarin and *RR* have vitamin K deficiency. This means that the heterozygote advantage depends on the environment in both cases—it disappears in the absence of malaria or warfarin. The more general principle, that selection can change in different environments, will be explored more fully in Chapter 6.

A rare example of heterozygote advantage in a natural population is the plumage polymorphism in common buzzards (*Buteo buteo*) in Europe (Kruger, Lindstrom, and Amost 2001). Buzzard feathers vary from dark brown to almost pure white, and much of this variation is due to a single additive locus with two alleles. Lifetime number of offspring fledged was 2.0 and 2.4 for the dark and light colored homozygotes, respectively, and twice as high (4.4) for the heterozygotes.

EXERCISE: Calculate the relative fitnesses and selection coefficients for the three buzzard genotypes. Place the genotypes on a graph based on the relative fitnesses like those in Figure 3.7. Assuming that the relative number of fledglings determines the frequencies of the three genotypes, estimate population mean fitness.

ANSWER: (D = dark allele, d = light)

Genotype	Absolute fitness (no. of fledglings)	Relative fitness (w)	Selection coefficient ($s = 1 - w$)
DD	2.0	2.0/4.4 = 0.45	0.55
Dd	4.4	4.4/4.4=1	0
dd	2.4	2.4/4.4 = 0.54	0.46

Buteo buteo

Frequency of DD (P) is 2/8.8 = 0.23,
H = 4.4/8.8 = 0.5, and Q = 2.4/8.8 = 0.27
(these sum to 1). From Equation 3.15 the population mean fitness is:

$$\bar{w} = 0.23 \times 0.45 + 0.5 \times 1 + 0.27 \times 0.54$$

$$\bar{w} = 0.75$$

Heterozygote advantage is often confused with **heterosis** (also called **hybrid vigor**), but these are distinct phenomena. Heterosis occurs when the F_1 generation has higher fitness than the parental strains or subpopulations that were **crossed** (mated) to produce them. Heterosis commonly occurs when inbred strains are crossed, but it can also occur when isolated natural subpopulations come together and interbreed. To understand why it occurs, consider what genotypic frequencies would be like in subpopulations that have undergone drift or in inbred lines in a laboratory. Inbred lines are created by propagating lines with very small numbers of breeding individuals in each generation, so in both cases, genetic drift will cause random fixation of alleles. (Inbred lines are often created without true inbreeding.) This will cause inbreeding depression regardless.

Inbreeding depression can be caused by the fixation of deleterious recessives as discussed previously, or it can be caused by overdominant loci in which homozygotes are less fit. Since genetic drift is random, different subpopulations or inbred lines will tend to be fixed for different alleles at many loci, so when they interbreed, the F_1 population will be highly heterozygous. The increase in fitness that results is called heterosis, which can be caused by masking deleterious recessives (Figure 3.7A or C) or by heterozygote advantage (Figure 3.9A). Thus heterosis is the opposite of the inbreeding depression that is caused by genetic drift (or the inbreeding depression caused by true inbreeding). Heterozygote advantage can be one cause of heterosis, but heterozygote advantage and heterosis are not the same thing. One key difference is that heterosis is usually due to a number of loci, whereas heterozygote advantage refers to effects at a single locus. These points are summarized in Table 3.3. Most evidence to date suggests that inbreeding depression and heterosis are mainly caused by deleterious recessives (Keller and Waller 2002).

To examine the effects of inbreeding depression in nature, Jiménez et al. (1994) established a laboratory population of white-footed mice (*Peromyscus leucopus*) from wild-caught animals. Mice were either outbred (F_{IS} = 0) or inbred (F_{IS} = 0.25) in the lab and were then marked and released back into the same field location. The inbred mice were significantly less likely to be recaptured, and there were no differences between inbred and outbred mice in dispersal, strongly suggesting that inbreeding depression reduced survivorship. This kind of study, called a **mark-recapture study,** is the most common way of measuring survivorship in natural animal populations.

Heterosis has been very important in plant breeding, and is one of the main reasons why the seed planted by farmers is often the F_1 hybrid of a cross between inbred lines. In nature, heterosis can occur when individuals from different subpopulations are crossed (mated), because genetic drift can cause random fixation of different alleles in different subpopulations,

TABLE 3.3 *Summary of fitness effects of high and low frequency of heterozygotes*

	Subpopulations are	
	Small and isolated	Crossed with each other
Genotypic frequencies	Highly homozygous	Highly heterozygous
Fitness is	Low	High
Fitness effects called	Inbreeding depression	Heterosis or hybrid vigor
Fitness effects caused by	Deleterious recessives expressed or loss of heterozygote advantage	Deleterious recessives masked or heterozygote advantage

Note: The two causes of fitness differences are distinct (see text).

including deleterious recessives. Evidence for heterosis was found in a study of an intertidal crustacean, *Tigriopus californicus* (Edmands 1999). Edmands crossed individuals from populations separated by distances ranging from 5 m to 2000 km, and in general found increases in hatching success and survival in the F_1 offspring compared to the average of the parental populations.

Frequency dependent selection

In some cases the fitnesses of each genotype do not have a single value as in Table 3.2 and as in the previous examples. An important example of this situation is **frequency dependent selection,** in which the fitness of each genotype depends on its frequency in the population. In positive frequency dependence, genotypic fitnesses increase as their frequencies increase. This can occur in defensive mimicry systems, in which the predator more readily learns to avoid the common phenotype and thus attacks the rare one at a higher rate. A good example of this is the Müllerian mimicry system in tropical *Heliconius* butterflies. Both *Heliconius erato* and *H. melpomene* are distasteful to their bird predators, and they have different wing-color patterns in different regions (Figure 3.10). Mallet and Barton (1989) captured and released marked *H. erato* butterflies of both morphs shown in Figure 3.10 into the habitats where the other morph was found, and also back into their home habitat as a control. The morphs that did not match the common type were recaptured less than half as often as the morphs that did match the common type, suggesting strong positive frequency dependent selection.

Figure 3.10 Mimetic wing patterns in *Heliconius* butterflies from Peru. The butterflies on the left are *H. erato*, and those on the right are *H. melpomene*. The top pattern is called "postman" and the bottom "rayed." (After Mallet and Barton 1989.)

With negative frequency dependence, fitness increases as a genotype becomes rarer, slowing and eventually reversing the downward trend in frequency of that genotype. Conversely, fitness declines with increasing frequency, making it difficult for one allele to spread to fixation in a population. Therefore, negative frequency dependence helps maintain polymorphism, and thus genetic variation, in a population. There are more examples of negative than positive frequency dependence, at least in part because positive frequency dependence causes rapid fixation and is therefore rarely evident in natural populations. Also, negative frequency dependence captures the attention of population geneticists because it maintains genetic variation.

The examples of disassortative mating given in Chapter 2—two sexes and self-incompatibility in plants—are both examples of negative frequency dependent selection. This is why sex ratios that deviate substantially from 50:50 are so rare in nature. As one sex becomes rare, individuals of the rare sex have higher mating success because there are more of the opposite sex available, and this increases the numbers of the rarer sex. Similarly, self-incompatibility loci often have many alleles at roughly equally low frequencies, because individuals with common alleles are incompatible with some or all of the other individuals with that same allele and are thus at a mating disadvantage.

Negative frequency dependence is also found in predator–prey and parasite–host interactions, because predators and parasites tend to adapt to the most common prey or host, increasing the fitness of rare prey or hosts. Likewise, prey and host defenses tend to adapt to the most common predator and parasite types, increasing the fitness of rare predator and parasite phenotypes. A remarkable example is given by a scale-eating cichlid fish in Lake Tanganyika, *Perissodus microlepis* (Hori 1993). These fish approach other fish from behind and feed on their scales. The scale-eaters have asymmetrical jaws, so that some individuals have mouths facing to the left (sinistral) and others to the right (dextral). Sinistral fish attack the right side of their prey and dextrals always attack the left side. When sinistrals are common, prey fish preferentially defend their right sides, increasing the success rate of the rarer dextral fish and vice versa. As a result, the frequency of sinistrals and dextrals cycle regularly around 0.5 with a period of five years.

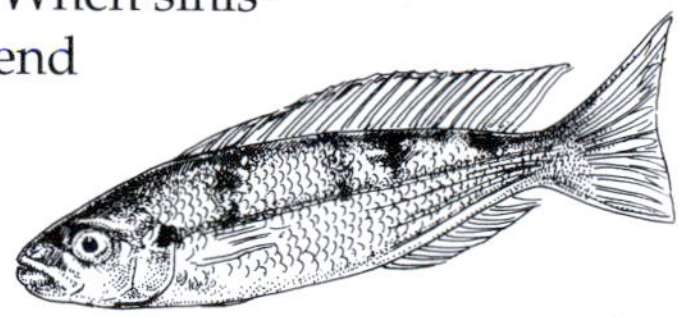

Perissodus microlepis

Population Genetics: Synthesis

Table 3.4 summarizes the effects of the four forces that change allele frequencies on genetic variation at two levels. Within subpopulations, we can measure variation as heterozygosity (H_I), whereas variation among subpop-

TABLE 3.4 *The effects on genetic variation of the four factors that change allele frequency*

	Level of variation[a]		
	Within subpopulations: Heterozygosity (H_I)	Among subpopulations: Differentiation (F_{ST})	Affect all loci equally?
Mutation	↑	↑	No
Gene flow (migration)	↑	↓	Yes
Drift	↓	↑	Yes
Selection	↑↓	↑↓	No

[a]Up arrows (↑) indicate an increase in variation; down arrows (↓) indicate a decrease in variation; dual arrows (↑↓) indicate both an increase and a decrease in variation.

ulations (differentiation) is measured with F_{ST}. Because mutation is the ultimate source of all genetic variation, it increases variation within subpopulations. Mutation also tends to increase differentiation, because the chance of the same random mutations occurring in different subpopulations is low. Recall that both these effects are slow relative to the other three forces due to the rarity of most mutations. Mutations are not equally likely at all loci.

As explained earlier (compare Figures 3.3 and 3.6), gene flow and genetic drift are opposite in their effects—gene flow tends to convert among-subpopulation variation (differentiation) to within-subpopulation variation, while genetic drift converts within-subpopulation variation to among-subpopulation variation. There are situations in which these effects on within-subpopulation variation would not be true, but they are less common. For example, under the island model, gene flow would decrease H_I if $p_i = 0.5$ and $\bar{p} \neq 0.5$, or have no effect if $p_i = \bar{p}$ (see Equation 3.3). Gene flow and genetic drift both affect all loci equally on average, because it is whole organisms with their entire genomes that migrate, not individual loci, and it is entire haploid genomes in the gametes that are sampled in low numbers to produce drift in small populations. One exception to this rule for migration is that if there are gene loci affecting the probability of migration, any loci linked to these loci would be affected more by migration than would the rest of the genome.

Natural selection is unique because it can have all possible effects on variation at both levels. Within subpopulations, selection can decrease variation if one homozygote is favored, or conversely increase or maintain variation if there is heterozygote advantage, negative frequency dependence, or disruptive selection (see Chapter 6). Current evidence suggests that neither heterozygote advantage nor disruptive selection is very common, so it is probably safe to say that selection usually decreases variation within subpopulations. If the environments that different subpopulations inhabit differ in some important factor (light, temperature, predation, population density, and so on), as will often be the case, then selection for **local adaptation** can occur, leading to an increase in differentiation. Conversely, any selection that does not differ between subpopulations, due to similar environments or fundamental features of a species, will lead to decreases in differentiation. Both these can be occurring at the same time because selection does not act on the entire genome. Selection acts on the phenotype, so it affects only variable gene loci that affect the given phenotypic trait under selection, as well as genes linked to these loci.

The fact that genetic drift and gene flow affect all loci whereas selection does not is paralleled in the two types of nonrandom mating we discussed in Chapter 2. Inbreeding affects all loci because related individuals are related across their genomes, whereas assortative mating affects only loci that affect the phenotypic traits that are used as cues for mating (plus linked loci). So the key to whether something affects the whole genome is usually whether or not it acts through the phenotype.

Interactions Between the Four Evolutionary Forces

We will now examine two or three of the evolutionary forces together, to explore how they interact in determining genetic variation within and among subpopulations. This involves relaxing more than one Hardy-Weinberg assumption at a time. In general, mutation and genetic drift are weaker than selection and migration in determining allele frequencies, except when selection is very weak (as in the case of a rare deleterious recessive) or when effective population size is very small. It typically takes very little migration or convergent selection to overcome differentiation due to genetic drift or mutation. Drift and mutation are more important in determining levels of molecular variation, such as DNA sequences or allozymes, than higher level phenotypic traits because, in general, selection will be weaker on molecular variation. Molecular variation may often be **neutral** or nearly neutral, meaning that this variation has little or no effect on fitness, whereas higher-level phenotypic traits will often have important fitness effects (see Chapter 6).

Mutation–selection balance

Recall from Figure 3.8 that selection acts to eliminate deleterious recessive alleles, but that this process is extremely slow when the alleles become rare. The slow process of elimination can be balanced by mutations creating new copies of the deleterious allele, creating an equilibrium. This **mutation–selection balance** can maintain the deleterious allele at low frequency. Two equations approximate this equilibrium frequency for two cases:

Case 1. When the harmful allele is completely recessive ($h = 0$):

$$\hat{q} \approx \sqrt{\frac{\mu}{s}} \quad \textbf{3.16}$$

Case 2. When the harmful allele is partially dominant ($h > 0$):

$$\hat{q} \approx \frac{\mu}{hs} \quad \textbf{3.17}$$

Note that q is the frequency of deleterious allele (this is a convention in most population genetics selection models), and that all terms in these equations refer to this allele. Usually this is also the recessive allele, but in rare cases the dominant allele can be deleterious; its frequency would still be denoted as q, which can be confusing. In both cases the equilibrium frequency of the deleterious allele ($\hat{q}$) is lower if the mutation rate to that allele (μ) is lower, or if selection against it (s) is stronger. The second case (Equation 3.17) is that depicted in Figure 3.7, that is, h is the degree of dominance of the disfavored allele. For a given ratio of mutation rate to selection (μ/s), the equilibrium frequency is higher the more recessive the allele is. For example, if μ/s is 0.01, then $\hat{q}$ is 0.1 when the allele is completely recessive (from Equation 3.16) and 0.01 when the allele is completely dominant (Equation 3.17, $h = 1$). This is because selection is more effective against the dominant allele (see Figure 3.8).

For example, Waser and Price (1981) studied a flower color polymorphism in Rocky Mountain populations of *Delphinium nelsonii*. Most plants had blue flowers, but white-flowered plants occurred at an average frequency of 7.4×10^{-4}. Pollinators visited the blue-flowered plants at a higher rate, resulting in an average seed production of 86.4 for blue-flowered plants versus 53.2 for white-flowered plants.

EXERCISE: Assuming that the flower-color polymorphism is due to a single locus with the blue allele completely dominant to the white, and also assuming that these populations are at equilibrium and approximately Hardy-Weinberg frequencies, what is the mutation rate to the white allele?

ANSWER: Relative fitness (w) of the white homozygote is 53.2/86.4 = 0.62, so the selection coefficient (s) against it is 0.38. Solving Equation 3.16 for μ gives $\mu = sq^2 = 0.38 \times 7.4 \times 10^{-4} = 2.8 \times 10^{-4}$. This calculation is based only on female fitness (seed production) and ignores male fitness (seeds sired), but male fitness is likely to be similarly affected by pollinator visitation.

Selection vs. migration

Selection for local adaptation in different subpopulations can cause differentiation if the selection coefficient *s* averaged across subpopulations is greater than the fraction of migrants each generation *m*. (This is a rough approximation; for more detail see Lenormand 2002.) The greater the difference between *s* and *m*, the greater the differentiation, but only if different alleles are favored in the different populations, as is implied by selection for local adaptation. Of course, this differentiation will only be at the loci affecting the traits selection acts upon.

An example of the balance between selection and migration comes from the phenomenon of industrial melanism mentioned in Chapter 2. As industrial pollution declined in the mid-1900s, soot became confined to the areas around some major industrial cities. This situation caused selection for melanic (dark) moths near the cities and for the lighter nonmelanic forms farther away where the air was cleaner. Local adaptation occurred in two species of moths, the peppered moth and the scalloped hazel moth, showing that selection was stronger than gene flow. However, the differentiation for the melanic alleles existed over shorter distances in the hazel moth, due to less gene flow in this species compared to the peppered moth (Figure 3.11). Therefore, adaptive local differentiation in the peppered moth was reduced by gene flow, an example of an **evolutionary constraint** (see Chapter 6 for a more detailed description of constraints).

Selection vs. genetic drift

Selection will primarily determine allele frequencies if $s > 1/(2N_e)$, otherwise the allelic variants are **nearly neutral** or **effectively neutral,** that is their frequencies are determined more by genetic drift than by selection. Therefore, the smaller the effective population size, the stronger selection has to be (larger selection coefficient) to counteract the effects of genetic drift. This means that genetic drift can also constrain adaptive evolution by decreasing the frequency of adaptive alleles and increasing the frequency of deleterious alleles. In the *Tribolium* example in Figure 3.5, with $N_a = 10$, one of the subpopulations fixed for the deleterious black allele.

Eckert, Manicacci, and Barrett (1996) provide an example of the balance between genetic drift and selection in the maintenance of a simple floral polymorphism in purple loosestrife (*Lythrum salicaria*). Purple loosestrife is a wetland plant native to Europe that became invasive after being introduced to North America. The three morphs are maintained by negative frequency dependence arising from disassortative mating, but morphs can be lost through genetic drift. Figure 3.12 shows the numbers of populations in different size classes that still have all three morphs in gray, and those that have lost one or two morphs through genetic drift in black. In France, where the

(A)

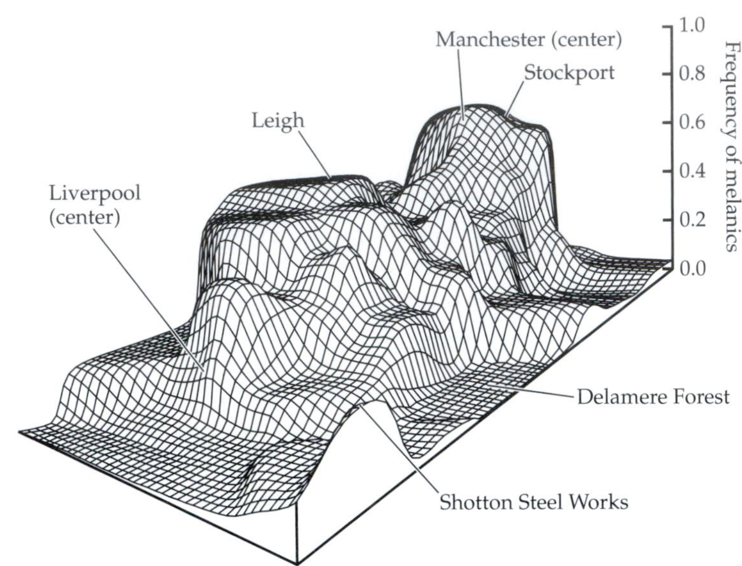
Manchester (center)
Stockport
Leigh
Liverpool (center)
Delamere Forest
Shotton Steel Works
1.0
0.8
0.6
0.4
0.2
0.0
Frequency of melanics

(B)

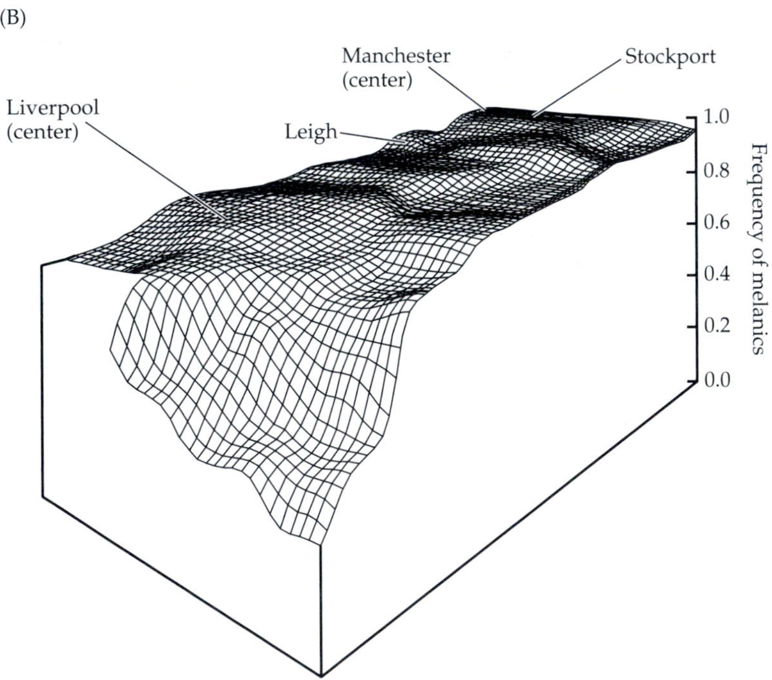
Manchester (center)
Stockport
Liverpool (center)
Leigh
1.0
0.8
0.6
0.4
0.2
0.0
Frequency of melanics

Figure 3.11 The balance between selection for local adaptation and gene flow in two moth species in England. The height of the surface represents the frequency of melanic forms at different points in northwest England. (A) Melanic frequency in scalloped hazel moths. (B) Peppered moths. (After Bishop and Cook 1975.)

plant is native, only the smallest populations have lost morphs because selection is more effective than drift in the larger populations and thus all three morphs are maintained. In Ontario, however, morphs have been lost even in populations between 200 and 500 individuals. The interpretation is that the introduced Ontario populations were founded by small numbers of individuals, and in some cases the founders lacked one or two of the morphs, or they were in such low frequency in the founders that they were lost through genetic drift before the population could increase in size enough for selection to be effective in maintaining them.

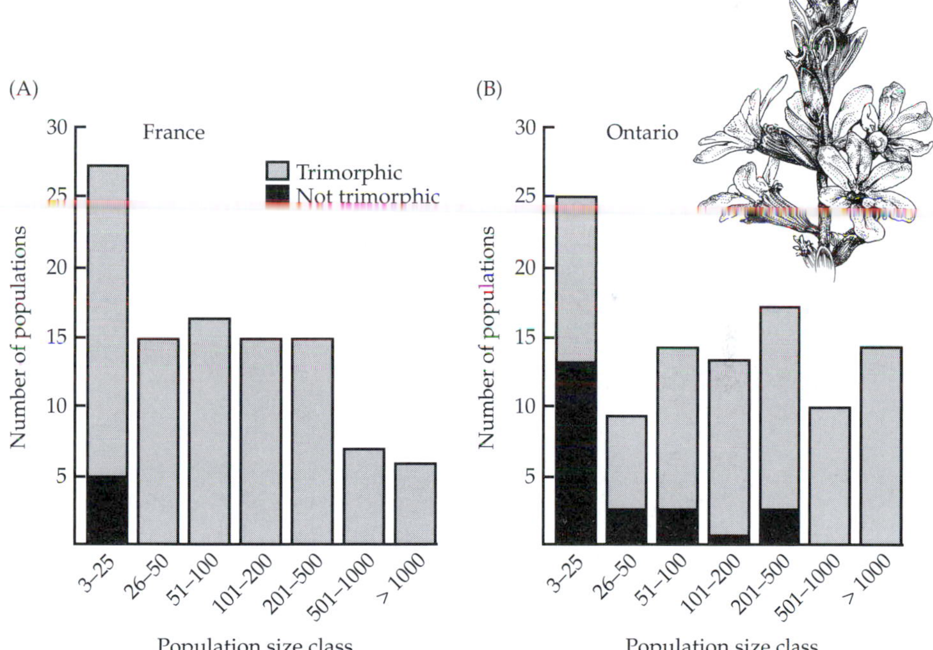

Figure 3.12 Numbers of populations of purple loosestrife, (*Lythrum salicaria*) with all three floral morphs (gray portions of bars) and those that have lost one or two morphs (black portions of bars), plotted versus population size. (A) Native populations in France. (B) Introduced populations in Ontario. (After Eckert, Manicacci, and Barrett 1996.)

Migration vs. genetic drift

Assuming the island model of migration with no mutation or selection, the level of differentiation at an equilibrium between migration and genetic drift is approximated by:

$$F_{ST} = \frac{1}{1+4N_e m} \qquad \textbf{3.18}$$

From this equation, we can see that as either N_e or m increases, differentiation decreases. This makes intuitive sense, because large N_e means less genetic drift to create differentiation, and lots of migration mixes alleles among subpopulations, reducing whatever differentiation genetic drift creates. The term $N_e m$ approximates the number of migrants per generation, since it is the product of the effective population size and the proportion of those individuals that are migrants. The decline in F_{ST} with increasing $N_e m$ is extremely rapid (Figure 3.13). Therefore, with only one migrant per generation, $F_{ST} = 0.2$, which means that this is enough gene flow to prevent very great differentiation due to genetic drift.

A surprising feature of this result is that the number of migrants needed to maintain a given level of differentiation is independent of subpopulation size. This is true because a smaller proportion of migrants is needed to coun-

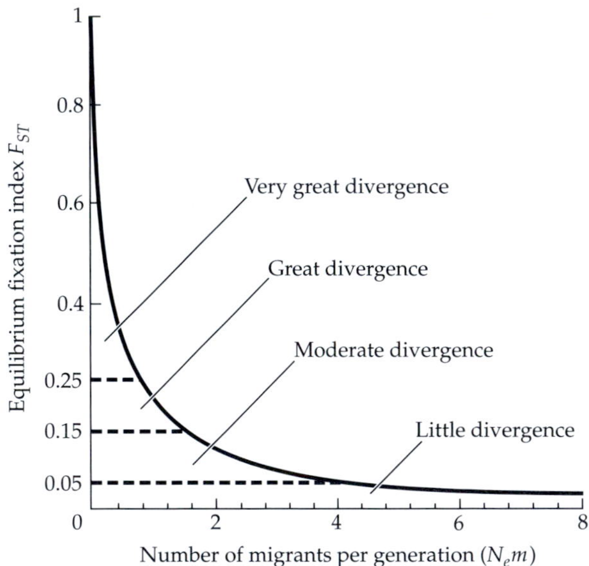

Figure 3.13 Theoretical equilibrium population differentiation (measured as F_{ST}) versus number of migrants per generation, assuming the island model of migration. The curve is from Equation 3.18.

teract the smaller amount of genetic drift in larger populations. For example, with a subpopulation effective size of 5, there is lots of genetic drift, and one migrant per generation is the high proportion ($m = 0.2$) needed to counteract this high level of drift. On the other hand, if $N_e = 100$, then one migrant per generation is only a small proportion ($m = 0.01$), but there is little genetic drift to counteract.

Methods of measuring gene flow

Both direct and indirect methods for measuring gene flow have been used. The direct methods usually involve marking animals and recapturing them to determine which have migrated, or observations of seed dispersal and pollinator movements in plants. Ideally such studies would also include measurements of mating success of the migrants in their new population, but this is rarely done. The directness of the estimates of gene flow derived from these methods is appealing, but these estimates have drawbacks. The studies required are extremely time consuming and difficult to do properly. An even bigger problem is that the studies encompass a short time scale, whereas gene flow may be episodic, occurring mostly after a major disturbance or other change in the habitat. These episodes could account for a large amount of gene flow, but could be missed by short-term direct studies. This problem is analogous to the effects of historical bottlenecks on effective population size discussed earlier.

For these reasons, indirect methods have been developed that are based on the drift–migration equilibrium discussed above. These methods are less time-consuming than direct methods and they include the effects of past gene flow events. The indirect methods also have serious problems of their own, however, so results should be interpreted with caution (Whitlock and McCauley 1999). To obtain an indirect estimate of gene flow, molecular markers are scored on a sample of individuals from a number of subpopulations, and these are used to estimate F_{ST}. This estimate of F_{ST} is then used to estimate the number of migrants per generation, $N_e m$, from Equation 3.18. This method assumes the island model of migration, and that no mutation or selection affects differentiation. The assumption of no mutation is likely to be safe due to the rarity of mutation, except perhaps at highly mutable markers like microsatellites. The assumption of no selection is more problematic, but many molecular markers, especially DNA markers, are likely neutral or nearly neutral so that selection will not be a problem in many cases. If multiple markers (loci) are used, and all give similar answers, then this suggests that selection is not affecting differentiation. An F_{ST} for one locus that differs substantially from the other loci suggests that selection is acting at that locus. If F_{ST} is greater than the other loci, this suggests selection for local adaptation at that locus, or one linked to it, is creating differentiation; and if F_{ST} is much lower at one locus, it suggests homogenizing selection across subpopulations is reducing differentiation.

For example, Paul Ehrlich and many students and colleagues have studied the ecological genetics of the butterfly *Euphydryas editha* for several decades. Slatkin (1987) used some of their allozyme data to estimate gene flow among 21 subpopulations (Table 3.5). The F_{ST} values for seven of the loci agree fairly closely, within the expected range of random error, and show low levels of differentiation (F_{ST} of 0.06 or less). This suggests overall high levels of gene flow among the subpopulations. The F_{ST} value for the *hk* locus is much higher, 0.29, showing very high differentiation; this suggests selection for local adaptation at *hk* or a locus closely linked to *hk*. Using Equation 3.18 but excluding *hk*, the average number of migrants per generation (N_em) was estimated as 7.8, consistent with the interpretation of high gene flow. However, Ehrlich and colleagues performed extensive mark–recapture studies in these populations, and they rarely recaptured them in subpopulations outside where they were marked. Their direct estimates of the number of migrants per generation from these studies was only 0.1. The large discrepancy between the direct and indirect estimates of gene flow suggests the possibility of a great deal of gene flow in the recent past, or that the indirect method is not accurately estimating gene flow.

Further examples of estimates of N_em and F_{ST} from a variety of organisms are given in Table 3.6.

EXERCISE: Calculate some of the N_em values in Table 3.5; make sure your values agree with those in the table.

ANSWER: To do this calculation, Equation 3.18 needs to be rearranged to give N_em in terms of F_{ST}:

$$N_em = \frac{1}{4F_{ST}} - \frac{1}{4} \qquad \textbf{3.19}$$

Selection, genetic drift, and gene flow: Wright's shifting balance theory

Sewall Wright proposed a comprehensive view of evolution, called the *shifting balance theory*, that works through an interaction between genetic drift, migration, and selection. Although the importance of the shifting balance process in nature remains controversial (Whitlock and Phillips 2000), it provides a useful framework for thinking about evolution. Central to the shifting balance is the **adaptive topography** or **adaptive landscape** (Figure 3.14),

TABLE 3.5 *Values of F_{ST} from eight allozyme loci and estimates of $N_e m$ estimated from the F_{ST} values using Equation 3.18*

Locus	F_{ST}	$N_e m$ (estimated)
pgm	0.028	8.7
pgi	0.052	4.6
hk	0.291	0.6
got	0.017	14.5
ak	0.062	3.8
bdh	0.034	7.1
α-gpdh	0.027	9.0
to	0.035	6.9

Source: Data from Slatkin 1987.

TABLE 3.6 *Estimates of Nm and F_{ST}*

Species	Type of organism	Estimated $N_e m$	Estimated F_{ST}
Stephanomeria exigua	Annual plant	1.4	0.152
Mytilus edulis	Mollusc	42.0	0.006
Drosophila willistoni	Insect	9.9	0.025
Drosophila pseudoobscura	Insect	1.0	0.200
Chanos chanos	Fish	4.2	0.056
Hyla regilla	Frog	1.4	0.152
Plethodon ouachitae	Salamander	2.1	0.106
Plethodon cinereus	Salamander	0.22	0.532
Plethodon dorsalis	Salamander	0.10	0.714
Batrachoseps pacifica ssp. 1	Salamander	0.64	0.281
Batrachoseps pacifica ssp. 2	Salamander	0.20	0.556
Batrachoseps campi	Salamander	0.16	0.610
Lacerta melisellensis	Lizard	1.9	0.116
Peromyscus californicus	Mouse	2.2	0.102
Peromyscus polionotus	Mouse	0.31	0.446
Thomomys bottae	Gopher	0.86	0.225

Source: Data from Slatkin 1985.

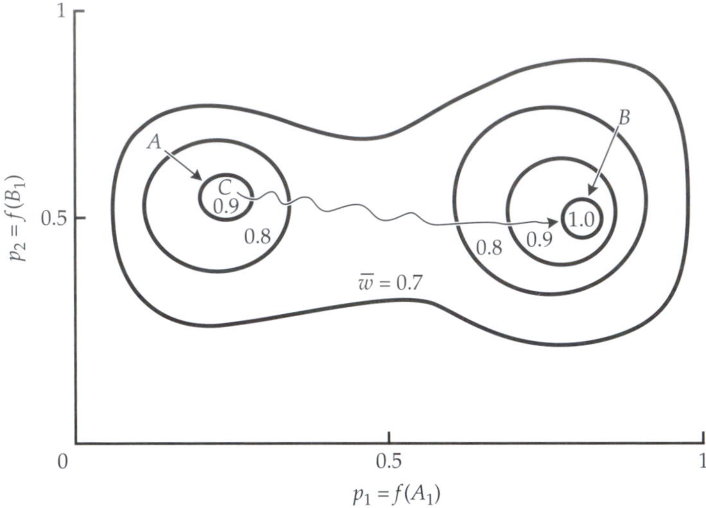

Figure 3.14 Hypothetical adaptive landscape. The axes represent the frequency (p) of one allele at two loci (A and B); each have two alleles (therefore the axes represent $1 - q$ as well). The contours are lines of equal population mean fitness, with one peak at $w = 0.9$ and a higher peak at $w = 1$. The initial positions of three populations are shown by the letters A–C, with their evolutionary trajectories depicted by the arrows.

in which the axes are allele frequencies at two loci and the surface depicted with the contour lines is fitness averaged over all individuals in the population. This is exactly the same type of plot as Figure 3.9C, except that the adaptive landscape is extended to two loci instead of only one. Each contour line represents combinations of allele frequencies at the two loci that have the same population mean fitness, just as a contour line on a topographic map represents locations of equal elevation in two-dimensional space. It is critical to stress that the adaptive topography we are discussing is at the population, not the individual, level—the axes are allele frequencies, not genotypes, and the contour lines show population mean fitness. Wright himself had at least two different conceptions of the adaptive landscape, which has contributed to confusion on this issue (Provine 1986, Chapter 9). Another common misconception is to think of the adaptive landscape as representing the spatial area in which the populations live. The "landscape" here has nothing to do with an actual landscape, because the axes are not space but rather allele frequencies. This is in marked contrast to Figure 3.11, in which the horizontal axes *do* represent geographical space and the contours depict frequency of the melanic morphs.

The shape of the topography is caused partly by **epistasis,** a form of gene action in which two or more loci interact with each other to determine the phenotype (fitness here). You can visualize epistasis for population mean fitness in Figure 3.14 by choosing one allele frequency at one of the loci (say, $p_2 = 0.5$) and noting how fitness varies at that one allele frequency with changes in allele frequency at the other locus. Epistasis is discussed further in the next chapter.

Figure 3.14 depicts two adaptive peaks, that is, joint allele frequencies at the two loci that lead to maximal population mean fitness. The one with mean fitness equal to 0.9 is a local adaptive peak, that is, a local fitness maximum but lower than the highest mean fitness. It is also a locally stable equilibrium under selection alone. The arrows show **evolutionary trajectories** of populations, that is, possible ways that the joint allele frequencies might evolve across generations. The arrows labeled *A* and *B* show the trajectories in large populations under the influence of selection alone; recall that selection by itself always increases population mean fitness. This means that with selection only, a population at the peak with mean fitness of 0.9 is stuck—selection cannot move the population through the valley of lower mean fitness toward the peak with mean fitness equal to 1. Wright's idea was that if populations were small enough, genetic drift would sometimes move the population down the slope and through a fitness valley onto the slope of a higher peak. Once there, selection can move the population up to the higher peak. These steps are depicted by the trajectory labeled *C*, and are the first two phases of the shifting balance process (genetic drift moving a population through a fitness valley and selection moving the population up to a new peak).

The third and final phase of the shifting balance process is called **interdemic selection.** Imagine that there was another population still at the lower peak after the first population moved to the higher peak. The first population has higher mean fitness and would therefore produce more offspring each generation. If there is migration between the populations, then this difference in productivity would mean that the net migration would be from the more fit population to the less fit population. This would tend to move the less fit population down the slope and through the valley, because the migrants from the more fit population shift the allele frequencies of the less fit population. Through this three-phase process, all populations can theoretically be brought to the highest peak.

PROBLEMS

3.1. The fly *Eurosta solidaginis* forms galls (enlarged areas) on goldenrod plants within which the fly larvae feed and develop. A study of allozyme frequencies in 21 subpopulations of *E. solidaginis* on two different species of goldenrod distributed from Minnesota to Maine (Waring, Abrahamson, and Howard 1990) produced the following estimates of F_{ST}:

Locus	F_{ST}
GAP-1	0.10
TPI-1	0.11
IDH-1	0.02
HBDH-1	0.62
GPD-1	0.11
PGM-1	0.14

a. Calculate the expected number of migrants per generation for each locus. What is your interpretation of these data? Do you think differentiation at these loci is caused solely by a balance between migration and genetic drift? Why or why not?

b. Most of the differentiation at *HBDH-1* shown in the data above occurs between the two species of host plants; nine of the subpopulations occurred on *Solidago altissima* and the other 12 subpopulations were on *S. gigantea*. The frequency of one of the two *HBDH* alleles is 0.84 on *Solidago altissima* and 0.13 on *S. gigantea*. How does this fact affect your interpretation of the results?

3.2 Levin (1978) studied allele frequencies at the 6-*pgd* allozyme locus in 73 subpopulations of the self-incompatible species *Phlox drummondii*. Of these 73, 66 were fixed for the *a* allele, with allele frequencies and observed heterozygosities at the other loci given below (Levin's original subpopulation numbering altered for simplicity):

Subpopulation	p	H_I
1–66	1	0
67	0.86	0.06
68	0.8	0.12
69	0.7	0.2
70	0.96	0.03
71	0.96	0.09
72	0.73	0.15
73	0.91	0.06

a. Calculate the three *F*-statistics for these data and check to make sure they have the correct mathematical relationship. Compare them with those calculated for the self-compatible *P. cuspidata* calculated in the text box (pp. 60–61). Do these comparisons fit your expectations based on the mating systems of these two species? Why or why not? If not, do you have a possible explanation for the lack of fit?

b. Calculate $N_e m$ from these data. If the migration rate m was found to be 0.1, what would your estimate of N_e be?

c. Now assume that N_e is very large, 6-*pgd* is neutral in these subpopulations, and the migration rate is still 0.1. What would the frequencies of the *a* allele in subpopulations 68 and 69 after 10 generations be? After 25 generations? What biological principle is illustrated by these results?

3.3 A rare triggerplant from Australia (*Stylidium coroniforme*) has only five known populations (Coates 1992). One of these populations has been monitored for several years, and over five years in the early 1980s, 2, 3, 25, 32, and 86 plants were recorded. Assuming that N_e in each year equals N_a in that year, estimate N_e as well as the average N_a over this period. What biological principle is illustrated by the difference between the two?

3.4 Due to suburban development, a large panmictic population of moles is split into a metapopulation of five subpopulations of 20 males and 20 females each, with no migration among subpopulations. The variance in reproductive success is equal to 2 in females but is 8 in males, due to a few dominant males fathering most of the offspring.

a. Estimate N_e within each of these subpopulations.

b. Calculate what F_{ST} will be at 5 and at 10 generations after the split if the subpopulations remain at the N_e you estimated in part (a). Also describe in words what has happened.

3.5 The sex ratio in fig wasps becomes even more female-biased as the number of females laying eggs in one fig declines. In figs with only one wasp mother, there were an average of 72 wasps per fig but only 10% were male. Calculate N_e in this case, and explain why this answer is different than the fig wasp example in the text (see pg. 62).

3.6 Selander and Yang (1969) trapped wild mice (*Mus musculus*) from four large chicken barns near Ramona, California, and performed electrophoresis on a number of loci. F_{ST} values for three of these loci (hexose-6-phosphate dehydrogenase, NADP-isocitrate dehydrogenase, and hemoglobin) were 0.10, 0.16, and 0.11, respectively, for an average $F_{ST} = 0.12$. Assuming that the four subpopulations in each barn were founded at the same time from the same large panmictic population, there is no migration among the barns, and the effective subpopulation size has been the same in each barn since they were founded, estimate how many years ago the subpopulations were established if $N_e = 20$. How long ago if $N_e = 100$? Explain the biological reason for the difference between the two.

3.7 Stanton, Snow, and Handel (1986) studied selection on a floral color polymorphism in wild radish (*Raphanus raphanistrum*). Pollinators preferred yellow-flowered plants, so average total fitness (estimated seed production and seed siring) was 88.3 for yellow-flowered plants and 56.6 for white-flowered plants. Floral color is determined by a single locus with the white allele completely dominant to the yellow allele.

a. Calculate relative fitness for the three genotypes and the selection coefficient. What is unusual about this case?

b. The frequency of yellow-flowered plants in this study population averaged 0.85. Assuming that this frequency results from an equilibrium between mutation and selection, and the population is in HWE, what is the mutation rate to the white allele? Do you think that the assumption of an equilibrium between mutation and selection is reasonable here? Why or why not? Give one alternative explanation for the frequency of yellow plants.

c. What would the mutation rate be if the white allele was completely recessive, assuming the same allele frequencies as above? What biological principle is illustrated by the difference among these two mutation rates?

SUGGESTED READINGS

Barrett, S. C. H., and D. Charlesworth. 1991. Effects of a change in the level of inbreeding on the genetic load. Nature 352:522-524. An experimental study of the effects of inbreeding and outcrossing on fitness.

Crow, J. F. 1991. Was Wright right? Science 253:973. Commentary on the Wade and Goodnight paper (see citation below).

Crow, J. F., and M. Kimura. 1970. *An Introduction to Population Genetics Theory*. Harper & Row, New York. The standard reference for the mathematical theory underlying population genetics.

Ford, E. B. 1975. *Ecological Genetics*. Chapman & Hall, London. A summary of classical work, mainly in Britain, on changes in allele frequencies for visible polymorphisms in nature.

Hartl, D. L. 2000. *A Primer of Population Genetics,* 3rd Edition. Sinauer Associates, Sunderland, MA. A more mathematical introduction to the field.

Hartl, D. L., and A. G. Clark. 1997. *Principles of Population Genetics,* 3rd Edition. Sinauer Associates, Sunderland, MA. A detailed and comprehensive treatment of population genetics.

Lewontin, R. C. 1974. *The Genetic Basis of Evolutionary Change*. Columbia University Press, New York. A survey of the important questions in population genetics by a leader in the field.

McCauley, D. E. 1994. Contrasting the distribution of chloroplast DNA and allozyme polymorphism among local populations of *Silene alba*: Implications for studies of gene flow in plants. Proc. Natl. Acad. Sci. USA 91:8127–8131. This study compares indirect estimates of gene flow using two different types of markers.

*Slatkin, M. 1987. Gene flow and the geographic structure of natural populations. Science 236:787–792. A review of the effects of gene flow in evolution and some methods for estimating gene flow.

Taylor, M. F. J., Y. Shen, and M. E. Kreitman. 1995. A population genetic test of selection at the molecular level. Science 270:1497–1499. This paper provides evidence for differential selection among populations at a single gene locus.

Wade, M. J., and C. J. Goodnight. 1991. Wright's shifting balance theory: An experimental study. Science 253:1015–1018. One of the few experimental tests of the shifting balance theory.

**Indicates a reference that is a suggested reading in the field and is also cited in this chapter.*

SUGGESTED READINGS QUESTIONS

The following questions are based on papers from the primary literature that address key concepts covered in this chapter. For the full citation for each paper please see Suggested Readings.

From Barrett, S. C. H. and D. Charlesworth. 1991. Effects of a change in the level of inbreeding on the genetic load.

1. What is genetic load? What theoretical prediction concerning genetic load are the authors testing?
2. What was the main important difference between the two populations used?
3. What was the experimental procedure imposed on the populations?
4. How did the results differ between the two populations? Does this difference support or refute the theoretical prediction referred to in question 1?
5. What specific comparison in the data enables them to distinguish between overdominance and deleterious recessives as a mechanism for inbreeding depression?

From McCauley, D. E. 1994. Contrasting the distribution of chloroplast DNA and allozyme polymorphism among local populations of *Silene alba*: Implications for studies of gene flow in plants.

1. What two sources of gene flow is McCauley trying to distinguish between?
2. What is the most important difference between the two types of genetic markers the author used, and how does this difference help him distinguish between the two sources of gene flow?
3. What analysis did he use to obtain information about gene flow from the genetic marker data? How does this analysis estimate gene flow? Is this a direct or indirect estimate?
4. What do his results suggest about the relative amounts of gene flow through the two sources?
5. What potentially confounding effects exist in this study? In other words, what besides gene flow could affect the results?

From Slatkin, M. 1987. Gene flow and the geographic structure of natural populations.

1. How do genetic drift, gene flow, and selection affect the level of genetic differentiation among populations (i.e., how do they affect differences in allele frequencies among populations)? Do they increase differentiation, decrease differentiation, or both?
2. What are the relative strengths of genetic drift, gene flow, and selection in determining genetic differentiation among populations? What variables need to be measured to determine this?
3. What equation estimates the amount of differentiation resulting from genetic drift and gene flow in combination? How does selection affect this estimate?
4. What are direct and indirect methods of estimating of gene flow? What are some advantages and disadvantages of each?

From Taylor, M. F. J., Y. Shen, and M. E. Kreitman. 1995. A population genetic test of selection at the molecular level.

1. Why was *Hpy* predicted to have higher levels of differentiation than *Hejs*?
2. What was the authors' evidence for geographic variability in selection for pyrethroid resistance?
3. Did the data support the prediction in question 1 above?

From Wade, M. J. and C. J. Goodnight. 1991. Wright's shifting balance theory: An experimental study.

1. What are the three phases of the shifting balance process?
2. How are the processes envisioned by Wright in Phase 1 potentially in conflict with Phases 2 and 3? Focus on the roles of genetic drift, migration, and selection.
3. Describe the main features of the experimental procedure used by Wade and Goodnight. What is the key difference between the experimental and the control arrays?
4. Which phase of the shifting balance process does this experiment test? Why did they focus on this phase? (See also the 1991 Crow suggested reading.)
5. What do their results suggest about the plausibility of this one phase of the shifting balance process?

CHAPTER REFERENCES

Bishop, J. A., and L. M. Cook. 1975. Moths, melanism, and clean air. Sci. Am. 232:90–99.

Buri, P. 1956. Gene frequency in small populations of mutant *Drosophila*. Evolution 10:367–402.

Caccone, A. 1985. Gene flow in cave arthropods: A qualitative and quantitative approach. Evolution 39:1223–1235.

Clutton-Brock, T. H., S. D. Albon, and F. E. Guinness. 1988. Reproductive success in male and female red deer. Pp. 325–343 in T. H. Clutton-Brock, ed. *Reproductive Success*. The University of Chicago Press, Chicago.

Coates, D. J. 1992. Genetic consequences of a bottleneck and spatial genetic structure in the triggerplant *Stylidium coroniforme* (Stylidiaceae). Heredity 69:512–520.

Conner, J. K., R. Davis, and S. Rush. 1995. The effect of wild radish floral morphology on pollination efficiency by four taxa of pollinators. Oecologia 104:234–245.

Conner, J. K., S. Rush, and P. Jennetten. 1996. Measurements of natural selection on floral traits in wild radish (*Raphanus raphanistrum*). I. Selection through lifetime female fitness. Evolution 50:1127–1136.

Eckert, C. G., D. Manicacci, and S. Barrett. 1996. Genetic drift and founder effect in native versus introducted populations of an invading plant, *Lythrum salicaria* (Lythraceae). Evolution 50:1512–1519.

Edmands, S. 1999. Heterosis and outbreeding depression in interpopulation crosses spanning a wide range of divergence. Evolution 53:1757–1768.

Falconer, D. S., and T. F. C. Mackay. 1996. *Introduction to Quantitative Genetics*. Longman, Harlow, UK.

Frank, S. A. 1985. Hierarchical selection theory and sex ratios. II. On applying the theory, and a test with fig wasps. Evolution 39:949–964.

Graur, D., and W.-H. Li. 2000. *Fundamentals of Molecular Evolution*, 2nd Edition. Sinauer Associates, Sunderland, MA.

Hori, M. 1993. Frequency-dependent natural selection in the handedness of scale-eating cichlid fish. Science 260:216–219.

Jiméñez, J. A., K. A. Hughes, G. Alaks, L. Graham, and R. C. Lacy. 1994. An experimental study of inbreeding depression in a natural habitat. Science 266:271–273.

Keller, L. F., and D. M. Waller. 2002. Inbreeding effects in wild populations. Trends in Ecology and Evolution 17:230–241.

Kruger, O., J. Lindstrom, and W. Amost. 2001. Maladaptive mate choice maintained by heterozygote advantage. Evolution 55:1207–1214.

Lenormand, T. 2002. Gene flow and the limits to natural selection. Trends in Ecology and Evolution 17:183–189.

Levin, D. A. 1978. Genetic variation in annual phlox: Self–compatible versus self-incompatible species. Evolution 32:245–263.

Mallet, J., and N. H. Barton. 1989. Strong natural selection in a warning-color hybrid zone. Evolution 43:421–431.

Mitton, J. B. 1997. *Selection in Natural Populations.* Oxford University Press, New York.

Provine, W. B. 1986. *Sewall Wright and Evolutionary Biology.* University of Chicago Press, Chicago.

Rank, N. E. 1992. A hierarchical analysis of genetic differentiation in a montane leaf beetle *Chrysomela aeneicollis* (Coleoptera: Chrysomelidae). Evolution 46:1097–1111.

Rich, S. S., A. E. Bell, and S. P. Wilson. 1979. Genetic drift in small populations of *Tribolium.* Evolution 33:579–584.

Selander, R. K., and S. Y. Yang. 1969. Protein polymorphism and genic heterozygosity in a wild population of the house mouse (*Mus musculus*). Genetics 63:653–667.

Slatkin, M. 1985. Rare alleles as indicators of gene flow. Evolution 39:53–65.

Stanton, M. L., A. A. Snow, and S. N. Handel. 1986. Floral evolution: attractiveness to pollinators increases male fitness. Science 232:1625–1627.

Templeton, A. R., and B. Read. 1994. Inbreeding: One word, several meanings, much confusion. Pp. 91–105 in V. Loeschske, J. Tomiuk and S. K. Jain, eds. *Conservation Genetics.* Birkhauser Verlag, Basel, Switzerland.

Waring, G. L., W. G. Abrahamson, and D. J. Howard. 1990. Genetic differentiation among host–associated populations of the gallmaker *Eurosta solidaginis* (Diptera:Tephritidae). Evolution 44:1648–1655.

Waser, N. M., and M. V. Price. 1981. Pollinator choice and stabilizing selection for flower color in *Delphinium nelsonii.* Evolution 35:376–390.

Weir, B. S. 1996. *Genetic Data Analysis II.* Sinauer Associates, Sunderland, MA.

Whitlock, M. C., and D. E. McCauley. 1999. Indirect measures of gene flow and migration: $F_{ST} \neq 1/(4Nm + 1)$. Heredity 82:117–125.

Whitlock, M. C., and P. C. Phillips. 2000. The exquisite corpse: A shifting view of the shifting balance. Trends in Ecology and Evolution 15:347–348.

4

Quantitative genetics I: Genetic variation

Mendelian Basis of Continuous Traits

The previous chapters have focused on the population genetics of single loci with only two alleles. If a large majority of variation in a phenotypic trait is determined by one locus, the result is a visible polymorphism like Mendel's pea traits or flower color in *Delphinium* (see Chapter 2). However, as noted in Chapter 2, most phenotypic traits do not fall into distinct categories, but rather are continuously distributed. Examples of continuously distributed traits are shown in Figure 4.1. These traits are called **quantitative** or **metric** traits because they need to be measured, not just scored as round or wrinkled or blue or white. In classical or statistical **quantitative genetics,** the phenotypes of individuals of known genetic relationship (usually parents and offspring or siblings) are measured, and the genetic and environmental sources of phenotypic variation are determined statistically. The genetic information in classical quantitative genetics is derived from the fact that the related individuals share a known proportion of their genes (as quantified by the coefficient of relatedness r_{IJ}; see Chapter 2). In the newer technique of **quantitative trait locus (QTL) mapping,** variation in genetic markers that are scattered throughout the genome is statistically related to phenotypic variation (see Chapter 5). In QTL mapping, the genetic information comes from the genetic map produced from the markers. Both techniques have strengths and weaknesses, and both are valuable for the study of natural populations. The key similarity is that they focus on continuously distributed *phenotypic* traits, whereas population genetics is much more concerned

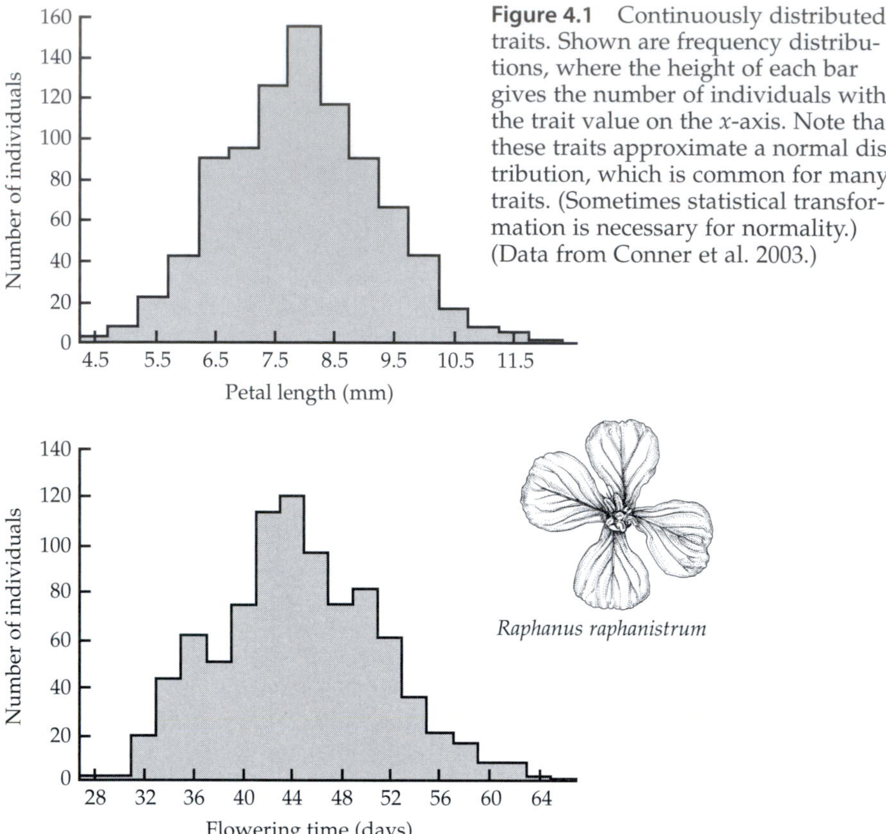

Figure 4.1 Continuously distributed traits. Shown are frequency distributions, where the height of each bar gives the number of individuals with the trait value on the x-axis. Note that these traits approximate a normal distribution, which is common for many traits. (Sometimes statistical transformation is necessary for normality.) (Data from Conner et al. 2003.)

with discrete *genotypes*. Population genetics techniques are applied directly to phenotypes only in the rare cases in which phenotypic variation is discrete, in other words, for visible polymorphisms. So for this chapter and the next two, the emphasis is shifted from genotypes to phenotypic traits.

In population genetics, the population is characterized with allele and genotype frequencies (Chapter 2). Because these are discrete, it is a simple matter to count individuals with each genotype and calculate the frequencies (proportions). In Chapter 3 we used F-statistics to describe how discrete variation is distributed within and among populations. With continuous phenotypic traits, in most cases the alleles that are present in the population or even the loci that affect the trait are unknown, so we need to use the statistical measures of **mean** and **variance** (and later covariance, discussed in Chapter 5) to characterize populations (Figure 4.2). Much of this chapter is devoted to relating these two methods by showing how the means and variances are determined by allele frequencies and the environment. In this way

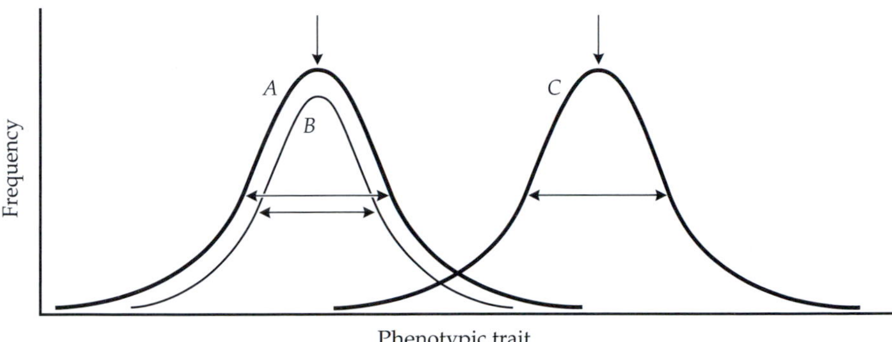

Figure 4.2 Three normal distributions (idealized versions of real data such as those in Figure 4.1), illustrating mean and variance. The mean (indicated by single-headed arrows) is another term for the average phenotype in the population; and the variance (indicated by double-headed arrows) is a measure of how variable the population is; in other words, the width of the distribution. Populations *A* and *B* have the same mean but different variances, while *A* and *C* have different means but the same variance. (See Appendix for the formulas for mean and variance.)

we will show how statistical quantitative genetics is derived from Mendelian genetics, and we set the stage for a discussion of QTL mapping in the next chapter. The last part of this chapter focuses on how genetic variance is measured. Recall that the reason we are interested in genetic variance is because it is the raw material for evolutionary change. In Chapters 5 and 6 we will combine the variance measures discussed in this chapter with measures of selection to understand rates of evolution of the mean phenotype.

Most phenotypic traits have a continuous distribution even though all genetic variation is discrete. For example, there are three distinct genotypes at a locus with two alleles. The continuous distribution of most traits occurs for two reasons—most traits have more than one gene locus affecting them, and most traits are also affected, sometimes to a large degree, by the environment. Remember that means, variances, and allele frequencies are all properties of the population, but the first step in relating these is to discuss what determines an individual's **phenotypic value (*P*),** which is simply the measurement of a given trait for that individual. For example, the phenotypic value for the petal length of a particular plant in Figure 4.1 might be 7.5 millimeters, and the same plant's *P* for flowering time might be 44 days. The phenotypic value of an individual is determined by the individual's genotype and the environment, which we will define as all nongenetic effects on the phenotype:

$$P = G + E \qquad \textbf{4.1}$$

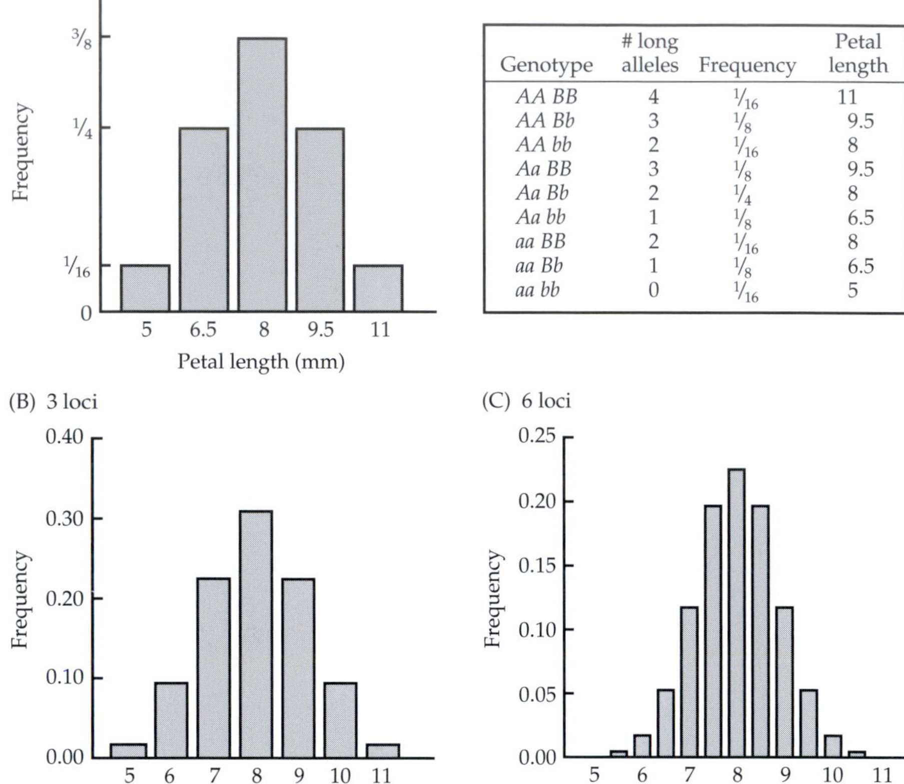

Genotype	# long alleles	Frequency	Petal length
AA BB	4	1/16	11
AA Bb	3	1/8	9.5
AA bb	2	1/16	8
Aa BB	3	1/8	9.5
Aa Bb	2	1/4	8
Aa bb	1	1/8	6.5
aa BB	2	1/16	8
aa Bb	1	1/8	6.5
aa bb	0	1/16	5

Figure 4.3 A hypothetical example (based on the real petal length data in Figure 4.1) showing genotypic values (along the *x*-axes). The three graphs show how increasing numbers of loci affecting a trait causes the trait distribution to become more continuous even in the absence of environmental deviations. In (A), there are two loci with two alleles each, which is the simplest case for a trait affected by more than one locus. The loci act additively (no dominance or epistasis), so each allele indicated with an uppercase letter adds 1.5 mm of petal length over the *aa bb* genotype, which has 5 mm petals. The table shows the frequency of each genotype under HWE and with $p = q = 0.5$ for both loci, and the graph shows the phenotypic distribution that results. (B) and (C) show the phenotypic distribution with 3 and 6 loci; in these cases each allele with an uppercase letter adds 1 and 0.5 mm, respectively, keeping all other conditions the same as in part (A).

where G is the genotypic value and E is the environmental deviation. The **genotypic value** is the phenotype produced by a given genotype averaged across environments. Thus it can only be measured by replicating clones or highly inbred lines across environments. However, genotypic value is a useful conceptual tool for outbred sexual species as well. The graphs in Figure 4.3 show the values of G for each genotype with 2, 3, or 6 loci; this shows how additional loci make the trait distribution more continuous even in the absence of any environmental effects.

Environmental factors themselves usually vary continuously—think of such factors as temperature, rainfall, sunlight, and prey availability. Therefore, the environment causes the phenotypic values produced by different individuals with the same genotype to deviate continuously from G; these deviations are represented by the E term in Equation 4.1. In other words, the **environmental deviation** is the difference between the phenotypic and genotypic values caused by the environment. These points are illustrated in Figure 4.4, which depicts a set of hypothetical results from rearing 12 individuals with the *AA BB* genotype from Figure 4.3A in random positions in the environment. The resulting distribution of environmental deviations reflects the most common assumptions in modeling environmental effects, that is, that E is normally distributed with mean = 0. Thus, the most common environments cause little or no deviation from the genotypic value of 11 for the *AA BB* genotype, and environments causing larger deviations are progressively rarer. Because the mean environmental deviation (E) is zero, when the means are substituted into Equation 4.1, we find that the mean phenotypic value equals the mean genotypic value— just a different

Environment	E	Phenotypic value of *AA BB*
1	−0.9	10.1
2	−0.6	10.4
3	−0.5	10.5
4	−0.4	10.6
5	−0.1	10.9
6	0	11.0
7	0.1	11.1
8	0.2	11.2
9	0.4	11.4
10	0.6	11.6
11	0.7	11.7
12	1.0	12.0

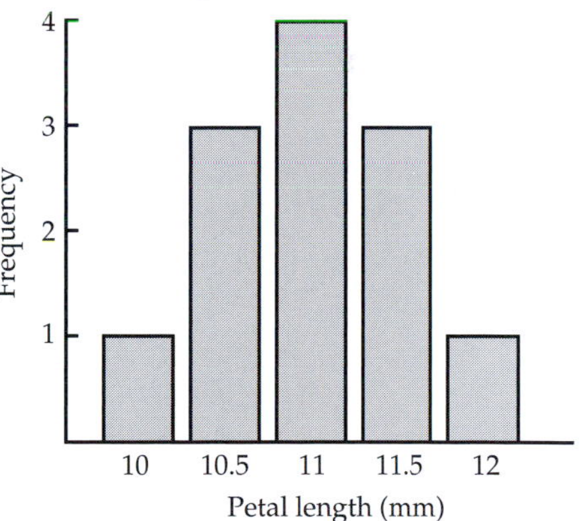

Figure 4.4 An illustration of the effects of hypothetical environmental deviations (E) on the phenotypes produced by the *AA BB* genotype in Figure 4.3A. The deviations are continuously and normally distributed with mean = 0. These deviations give rise to the same distribution of phenotypic values (P) in the graph, except that the mean = 11 mm, which equals G for the *AA BB* genotype.

way of stating our definition of *G* as the phenotypic value produced by a genotype averaged across environments.

We can combine Figures 4.3A and 4.4 by adding a normal curve representing the environmental deviations in Figure 4.4 over the bar for each genotypic value in Figure 4.3A. The result is shown in Figure 4.5. You can see that the curves overlap, so for example, some individuals with the *Aa Bb* genotype are actually taller than some *AA Bb* individuals due to the environmental deviations. This overlap further smoothes out the differences between genotypes, producing the dashed normal curve overlaying the entire histogram. Therefore, a continuous phenotype can result even if the trait is **oligogenic,** that is, affected by only a few loci (two in our hypothetical case). Traits that are affected by many gene loci are often called **polygenic** traits. Note that by drawing the same curves over each bar, we are assuming that the environment has the same effect (i.e., the same *E*'s as in Figure 4.4) on each genotype. This is rarely the case, so we will relax this assumption when we discuss genotype-by-environment interaction in the next chapter.

The distinction between discrete polymorphic traits that show simple Mendelian inheritance and quantitative traits is in the magnitude of the effect of a single locus on the trait relative to other sources of variation (i.e., other loci and the environment). A simple Mendelian polymorphism results when a single **major gene** is responsible for the majority of phenotypic variation in the trait, and a quantitative trait results when the variation is

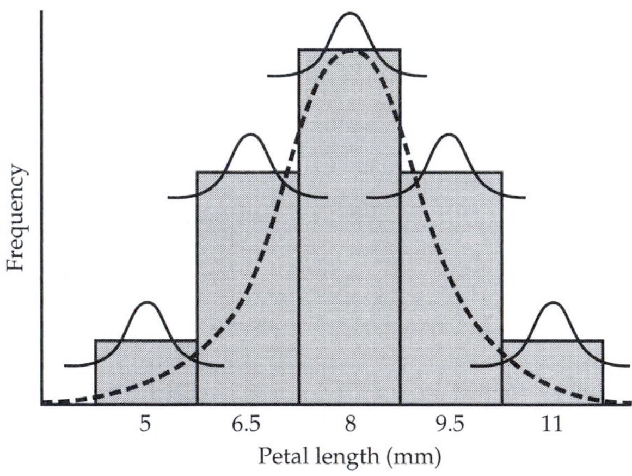

Figure 4.5 Combining *G* with *E* to show how a continuous phenotypic distribution can result from only two loci with two alleles each. The value on the *x*-axis for each bar is *G*, as in Figure 4.3A, and the small normal curve over each bar represents the distribution of phenotypes produced by each genotype when the environmental deviation (*E*, from Figure 4.4) is included. Because the small distributions overlap, the entire distribution approximates a smooth normal curve (dashed).

divided among several to many gene loci and the environment. Genes that contribute a small amount of variation are called **minor genes,** but in reality there is a continuum of effects, from very small to large. Only with the advent of QTL mapping are we beginning to get an idea of the number of gene loci affecting quantitative traits and the distribution of effects of those loci on phenotypic variation. *Statistical quantitative genetics* treats the genome as a "black box," providing no information on individual loci, but rather abstracting the effects of genes as statistical parameters such as variance and covariance (see the discussion later in this chapter). Even so, statistical quantitative genetics is very useful for answering a variety of evolutionary questions and will be the method used to study many questions and organisms until QTL mapping becomes simpler, faster, and cheaper.

Different types of gene action

Because quantitative genetics is focused on the phenotype, we need to be more explicit about different modes of gene action, or the manner in which genotypes affect phenotypes. Additive gene action is aptly named because one can merely add up the effects of each allele in a genotype to determine the total effect on the phenotype. Hypothetical examples of additive gene action were given in Figure 4.3. Note that petal length in those examples is determined simply by the number of uppercase letter alleles present in the genotypes. The phenotypic effect of each allele is not affected by which other allele is present at the same locus, nor by which alleles are present at the other loci. Additivity does not imply equal effects of all alleles at a locus or all loci affecting a trait; there can be major and minor genes all interacting additively in their effects on a given phenotypic trait. The continuum of effects of alleles and loci can occur with additivity or any other mode of gene action. Two other modes of gene action, **dominance** and **epistasis,** are characterized by interactions between alleles at the same locus and different loci, respectively (Table 4.1).

When genes act in a dominant fashion, the interaction between alleles at one locus means that the diploid genotype at each locus needs to be consid-

TABLE 4.1 *Summary of how interactions among alleles at different levels (within or between loci) define different types of gene action*

	Interaction among alleles	
	No interaction	Interaction
Within a locus	Additive	Dominance
Between loci	Additive	Epistasis

ered as a whole to determine the phenotypic effect (see Figure 3.7). An important general point about gene action is that it is specific for a given locus and also for a given phenotypic trait, so that the degree of dominance or epistasis for a given locus can vary across traits at different levels of the phenotypic hierarchy (see Chapter 1). As a hypothetical example, suppose that predation on a species of shrimp by fish depends on shrimp size because, above a certain threshold size, the shrimp are too large for the fish to handle. If a locus has additive effects on shrimp body size, then this same locus could show dominance for a different trait, predation resistance, if heterozygotes at this locus are above the threshold size. This is because heterozygotes would have the same low predation risk as the large-body-size allele homozygotes. Therefore, this hypothetical locus would be additive for body size but dominant for predation resistance.

Epistasis is an interaction between alleles at different loci; hence the phenotypic effect associated with a particular genotype depends on which alleles are present at another locus. An example comes from butterfly wing genes (Table 4.2; Smith 1980), which show both dominance and epistasis. At the *B* locus, the *B* allele is completely dominant to *b*. Note that in individuals with the *BB* and *Bb* genotypes at the *B* locus, there is **underdominance** at the *C* locus (i.e., the heterozygote has the smallest wings), whereas in individuals with the *bb* genotype there is overdominance at the *C* locus (*Cc* has the biggest wings).

How can we quantify gene action?

Before we can show how population means and variances are affected by Mendelian genetics, we need measures that quantify gene action. We will ignore epistasis for simplicity. Previously we discussed the continuum of magnitudes of gene effects; the magnitude of the additive and dominant phenotypic effects of the alleles at a particular locus can be quantified as *a* and *d* respectively (Figure 4.6). These values are of only conceptual interest

TABLE 4.2 *Mean forewing length (mm) of female* Danaus *butterflies of differing genotypes at two loci (sample sizes in parentheses)*

	BB and *Bb*	*bb*
CC	41.91 (46)	40.96 (119)
Cc	40.81 (113)	42.13 (32)
cc	40.94 (150)	41.62 (21)

Note: These are estimates of genotypic values (*G*) because they are the averages of a number of individuals of the same genotype.

Danaus chryssipus

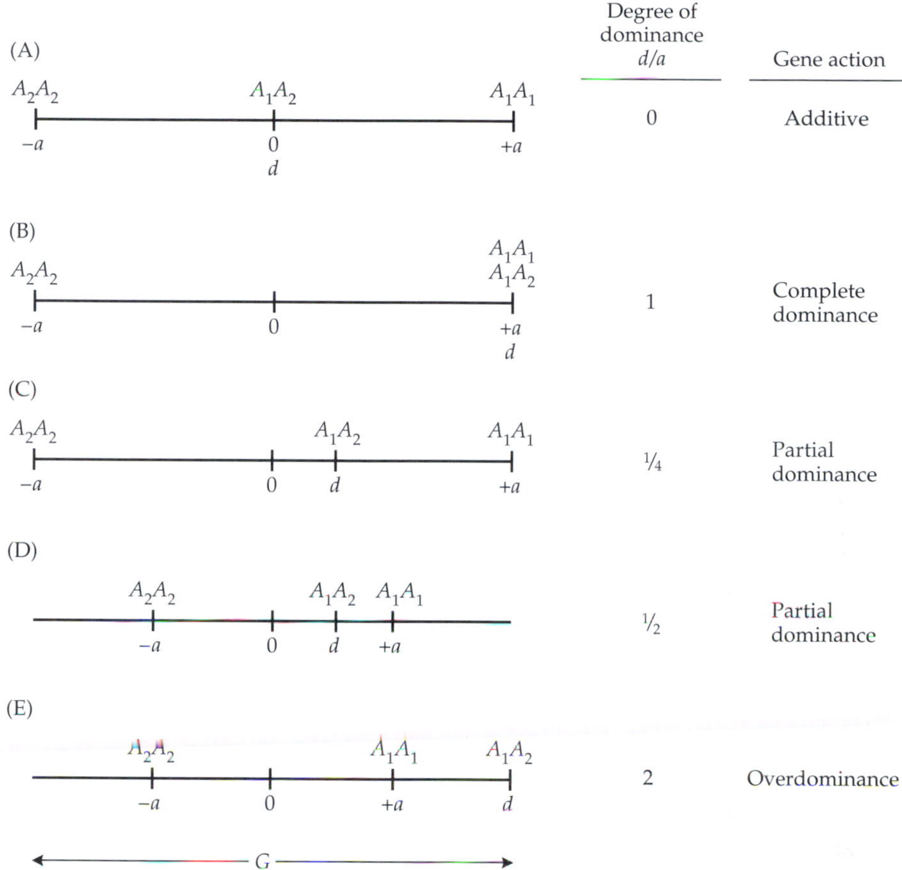

Figure 4.6 Diagram showing gene action quantified using *a* and *d*. The horizontal scale represents genotypic values.

in statistical quantitative genetics. They cannot be measured using these techniques, but they can be estimated using QTL mapping and so are of increasing practical interest (see Chapter 5). Figure 4.6 is similar to Figure 3.7 except that the horizontal axis in Figure 3.7 denotes fitness, whereas in Figure 4.6 the horizontal axis denotes *G* (or mean *P*) for any phenotypic trait.

In the scheme shown in Figure 4.6, the midpoint between the two homozygotes is set to zero, genotypic values (*G*) for the two homozygotes are $+a$ and $-a$, and *G* for the heterozygote is *d* (*d* quantifies dominance as does *h* in Figure 3.7, but they are not the same). Defining genotypic values in this way as $+a$, $-a$, and *d* greatly simplifies the formulas for the expected means and variances in a population. Figure 4.6A shows the additive case,

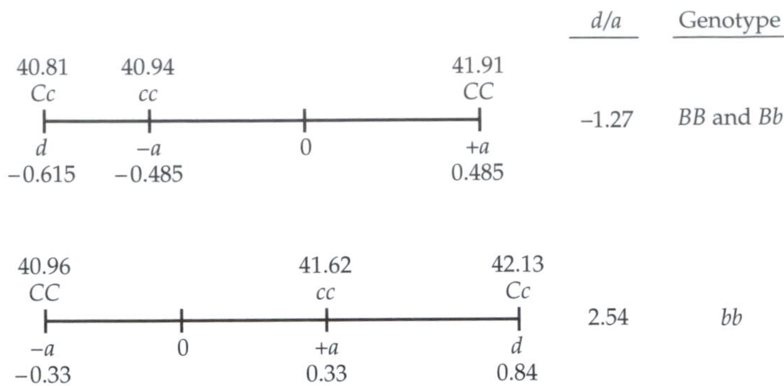

Figure 4.7 The effects of the genotypes at the *C* locus on wing length in *Danaus* butterflies (data from Table 4.2). For each line graph, the absolute values of *G* are above the graph and *a* and *d* are below. (Recall that the latter values are *G* expressed as a deviation from the midpoint between the two homozygotes.)

4.6B complete dominance, and 4.6C an instance of partial dominance. From this we can see that the degree of dominance can be expressed as *d/a*, which equals 0, 1, and $\frac{1}{4}$ in these three cases, respectively. Figure 4.6D depicts a locus with smaller *a*—that is, a smaller effect on genotypic values (*G*)—and thus a smaller effect on the phenotype (*P*) than the loci shown in parts A–C. This case illustrates that *a* is a measure of the magnitude of the effect of the locus on the phenotype; in other words, it shows whether the locus is a major or minor gene. Note that the absolute value of *d* is the same in (C) and (D), but since *a* is smaller in (D) the degree of dominance (*d/a*) is greater in (D) ($\frac{1}{2}$ compared to $\frac{1}{4}$) due to the smaller overall effect of the locus in (D). Finally, Figure 4.6E represents a locus with overdominance where $d/a > 1$.

In Figure 4.7 we illustrate the use of *a* and *d* with the data on *Danaus* butterfly wing lengths in Table 4.2. Since there is epistasis, the values of *a* and *d* for the C locus differ depending on the genotype at the *B* locus. The C locus has a larger additive effect on wing length with the *BB* or *Bb* genotypes ($a = 0.48$) than with the *bb* genotype ($a = 0.33$). However, since *d/a* is greater with *bb*, the overall effect of the *C* locus is slightly greater with *bb* than with *BB* or *Bb* (the spread between genotypes is greater). Note that *d* is negative with underdominance.

Population mean

We are now in a position to show how the population mean for a given phenotypic trait is affected by Mendelian genetics, that is, by allele frequencies and gene action. It is beyond the scope of this book to attempt this for multiple loci, so we will go back to the simplest case of one locus with two alleles. This is adequate to demonstrate the main concepts, and in the absence of epistasis, the results for each locus can simply be summed to obtain the effects of all loci on the mean of a phenotypic trait.

The derivation of the mean is shown in Table 4.3. The usual way of calculating a mean, that is, by summing values and dividing by the number of individuals summed, is equivalent to the method shown in the table when there are multiple individuals with the same value. In this method, the value for each class (the three genotypes here) is multiplied by its frequency, and then these products are summed to give the mean. The frequencies are the Hardy-Weinberg proportions, while the genotypic values are expressed in terms of a and d. The summation gives the equation for the mean:

$$\bar{G} = \bar{P} = a(p - q) + 2pqd \qquad \textbf{4.2}$$

This equation is important because it shows how population means are determined by the allele frequencies, the magnitude of the additive effect, and the degree of dominance. The first term represents the effects of the homozygotes and shows that, as a increases, the mean increases if $p > q$ and decreases if $p < q$. (Recall that G for the aa homozygote is $-a$.) The second term describes the effect of the heterozygotes and equals the frequency of heterozygotes times d (which equals G for the heterozygotes). Again, in the absence of epistasis (we are assuming it to be absent), these terms can just be summed over all loci affecting the trait. Equation 4.2 is also important because it shows how changes in allele frequency affect the mean phenotype, which is how phenotypic evolution occurs (see Chapters 5 and 6). Finally, note that the mean in Equation 4.2 is expressed as a deviation from the midpoint of the two homozygotes (e.g., the numbers below the lines in Figure 4.7), not on an absolute scale (e.g., the numbers above the lines in Figure 4.7).

TABLE 4.3 *Derivation of the equation for genotypic mean*

Genotype	Frequency	Genotypic value	Product
AA	p^2	$+a$	p^2a
Aa	$2pq$	d	$2pqd$
aa	q^2	$-a$	$-q^2a$
			Sum of products = $a(p - q) + 2pqd$

Note: To simplify the sum of the products, use $p^2 - q^2 = (p + q)(p - q) = p - q$ because $p + q = 1$.

Population variance

In addition to the mean, we also need to know the variance in order to characterize continuous phenotypic traits in a population. Variance is absolutely critical, because as noted in earlier chapters, genetic variation is the raw material for evolutionary change. Variance is also the fundamental measure of variation in statistics, upon which many other statistical measures and tests are based (see also Figure 4.2 and Appendix). The variance is estimated as:

$$V_x = \frac{\sum_{i-1}^{n} (x_i - \bar{x})^2}{n-1} \qquad \textbf{4.3}$$

The numerator of this formula is the **sum of squares (SS),** or the sum over all individuals of the squared deviations from the mean. Therefore, if there are many individuals with values far from the mean in a population (i.e., curves *A* and *C* in Figure 4.2), then the sum of the deviations will be large and thus the variance will be as well. If most individuals have values close to the mean (e.g., curve *B* in Figure 4.2) then the deviations and the variance will be small. The denominator is the number of individuals in the population minus one (the degrees of freedom; see Chapter 2), which makes the variance an average squared deviation from the mean. This is why variance is sometimes called a **mean square (MS)** (see Box 4.2).

If the phenotypic values for a population are used in Equation 4.3, then this is the **phenotypic variance (V_P)** for that trait. We can partition this total variance in the population into additive components representing various genetic and environmental causes, assuming that there is no correlation or interaction between the genotypes and the environment. These assumptions can usually be met with proper experimental design and analysis, but the study of genotype-by-environment interactions is a very important endeavor in its own right (see Chapter 5). The simplest partition is:

$$V_P = V_G + V_E \qquad \textbf{4.4}$$

where V_G is the **genotypic variance** and V_E is the **environmental variance.** This equation closely parallels Equation 4.1 because these are the variances across the population of the individual phenotypic values, genotypic values, and environmental deviations, respectively. This partitioning is most useful for clonal or highly self-fertilizing organisms because the parental diploid genotypes are recreated in the offspring. It is less useful for sexually outcrossing species in which novel genotypes are created in each offspring by a random combination of one allele from each parent at each locus. Therefore, for these species we need to further partition V_G:

$$V_G = V_A + V_D + V_I \qquad \textbf{4.5}$$

where V_A is the **additive genetic variance,** V_D is the **dominance variance,** and V_I is the **interaction** or **epistatic variance.** (The latter two are collectively referred to as **nonadditive genetic variance.**) As the names imply, these components of genetic variation are caused by the three modes of gene action. The additive genetic variance is the most important for sexually reproducing species because only the additive effects of genes are transmitted directly from parents to offspring. The dominance effects are not transmitted directly because only one allele at each locus is inherited from each parent and it is the combination of parental alleles at each locus that determines the dominance relationships in the offspring. Similarly, independent assortment of alleles at different loci produces new combinations of alleles that determine the epistatic effects.

Because of direct transmission of additive effects, V_A is most important in determining changes in mean phenotypic value across generations in sexual species—which is the definition of phenotypic evolution. It also means that V_A is the easiest of the genetic components of variance to estimate using the resemblance between relatives because this resemblance is caused primarily by additive variation. This principle is the basis of the statistical quantitative genetic methods that we will examine later in this chapter and in the next. The effects of dominance and epistasis cause offspring phenotypes to deviate from the average of their parents, whereas under completely additive effects (including no environmental effects), the mean of the offspring equals the mean of the parents. As an example, consider a cross between two *Danaus* butterflies from Table 4.2, one with the *BB CC* genotype and the other with the *bb cc* genotype. Under complete additivity, the offspring (all *Bb Cc*) would have genotypic values equal to the average of the parents, that is, (41.91 + 41.62)/2 = 41.76. However, due to both underdominance and epistasis in this case, the double heterozygote offspring have G = 40.81, less than either parent.

The equation for additive variance in terms of allele frequencies and gene action is:

$$V_A = 2pq[a + d(q - p)]^2 \quad \textbf{4.6}$$

As in the equation for the population mean (Equation 4.2), this is valid for one locus only. Recall from Figure 2.10 that the first term, $2pq$, is maximum at $p = q = 0.5$. This means that genetic variance tends to be high at intermediate allele frequencies. This observation makes sense, because if one allele is rare, most individuals are homozygous for the other allele, and therefore there is little genetic variance in the population. If there is no dominance, $d = 0$, then the equation reduces to:

$$V_A = 2pqa^2 \quad \textbf{4.7}$$

which shows that purely additive variance is maximized at $p = q = 0.5$ (Figure 4.8A). When there is complete dominance ($a = d$), the maximum

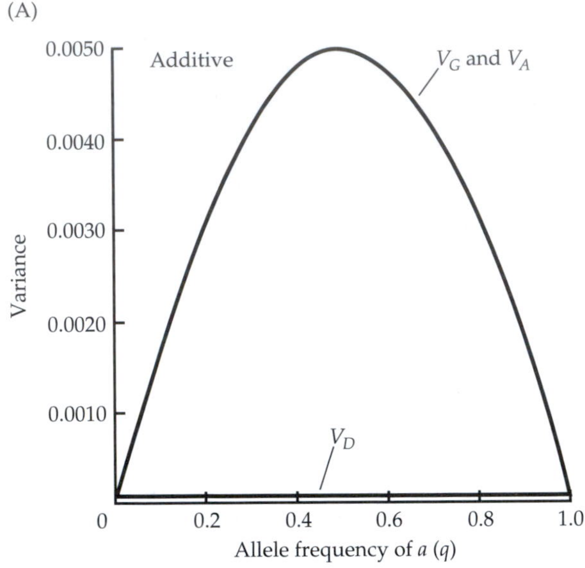

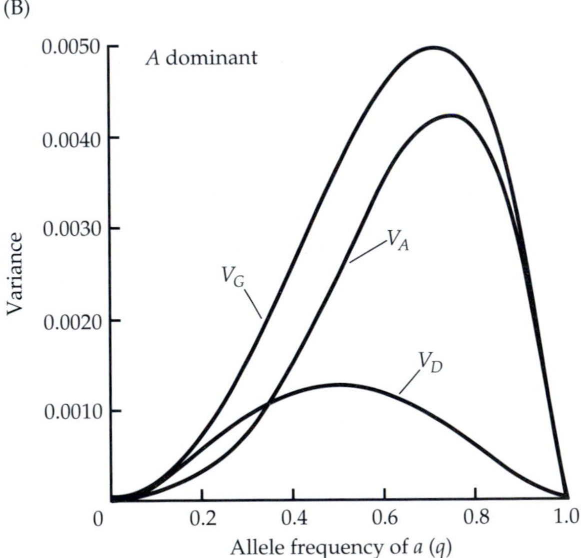

Figure 4.8 Genotypic variance (V_G), additive genetic variance (V_A), and dominance variance (V_D) for a single locus with two alleles in a population in HWE. Note that the x-axis denotes the frequency of the a allele, which is recessive in (B). Because this is a single locus there is no epistatic variance, and therefore $V_G = V_A + V_D$. (A) A completely additive locus, $a = 0.1$, $d = 0$. (B) Complete dominance, $a = d = 0.0707$. (From Equations 4.6–4.8.)

additive genetic variance (V_A) occurs when the recessive allele is more common ($q = 0.75$), making the $d(q - p)$ term in Equation 4.6 large and positive (Figure 4.8B). Intuitively, this occurs because with dominance and equal allele frequencies ($p = q = 0.5$), 75% of the individuals in the population have the dominant phenotype, so less genetic variation is phenotypically expressed. As q becomes larger than 0.75, the additive variance decreases because the $2pq$ term decreases faster then the $d(q - p)$ term increases. Note that dominance variance does peak at $p = q = 0.5$ in Figure 4.8B; this is because the equation for dominance variance is similar to Equation 4.7 in that the allele frequencies appear only in the $2pq$ term:

$$V_D = (2pqd)^2 \qquad \mathbf{4.8}$$

Finally, note that all these equations for components of genetic variance have a squared term because variance is defined as the squared deviation from the mean (Equation 4.3). The squared terms are included to prevent them from having negative values (because negative variability is meaningless). In practice, however, empirical estimates of variances can sometimes be negative due to random error.

Breeding value

The genotypic value is a property of the genotype, but if we want to understand the evolution of phenotypic traits, we need to understand how the trait is inherited, that is, how it is transmitted from parents to offspring. In a sexually reproducing organism, genotypes are not passed on from parents to offspring, but rather are created anew in each offspring by combining an allele from each parent at each locus. The **breeding value** can be defined as the effect of an individual's genes on the value of the trait in its offspring; this effect is caused by the additive effect of genes. The breeding value is sometimes called the *additive genotype*, and the variance of these breeding values is V_A. Thus, in sexual species, the breeding values are more important than the genotypic values (G). Breeding values are being used increasingly in ecological genetics for at least three reasons. There are improved methods for estimating breeding values, they provide a useful method of estimating genetic correlations (see Chapter 5), and they can be used to reduce bias in measuring selection (see Chapter 6). The improved method of estimation is called *best linear unbiased prediction* (BLUP; for details see Chapter 6 in Littell et al. 1996).

The name *breeding value* comes from its roots in animal breeding. For example, if a given bull has alleles that tend to increase meat production in its offspring, then it has a high value for use in breeding to increase meat production in the next generation. The breeding value for a given male is usually estimated as two times the deviation of the mean of his offspring from the population mean, as measured in a paternal half-sibling mating design (described later in this chapter). Therefore, if a bull's offspring tend to have higher meat production than average, then the parental bull has a positive breeding value for this trait. Sometimes half-sibling family means (that is, the mean of all the offspring of one sire) are used in place of breeding values.

The breeding value is two times the deviation from the mean because the male only contributes half of the offspring's alleles. Because in the half sibling design the male is mated to several randomly chosen females, the breeding value represents the effect of the paternal alleles in combination with a random sample of other alleles in the population. This emphasizes the point that breeding value, like all quantitative genetic parameters, depends on population allele frequencies. If the hypothetical bull in our example was introduced into a population of higher meat producers (i.e., higher frequencies of high-meat-production alleles), he could have a negative breeding value; that is, his offspring could be below the population mean. One can also measure a breeding value for traits that the male does not possess (for example, milk production in a bull) because it is based on measurements of offspring only.

Heritability

Heritability is the most widely estimated and discussed quantity in quantitative genetics, whether in plant or animal breeding or when applied to natural populations. Heritability is important because it partly determines how rapidly the mean phenotype evolves in response to artificial or natural selection. Heritability is the proportion of the total phenotypic variance that is due to genetic causes; in other words, heritability measures the relative importance of genetic variance in determining phenotypic variance. There are two types of heritability, broad-sense heritability and narrow-sense heritability. **Broad-sense heritability** is based on the genotypic variance:

$$h_B^2 = \frac{V_G}{V_P} \quad \textbf{4.9}$$

Therefore, broad-sense heritability measures the extent to which phenotypic variation is determined by genotypic variation. Broad-sense heritability includes the effects of dominance and epistatic variance and therefore is most useful in clonal or highly self-fertilizing species in which genotypes of the offspring are virtually identical to those in the parents; in these species it is the broad-sense heritability that affects evolutionary rates. In outbreeding species, evolutionary rates are affected by **narrow-sense heritability,** the proportion of total phenotypic variance that is determined by the additive genetic variance:

$$h_N^2 = \frac{V_A}{V_P} \quad \textbf{4.10}$$

Since in this book we focus on sexual species, from here on when the term *heritability* is used without qualification it will mean narrow-sense heritability. The symbol h^2 derives from Sewall Wright's definition of h as the ratio of the additive and phenotypic standard deviations. (A **standard deviation** is the square root of a variance.) The ratio of standard deviations is now rarely used. The square in the symbol h^2 can serve as a reminder that heritability is a ratio of variances because, as we have noted, a variance is based on squared deviations from the mean.

Heritability is often misinterpreted as the degree of genetic determination, or the extent to which the phenotype is determined by the genotype, or the extent to which the phenotype is determined by genes inherited from the parents. These interpretations are all incorrect, because there can be many loci that affect the trait but are fixed and therefore do not contribute to genetic variation. Note that in the equations for additive and dominance variance above (Equations 4.6–4.8), when one allele is fixed, then the frequency of the other allele is zero and the entire equation equals zero. However, the equation for population mean (Equation 4.2) does *not* equal zero when one allele is fixed. Thus, a locus can still affect a trait mean even when it is not variable. For example, the fact that humans have two eyes is undoubtedly genetically determined, but the heritability of eye number is essentially zero. It is extremely important to remember that heritability refers to *variation* in the population because all of the descriptions and definitions above are based entirely on variance.

Heritability measures the extent to which resemblance among relatives, in comparison to unrelated individuals of the same species, is due to heredity. Obviously, all members of a species resemble each other more than they resemble another species; this is due to loci that are fixed throughout the species, or loci for which few or no alleles are shared with related species. The closer resemblance of relatives *within* the species or population is determined by additive genetic variation (and often the environment).

There are a number of other misconceptions relating to heritability, even among biologists, so it is important to stress the following points. First, there is not one unique heritability for a given trait in a given species because heritability can and often does differ among populations and among environments. It can differ among populations because V_A depends on allele frequencies (see Equation 4.6), so if populations are differentiated for loci affecting the trait of interest, heritability may differ. For example, Mitchell-Olds (1986) found significantly different heritabilities for germination time and two measures of size between a pair of Wisconsin populations of jewelweed (*Impatiens capensis*) grown

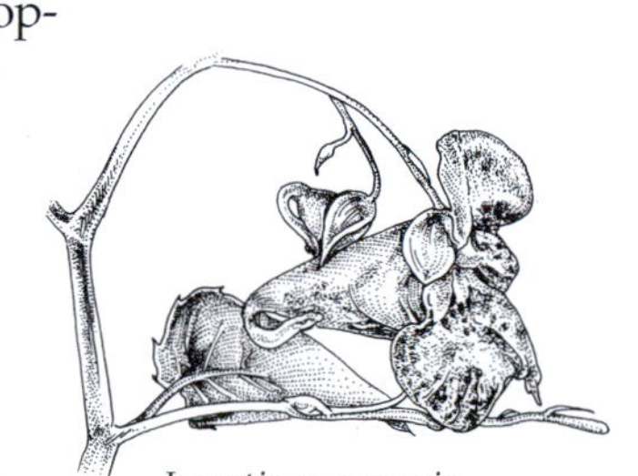

Impatiens capensis

in a common environment. However, such clear examples of differences in heritability among natural populations reared in a common environment are rare.

Heritability can differ among environments because V_E is part of V_P (see Equation 4.4). This can be seen more clearly when the equation for heritability is rewritten with the components of V_P written out:

$$h^2 = \frac{V_A}{V_A + V_D + V_I + V_E} \quad \mathbf{4.11}$$

Therefore, as V_E increases, heritability decreases, because a smaller proportion of the phenotypic variance is additive genetic. For example, the heritability of wing width in male *Drosophila melanogaster* was much greater under normal as compared to stressful conditions ($h^2 = 0.69$ vs. 0.09, respectively) even though V_A was slightly greater under stress (Hoffmann and Schiffer 1998). This lower heritability was caused by a much greater environmental variance under stress ($V_E = 9.2$ vs. 0.9 under stress and control conditions, respectively). The numerator of heritability can also be affected by the environment because expression of genetic variance can be affected by the environment; this is one type of genotype-by-environment interaction and is discussed further in the next chapter. In terms of Equation 4.6, the environment can affect the magnitudes of a and d, altering V_A without changes in allele frequencies.

Heritability is dimensionless because it is a ratio of variances of the same trait measured with the same units. Because the numerator is contained in the denominator, heritability theoretically varies from 0 to 1 although random error can cause estimates of heritability to sometimes be outside these bounds. A second common misconception is that a heritability of 0 means that the trait is not affected by genes. The reason this is not true is due to the effects of fixed loci mentioned earlier; in fact, all phenotypic traits have some genetic component because organisms are built according to the instructions in genes. The confusion disappears if one focuses on variation. A heritability of 0 does mean that there is no additive variance in that population in that particular environment, in which case the phenotypic variance is due entirely to environmental and nonadditive genetic variance.

A final misconception is that a heritability of 1 means that the environment cannot affect a trait. A heritability of 1 means only that the environmental variation affecting that particular population does not affect the phenotypic variance. If the environment changes, this could still change the mean and variance for the trait.

Because heritability is dimensionless, it is useful for comparison across species and traits that are measured on very different scales, and it is therefore widely estimated and reported (Table 4.4). However, as we have discussed, heritability can change with changes in either the additive variance or in the environmental or nonadditive variance. Therefore, comparing her-

TABLE 4.4 *Examples of heritability in natural populations*

Organism	Trait	h^2	Reference
Guppy	Orange spot size	1.08[a]	Houde 1992
Garter snake	Chemoreceptive response	0.32	Arnold 1981
Red deer	Female fecundity	0.46[b]	Kruuk et al. 2000
Meadow vole	Growth rate	0.54	Boonstra and Boag 1987
Darwin's finch	Bill length	0.65[b]	Boag 1983
Collared flycatcher	Male lifespan	0.15[b]	Merilä and Sheldon 2000
Seaweed fly	Male body size	0.70	Wilcockson et al. 1995
Cricket	Development time	0.32[b]	Simons and Roff 1994
Milkweed bug	Flight duration	0.20	Caldwell and Hegmann 1969
Scarlet gilia	Corolla width	0.29[b]	Campbell 1996
Jewelweed	Height	0.08[b]	Bennington and McGraw 1996

[a]Heritability estimates greater than 1 or less than 0 are sometimes obtained due to random error.
[b]Estimated in the field.

itabilities can be confusing and potentially misleading if the focus of interest is on additive variance. Unfortunately, directly comparing variances is difficult because they are not dimensionless and therefore vary with the scale of the trait or organism being measured.

Estimating Additive Variance and Heritability

Recall that additive genetic variance causes resemblance among relatives over and above the resemblance among unrelated members of the same population. Statistical quantitative genetics uses this fact to separate V_A from the nonadditive genetic variance and V_E. A common misconception about estimating heritability is that one can separate V_G from V_E by eliminating V_E through raising the organisms in a controlled laboratory environment. However, no matter how excellent the lab facilities and protocols are, there will inevitably be at least small differences among individuals in temperature, light, food, water, and so on. Environmental variance can certainly be reduced, but not eliminated. In addition, an estimate of heritability under highly controlled conditions may not accurately reflect heritability values in the field where evolution is occurring. This is because V_E is in the denominator of the heritability (see Equation 4.11), and so by reducing V_E artificially one overestimates heritability in the field. If there is genotype-by-environment interaction (see Chapter 5), then this is another reason why the laboratory estimate of V_A will not accurately reflect V_A in the field. For example, Conner, Franks, and Stewart (2003) reported greatly decreased heritability estimates for six floral traits in the field compared to the greenhouse for the same natural population of wild radish. These reduced heritabilities were due both to increased V_E and decreased V_A in the field relative to the greenhouse, a pattern matched by most other studies of this type.

In addition to separating V_G from V_E, additive and nonadditive variance also need to be separated to understand phenotypic evolution in outbreeding species. To make this separation, variances and heritabilities are estimated using phenotypic measurements on individuals of known genetic relationship, typically either parents and offspring or siblings. We will discuss each of these in turn.

Offspring–parent regression

To estimate heritabilities using offspring–parent regression, one first needs an adequate number of monogamously mated pairs for the parental generation. Fifteen pairs is a minimum, and 30 to 50 pairs are usually necessary for reasonably precise estimates. The procedure is to measure the trait(s) of interest on one or typically both of the parents and rear their offspring. When the offspring reach the same developmental stage at which the parents were measured, the same trait(s) are measured on the offspring and the average is taken over all the offspring measured in each family. This offspring average is then **regressed** on the measurements of the fathers, mothers, and/or the average of the two parents (called the **midparent;** Figure 4.9).

In an offspring–parent regression such as that in Figure 4.9, each family is represented by one point. Therefore, each of the 13 points in Figure 4.9 represents the average tarsus length of the two parents on the x-axis and the average tarsus length of all the offspring of those two parents on the y-axis; two to four offspring per family is typical. **Linear regression** (see Box 4.1) is a statistical technique of drawing the best-fitting straight line through these points, which produces an equation for the line in the familiar form:

$$y = a + bx \qquad \textbf{4.12}$$

In this equation, b represents the slope, or the change in y per unit change in x ($\Delta y/\Delta x$). The slope of offspring on midparent is the estimate of heritability. Therefore, the steeper the slope, the more closely the offspring

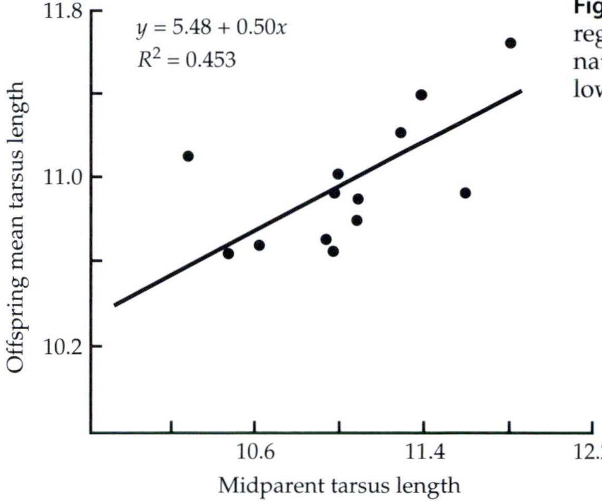

Figure 4.9 Offspring–parent regression of tarsus length in a natural population of tree swallows. (From Wiggins 1989.)

resemble their parents, and the higher the proportion of phenotypic variance that is additive genetic variance, that is, variation passed from parents to offspring. For example, if the slope is 1, then for an increase of one unit of midparent phenotypic value, one observes an increase of one unit in the offspring. In other words, when expressed as a deviation from the mean within each generation, the average offspring in a family has exactly the same phenotypic value as the average parent. A slope of 1 also means that the spread of points along the x-axis is the same as the spread of points along the y-axis; in other words, all of the phenotypic variance in the parents is passed on to the offspring because it is all additive genetic variance.

The slope of 0.5 in Figure 4.9 means that a one unit increase in the midparent value produces only half a unit increase in the offspring—that is, only half of the phenotypic variance is additive genetic variance and therefore transmitted. What happens to the other half of the phenotypic variance? It is caused by nonadditive genetic and environmental variance, so it does not contribute to the resemblance between parents and offspring. Does this imply that the total phenotypic variance in the offspring is half that of the parental generation? No, because the environmental and nonadditive variance is expressed in the variance among the full-siblings within each family (see the section on sibling analysis), and this variance is removed from the analysis by using a single average for the offspring from each family rather than the individual offspring values.

BOX 4.1 *Regression Details*

In any regression analysis, there is a standard error (SE) for the slope, a *P*-value testing the hypothesis that the slope is zero, and an R^2 value, which is the proportion of the variance in the dependent (*y*) variable that is explained by the variance in the independent (*x*) variable. As the scatter of points around the regression line increases, the R^2 decreases. In a simple linear regression with only one *x* variable, these three are very closely related to each other, and quantify how confident we are in the estimate of the slope.

In an offspring–parent regression, increasing the number of offspring included in each family will increase the precision of the estimate of the average trait value for that family. Because this increase in family size decreases random error in the *y*-variable, it will tend to decrease the standard error and the *p*-value for the slope, and increase the R^2. This means that we have more confidence in our estimate of the slope, and therefore our estimate of heritability, but the increased R^2 does *not* mean an increased heritability. The increased number of offspring per family will not have a systematic effect on the estimate of the slope (h^2), only on our confidence in the estimate. Increasing the number of points in the regression, that is, the number of families, has a similar effect; it will decrease the standard error and the *P*-value, but will not cause a predictable change in the slope. Increasing the number of families should not cause a predictable change in the R^2 value either; it does not affect the accuracy of estimation of each family mean and therefore does not affect the scatter of points around the regression line.

For example, the two offspring–parent regressions in Figure 4.10 were calculated from the same data, with the average of only two offspring per family used in Figure 4.10A and the mean of all four offspring that were measured in each family in Figure 4.10B. The two regressions have very similar slopes, therefore they make very similar estimates of heritability. (These are not statistically different, as indicated by the size of the standard errors.) The difference is that the scatter around the regression line is greater in Figure 4.10A (reflected in the higher SE and lower R^2 compared to Figure 4.10B), and so we have less confidence in this estimate. In Figure 4.10B we have greater confidence in our heritability estimate (lower SE) due to less scatter around the regression line. The lower scatter in turn leads to a higher R^2 because each family mean was estimated with less random error.

Figure 4.10 Offspring–parent regressions estimating heritability for pistil length in wild radish flowers. The slope of each fitted regression line is given with the standard error of the slope in parentheses. Four offspring were measured in each of 50 families; in (A), only two offspring per family were used to calculate the offspring mean, whereas all four were used in (B). Note that the increased number of offspring per family does not increase the heritability estimate, but does decrease the scatter around the regression line; this decrease is reflected in the lower standard error and the higher R^2. (Data from Conner and Via 1993.) ▶

Box 4.1 continued

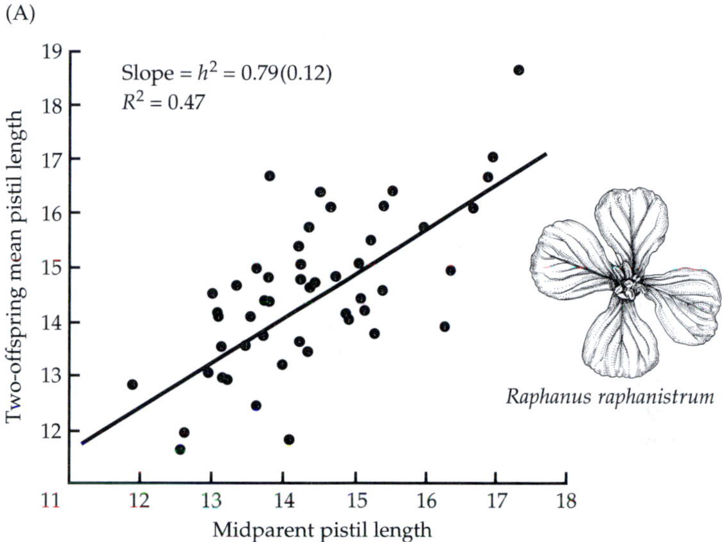

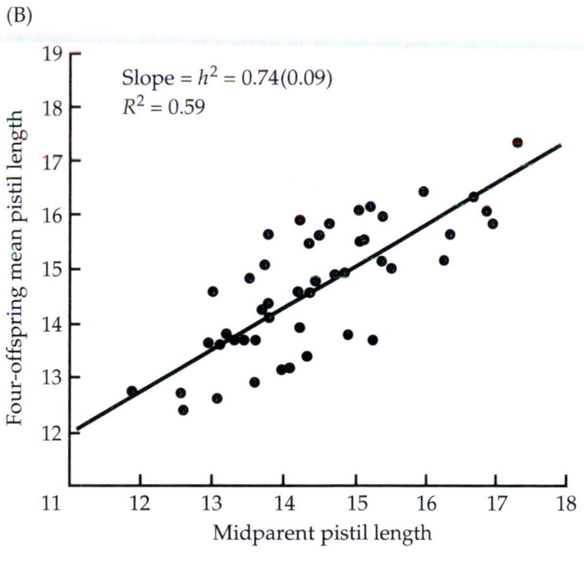

A few final notes on offspring–parent regression. The first is that the statistical technique of regression, now widely used throughout science, business, and elsewhere, was developed by Francis Galton in the late 1800s for the purpose of examining inheritance (Provine 1971). The name *regression* comes from Galton's observation that the offspring–parent regression was always less than 1 because heritabilities are rarely, if ever, 1. This means that the average offspring "regress" toward the parental mean; that is, there is less variance among the offspring means compared to the parental means, as noted above. Second, if the phenotypic value for only the mother or father is used on the *x*-axis, then the value of the slope is half the heritability because each parent contributes only half of the nuclear genes in the offspring. Third, to estimate the additive genetic variance (V_A) rather than the heritability using offspring–parent regression, one uses the offspring-parent *covariance* (not regression slope; see Chapter 5) multiplied by 2, regardless of whether the phenotypic values of one parent or the mean of both parents (midparent) is used.

Maternal and paternal effects

Our discussion of offspring–parent regression included the unstated assumption that all of the resemblance between offspring and parents was due to genetics. This assumption is often not valid in species in which there is parental care of any kind, including provisioning or teaching the offspring. If the amount or quality of parental care depends on the value of the phenotypic trait of interest in the parents, and in turn the amount or quality of care partly determines the phenotypic value of the same trait in the offspring, then this gives rise to a nongenetic resemblance between parents and offspring.

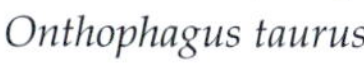
Onthophagus taurus

The most important cause of nongenetic resemblance between parents and offspring is a **maternal effect,** because in most plants and animals mothers have profound effects on their offspring through provisioning of the seed or egg. In mammals, effects occur through gestation and lactation, and in many animal species, through direct care of the young (Mousseau and Fox 1998). In species in which fathers contribute to offspring care there can be **paternal effects** as well. For example, larger male and female dung beetles (*Onthophagus taurus*) provide larger amounts of dung to their offspring than do smaller beetles, and this increased parental provisioning produces larger offspring (Hunt and Simmons 2000). There are also maternal genetic effects that do not contribute to additive genetic variation because it is the mother that transmits most or all of the nonnuclear genes in the mitochondria and chloroplasts (Kirkpatrick and Lande 1989).

The best way to obtain estimates of additive variance and heritability that are not contaminated by maternal effects is to use data on fathers only (e.g., do an offspring–father regression). This works extremely well in the vast majority of plants, arthropods, and reptiles in which the fathers contribute only sperm to the offspring. If the offspring–mother regression slope is greater than the offspring–father slope, then this suggests the presence of a maternal effect. In other animals, paternal care (care by fathers) is common, so paternal effects are possible. Paternal care is particularly common in birds, so the technique of **cross-fostering** is often used. In a cross-fostering experiment, eggs are switched between nests as soon as possible after egg laying, so that some or all of the eggs are reared (*fostered*) by animals that are not the genetic parents. After the offspring mature and are measured, their phenotypic values can be regressed on those of their genetic parents. This approach gives a good estimate of heritability because the genetic parents did not care for the offspring. The offspring values can also be regressed on the foster parents, which gives a good estimate of the parental effects. Cross-fostering does not wholly eliminate parental effects because there could be effects of egg provisioning; since these are through the mother, comparison of the offspring–mother and offspring–father regression slopes provides an estimate of the strength of egg maternal effects. For example, Wiggins (1989) reported a significant regression of offspring on their true genetic parents for tarsus length (see Figure 4.9), but he found no significant offspring–foster parent regression.

Sibling analyses

Just as with offspring–parent regression, **sibling analysis** is used to estimate additive genetic variance and separate it from environmental and nonadditive genetic variance. To do this, a large number of families of known parentage need to be raised, and a reasonable number of siblings from each family need to be placed in randomized positions in the laboratory, field, or greenhouse so that, on average, there are no environmental differences among the families. With this randomization, the environmental variance is distributed among siblings within families, and differences among family means are genetic. The analysis of variance (Box 4.2) tests whether there is significantly greater variance among family means than within families, thereby testing for significant genetic variation.

The simplest sibling analysis is based on full-sibling families, the same as those used in offspring–parent regression. The difference here is that data on the parental phenotypes are not used, but rather among-family variance is tested using a simple one-way ANOVA. However, as explained in Equation 4.17, this measure of variance is not a good estimate of V_A because the esti-

BOX 4.2 *Analysis of Variance (ANOVA)*

Just as we needed to understand the statistical technique of regression to interpret the results of an offspring-parent regression, we now need to understand the basic concepts of **analysis of variance** (ANOVA), the technique that Sir Ronald Fisher devised to do quantitative genetics on groups of siblings. ANOVA is a way of testing whether the means of a number of different groups differ in some measurement, or in other words, whether there is significant variance among the means. ANOVA and regression are closely related techniques, and both have continuous dependent variables. The difference is that the independent variable is continuous in regression (e.g., a quantitative trait) and it is categorical or discrete in ANOVA (e.g., different treatment groups or families). The simplest case is when there are only two groups, and then the ANOVA is equivalent to a *t*-test. For example, Figure 4.11 shows the distribution of adult dry weights for male and female flour beetles (*Tribolium castaneum*). Note that the mean for females is greater than that for males, but that the distributions overlap broadly. The ANOVA tests whether this difference in means is due only to random sampling error, or whether the difference is real. The test statistic for ANOVA has an *F*-distribution, and the *P*-value gives the probability that the difference in means could be due to chance alone; because this chance is less than 1 in 10,000 in this case, we can safely conclude that females really are heavier than males in this population, even with the large overlap in the distributions.

The less overlap there is between the two distributions, the more confident we can be of a difference between means, and the *P*-value will be smaller. Figure 4.12B and C show two ways that the overlap between two distributions

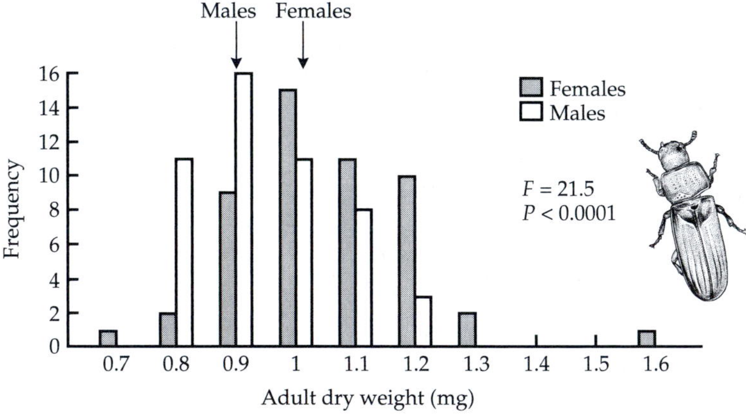

Figure 4.11 Adult dry weights in male vs. female flour beetles (*Tribolium castaneum*). Arrows show the means for males and females, and the *F*- and *P*-values are from the ANOVA that tests whether these means are different from each other. (Data from Conner and Via 1992.)

Box 4.2 continued

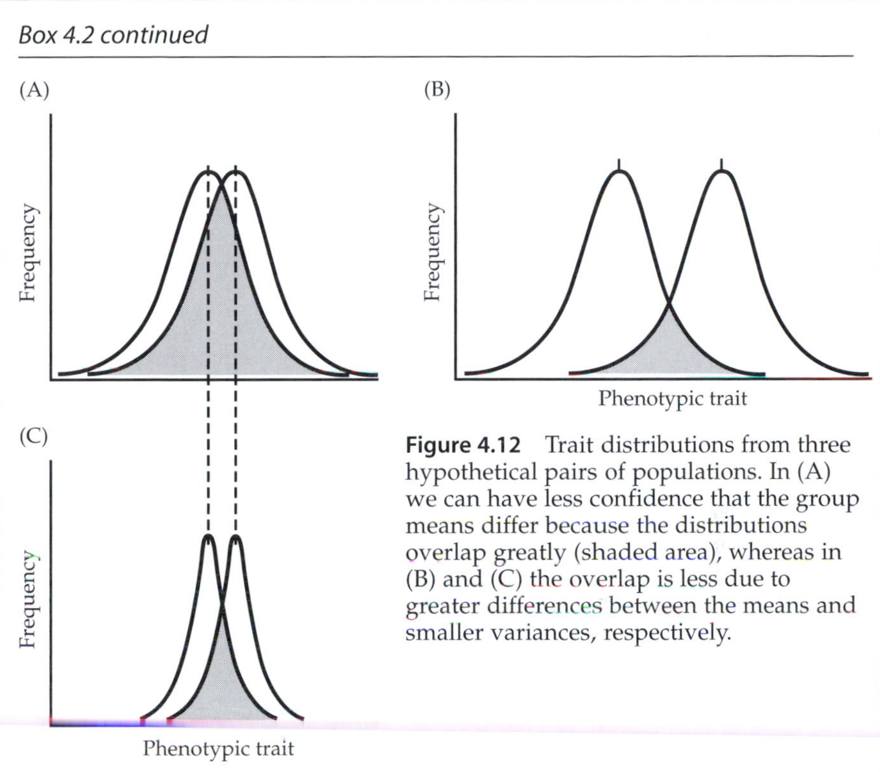

Figure 4.12 Trait distributions from three hypothetical pairs of populations. In (A) we can have less confidence that the group means differ because the distributions overlap greatly (shaded area), whereas in (B) and (C) the overlap is less due to greater differences between the means and smaller variances, respectively.

can be reduced relative to the largely overlapping distributions in Figure 4.12A. In Figure 4.12B the variances are the same as in (A), but the means are farther apart, and in Figure 4.12C the difference in means is the same as in (A) but the variances within each group are smaller. The key to our confidence that there is a real difference between the means, then, depends on the magnitude of the difference between the group means relative to the variance within the groups.

ANOVA is usually applied when there are more than two groups being compared. In quantitative genetics these groups are typically families (Figure 4.13). How can we express the differences among several to many group means? We could follow the approach depicted in Figure 4.12 and calculate the differences between all possible pairs of families, but with more than a few groups this becomes extremely inefficient. It is much simpler to calculate an average deviation of each family mean from a single standard, and for this statisticians use the grand mean or overall mean:

$$V_{means} = MS = \frac{\sum_{i=1}^{n}\left(\overline{Y}_i - \overline{\overline{Y}}\right)^2}{n-1} = \frac{SS}{d.f.} \qquad \textbf{4.13}$$

Here $\overline{Y}_i$ represents the mean value of group *i*, *n* is the total number of groups,

Box 4.2 continued

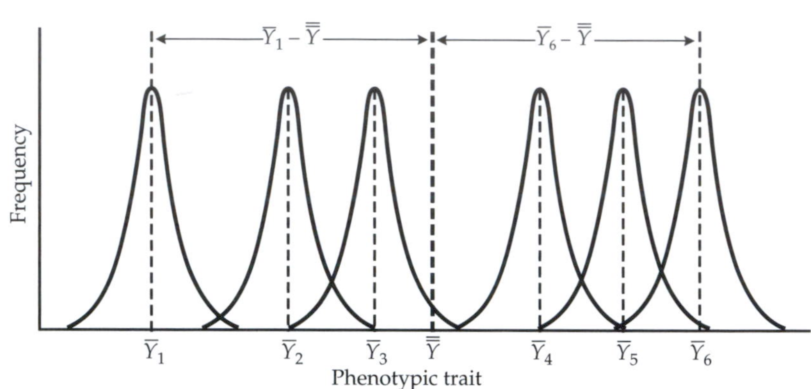

Figure 4.13 A hypothetical plot showing the phenotypic distributions of six families with their respective means, as well as the grand mean of all individuals $\bar{\bar{Y}}$. The deviations of all six family means from the grand mean are squared and summed to calculate the sum of squares (*SS*); the deviations of two families (1 and 6) are shown as examples. Note that the deviations for families 1–3 are negative, but the squaring makes all of their contributions to the sum of squares positive.

and $\bar{\bar{Y}}$ is the grand mean, or the mean of all the group means $\bar{Y}_i$. This equation should remind you of Equation 4.3 for a variance, because that is exactly what it is; it is the variance among group means instead of the variance among individuals. So instead of testing how different the means of two groups are relative to the variance within the groups, ANOVA with many groups tests whether the variance among group means is significantly greater than expected due to chance alone, relative to the variance among individuals within groups. This is why it is called *analysis of variance*. We use *Y* instead of *X* to remind us that the measurements are the dependent variable in the ANOVA. Recall that the numerator of the formula for a variance is usually called the *sum of squares (SS)*, and the entire ratio (the variance among means) is usually called the *mean square (MS)*.

The results of analysis of variance are usually presented in an ANOVA table (Table 4.5). The source of variance is rarely spelled out as it is here, so it is important to keep in mind what exactly each factor (a row in the table) represents. It is also important that the degrees of freedom and either the *SS* or the *MS* (*SS*/*d.f.*; see Equation 4.13) or both be presented. In the table there are six families with 10 offspring in each family, so the degrees of freedom (*d.f.*) at each level are one less than these values. The ratio of the *MS* is a measure of whether

Box 4.2 continued

there is more variance among groups (families here) than expected by chance, given the variance within groups. The *MS* ratio is the test statistic, and it has an *F*-distribution. The *P*-value tells the probability that the greater variance among groups compared to within groups could be due to chance alone, so the small value of *P* in Table 4.5 gives us a high degree of confidence that the family means differ.

TABLE 4.5 *Hypothetical ANOVA table for the data shown in Figure 4.13*

Factor	Source of variance	*SS*	*d.f.*	*MS*	*F*	*P*
Family	Among families	500	5	100	10	<0.005
Error	Within families	90	9	10		

mate also includes dominance variance and a different kind of maternal effect.

A better way to estimate additive variance and heritability with siblings is with a **nested paternal half-sibling** design (Figure 4.14). In this design, a

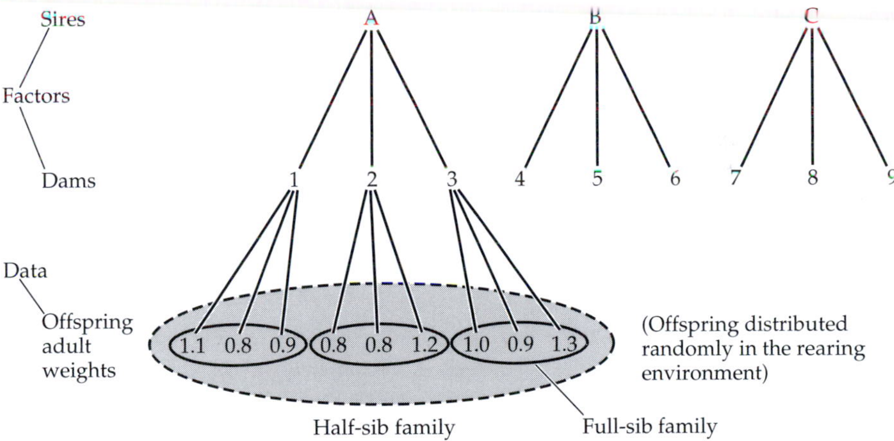

Figure 4.14 Diagram of a nested paternal half-sibling design. Only three sires are shown for clarity, although 30 or more sires are necessary for precise estimates of additive variance. Each sire is mated to three dams in this example, and three offspring from each dam are raised and measured (only the offspring of one sire are shown, for clarity). In the ANOVA, sire and dam nested within sire are the independent variables or factors, and the phenotypic measurements of the offspring are the data used for the dependent variable.

random sample of males (often called **sires**) are each mated to a unique set of randomly chosen females (called **dams**). A sample of offspring from each dam (three in Figure 4.14) is raised and the phenotypic traits of interest measured. Thus, the offspring of each dam represent full-sibling families (because they have the same mother and father), and these are nested within half-sibling families; that is, all the offspring of the same father. The latter are called half-sibling families because the offspring of different mothers share the same father. Nesting means that there are different dams mated to each sire, and this creates the hierarchical structure shown in Figure 4.14. As we will see below, the variance among half-sibling families is used to estimate V_A; significant variance among sires in the ANOVA is good evidence for additive variance in the population. The reason paternal half-sibling families are used is to eliminate maternal effects from the estimate, but just as in offspring parent regression, if there is paternal care then this estimate can be compromised by paternal effects. It is important that offspring be randomized across the environment they are reared in, so that on average the environments experienced by the offspring of each sire do not differ. This is why the environmental variance is not present in the variance among half-sibling family means.

The key parts of a nested ANOVA table on half-sibling data are shown in Table 4.6. The variance or mean squares (*MS*) is partitioned into three sources. The variance among sires is found by taking the mean of all offspring of each sire and calculating the variance of these sire means as in Equation 4.13. Similarly, the variance among dams is estimated by taking the average value for each full-sibling family (the offspring of each dam) and calculating the variance of these means, after correcting for differences among sire means. This is referred to as the variance among dams within sires, because each group of dams was mated to a different sire. The average variance among the offspring within each full-sibling family is referred to as the *error variance*. The column labeled expected mean squares shows the theoretical composition of each *MS* (σ^2 is the symbol used for the theoretical expectation of a variance). Note that each *MS* contains the variance from all levels below it in the nested

TABLE 4.6 *Nested analysis of variance for half-sibling data*

Title	Source of variance	*d.f.*	Expected *MS*
Sire	Among sires	$s–1$	$\sigma^2_w + k\sigma^2_D + dk\sigma^2 s$
Dam (sire)	Among dams (within sires)	$s(d – 1)$	$\sigma^2_w + k\sigma^2_D$
Error	Within full-sibling families	$sd(k – 1)$	σ^2_w

Note: The expected mean squares (*MS*) show the theoretical composition of the variance at each level. s = number of sires; d = number of dams per sire; k = number of offspring per dam.

hierarchy—the error variance includes only the average variance within full-sibling families, the dam within sire variance includes the error variance plus the variance among full-sibling families times the number of offspring/dam (k), and the sire MS includes these plus the variance among half-sibling families times k and the number of dams/sire (d).

Recall that the F-tests in ANOVA are constructed as a ratio of variances (MS). For example, the F-value testing for significant variance among sires (and thus for significant V_A) is the sire MS divided by the dam MS. Because the MS for each level in the hierarchy includes variance from all lower levels, the MS do not give an estimate of the variance at each level alone, except in the case of the error variance for which there are no lower levels. The expected MS are used to find the **variance components** at each level from the MS by subtracting the lower level MS and dividing by the coefficients. For example, the sire variance component is estimated as:

$$\sigma_S^2 = \frac{MS_S - MS_D}{dk} \qquad \textbf{4.14}$$

where MS_S and MS_D refer to the sire and dam mean squares, respectively.

EXERCISE: Derive the formula for the dam variance components from the expected MS in Table 4.6.

ANSWER: $$\sigma_D^2 = \frac{MS_D - MS_W}{k} \qquad \textbf{4.15}$$

These sire, dam, and error variance components are **observational variance components** (i.e., variance components due to the different factors in the mating design). The observational variance components sum to the total phenotypic variance, V_P. From these we can obtain the **causal variance components**—that is, those due to the underlying genetic and environmental sources of variance:

$$\sigma_S^2 = \tfrac{1}{4} V_A \qquad \textbf{4.16}$$

$$\sigma_D^2 = \tfrac{1}{4} V_A + \tfrac{1}{4} V_D + V_{Ec} \qquad \textbf{4.17}$$

Thus, the estimate of V_A from a nested paternal half-sibling design is four times the sire variance component. (The reasoning behind the fractions assigned to each causal component is beyond the scope of this book; see the text by Falconer and Mackay in the Suggested Readings list). The sire observational component includes only V_A because it estimates the effects of each male's genes, averaged over several mothers, so maternal effects cancel out. V_E cancels out also because the offspring are raised in randomized positions across the environment. There is no dominance variance, V_D, in the sire variance because the effects of only one of the parents (the sire) is included, whereas dominance depends on the alleles contributed by both parents at each locus. The dam observed variance component includes three causal components: V_A, V_D, and V_{Ec} (see Equation 4.17). V_D is included because these are full-sibling families, and thus the effects of both parents are included. V_{Ec} is the variance among dams due to differences among the full-sibling families in the common rearing environments shared within each family. This is a different kind of maternal effect than discussed above. With offspring–parent regression, the maternal effects are those that create similarities between mothers and offspring for the trait of interest. With sibling analysis the parental measurements are not included, so V_{Ec} measures a maternal or other environmental effect that causes similarities among the offspring within full-sibling families (and thus differences among the families) over and above the genetic similarities. For example, V_{Ec} might be caused by differences in milk production among mothers that lead to similarities in offspring body size within, and differences among, full-sibling families. As a final note, σ_S^2 and σ_D^2 both also include some epistatic variance, V_I, but these components are small, especially in the case of the sire variance, and are generally ignored.

So with traditional ANOVA analyses, there are three steps: first the *MS* is calculated, then the observational (sire, dam) variance components are calculated from the *MS*, and then the causal (additive, dominance, environmental) variances are estimated from the observed variance components. This was the procedure used for almost all half-sibling analyses done before the year 2000. Now these analyses are usually done using the newer statistical techniques of **maximum likelihood (ML)** or, more often, **restricted maximum likelihood (REML),** which generally produce the observational variance components directly. This eliminates the first step, and only requires the calculation of causal components from the observational components. The major advantage of the maximum likelihood technique is that it is less sensitive to **unbalanced data,** where there are unequal numbers of dams mated to each sire or unequal numbers of offspring per dam. Unbalance is common in experiments due to uneven mortality of individuals.

Futuyma, Keese, and Scheffer (1993) used nested half-sibling designs to test whether populations of a beetle, *Ophraella communa*, that feeds on the leaves of only one plant species, show genetic variation for feeding ability on the hosts of other *Ophraella* species. There were 55 sires each mated to two

TABLE 4.7 *ANOVA table from a half-sibling design testing for genetic variation in the amount of marsh elder* (Iva frutescens) *leaves consumed by* Ophraella *larvae*

	d.f.	*MS*	*F*	*P*
Sire	54	1.42	1.7	<0.05
Dam (sire)	44	0.83	2.1	0.0003
Error	193	0.39		

Note: $s = 55$; $d = 101/55 = 1.84$; $k = 2.91$.

Ophraella communa

dams, and three offspring from each dam were tested on each plant species. The final data were slightly unbalanced, because nine dams produced no offspring and for some fewer than three larvae per plant were successfully tested. The results of larval feeding trials for one of the alternative host plants, marsh elder (*Iva frutescens*), are shown in Table 4.7.

The sire variance component (see Equation 4.14) from these data is:

$$\sigma_S^2 = \frac{MS_S - MS_D}{dk} = \frac{1.42 - 0.83}{1.84 \times 2.91} = 0.11$$

Therefore, V_A is 0.44, because it is four times the sire variance component.

The dam variance component (see Equation 4.15) is:

$$\sigma_D^2 = \frac{MS_D - MS_W}{k} = \frac{0.83 - 0.39}{2.91} = 0.15$$

The dam variance component is slightly larger than the sire variance component, suggesting the presence of dominance variance or common environment effects (compare Equation 4.16–4.17). The sum of the three observational variance components is an estimate of the total phenotypic variance, V_P, which here is 0.11 + 0.15 + 0.39 = 0.65. The narrow–sense heritability is then 0.44/0.65 = 0.68.

The ANOVA shows significant variance among the offspring of different sires, which is good evidence for significant additive genetic variance. In addition, the heritability estimate of 0.68 is substantial. This means that these beetles have the genetic variation necessary for the evolution of increased feeding ability on the alternate host, perhaps facilitating a shift to that host in the future or indicative of a recent ancestor that fed on marsh elder.

EXERCISE: Table 4.8 shows the ANOVA table for the amount of leaf consumed by *Ophraella* larvae feeding on a different host plant. Calculate V_A and h^2 for these data. Is there evidence for dominance variance or common environmental effects? Are the degrees of freedom correct based on the formulas in Table 4.6?

TABLE 4.8 *ANOVA table from a half-sibling design testing for genetic variation in the amount of boneset* (Eupatorium perfoliatum) *leaves consumed by* Ophraella *larvae*

	d.f.	*MS*	*F*	*P*
Sire	54	1.04	2.0	<0.025
Dam (sire)	44	0.52	1.7	0.007
Error	198	0.30		

Note: $s = 55$, $d = 1.84$, $k = 2.96$.
Source: Data from Table 3 of Futuyma et al. 1993.

ANSWER: $V_A = 0.38$; $h^2 = 0.81$. No evidence for dominance or common environmental effects because the sire variance component (0.095) is actually greater than the dam variance component (0.07). This is theoretically impossible because variance components cannot be negative and the dam variance component contains more causal components than the sire variance component (see Equations 4.16 and 4.17). The theoretically impossible result is due to estimation error, which is always present. The degrees of freedom seem correct, except that the dam *d.f.* should be 46 based on 55 sires and 101 total dams. It is important to pay attention to the degrees of freedom in an ANOVA table, because a lack of a match to the values expected from the formulas in Table 4.6 is a good clue that there is a mistake or problem with the analysis. In this case, the difference is small and therefore likely to be inconsequential to the interpretation of the results.

If fathers contribute more than merely sperm to their offspring, then differences among fathers in paternal care can cause V_{Ec} to be present in the sire variance component. This can be reduced or eliminated by splitting up the offspring and having them reared by a number of different fathers and mothers, similar to cross-fostering in offspring–parent regression. However, also similar to cross-fostering, this action does not remove any effects that occur before the offspring can be transferred. For example, in some insects, males provide a nuptial gift of food or transfer nutrients and other substances during mating. If the female uses these to provision the eggs, then there could be V_{Ec} in the sire variance component that would be very difficult to control for. It is not known how important these effects are in nature.

Comparison of methods

There are a number of factors to consider when deciding whether to use offspring–parent regression, full-sibling analysis, or half-sibling analysis to estimate genetic variance and heritability. In natural populations, often the overriding factor is feasibility—there are many animals and some plants for which the nested half-sibling mating design is extremely difficult or impossible to implement. This is one reason why offspring–parent regression is very commonly used for birds. The other advantage is that offspring–parent regression can be conducted in unmanipulated natural populations of monogamously breeding animals. Ideally, molecular markers should be used to determine paternity of the offspring because in behaviorally monogamous birds, mating outside the pair can be common ("extra-pair copulations" or EPCs; Westneat, Sherman, and Morton 1990). Even maternity needs to be checked in some bird species in which females lay eggs in other female's nests ("egg-dumping"). If there are offspring included in the regression that are not genetically related to the parents they are being regressed on, then the heritability estimate will be biased downward (that is, lower than the actual heritability) because the genetic relatedness is less than is assumed by the analysis.

Another advantage of offspring–parent regression is that it is simpler statistically, but this simplicity comes at a price in flexibility and sometimes bias. The bias can occur if there is genotype-by-environment interaction (G × E; see Chapter 5), that is, if genotypes differ in their phenotypic response to the environment. This interaction can bias offspring–parent regressions because offspring and parents develop at different times, and therefore there is no way to make their environments exactly the same. The price in flexibility comes when genotype-by-environment interaction is itself the object of study. The usual technique (see Chapter 5) is to split full-sibling families (often nested within half-sibling families) and raise the offspring in different environments. This cannot be done with offspring–parent regression because there are only two parents.

Other issues in choosing a technique are precision and bias. The closer the genetic relationship between the relatives being measured, the higher the precision. Therefore, full-sibling and offspring–parent regression have better precision than half-sibling because the former are based on a coefficient of relatedness (r_{IJ}; see Chapter 2), of 0.5 versus 0.25 for the latter.

TABLE 4.9 *Summary of advantages and disadvantages of different methods of estimating additive genetic variance in outbreeding species*

Method	Advantages	Disadvantages
Offspring–parent	• Simple • Precise • Can use unmanipulated populations	• Environmental differences between parents and offspring • Biased by maternal effects (except father-offspring) • Cannot test for G × E
Full-sibling	• Precise	• High bias from maternal effects and dominance
Half-sibling	• Low bias • Can include other factors, e.g., different environments	• Imprecise • Complex • Large sample sizes needed

Bias is usually a more important issue than precision; it comes mainly from maternal effects (including common environment) and dominance. The former affects both offspring–mother and offspring–midparent regressions, and both can bias full-sibling analyses. Therefore, offspring–father regression and paternal half-sibling analysis are usually the least biased, but can be biased if there is paternal care. Typically, however, half-sibling designs are used in organisms like plants and insects in which paternal investment is rare or nonexistent. Note that half-sibling mating designs can be difficult to accomplish with vertebrates, so this design is usually used for plants and insects.

Because full-sibling designs have two sources of bias, they are definitely less useful for sexually reproducing organisms. They are useful for highly selfing plants, however; selfers produce progeny with diploid genotypes very similar to their own, so that V_D can contribute directly to evolutionary change. Similarly, for clonal organisms the genetic variance among clones is the relevant determinant of the rate of phenotypic evolution. Maternal effects can still be a source of bias for full-sibling analyses of selfers or clonal analyses.

Full-sibling designs have been commonly used for natural populations of outbreeding species when other designs were too difficult to implement. The variance among full-sibling families provides an upper bound to V_A, so perhaps this is better than nothing. However, because of dominance and common environment, often there will be significant full-sibling variance even when there is no significant V_A. For example, Agrawal et al. (2002) found highly significant variance among full sibling families for leaf trichome (hairs that deter herbivores) density in wild radish, but no evidence for heritability or additive variance.

In quantitative genetic experiments of any design, the number of families that the estimate of V_A is based upon is of prime importance to the power of the analysis. Twenty families should be considered a minimum, and 50 or more is necessary for reasonable statistical power. In particular, a finding of no significant additive variance with less than 50 families should be interpreted with caution because the lack of significance could easily be due to a lack of statistical power rather than a real lack of additive variance.

PROBLEMS

4.1. Sewall Wright studied the genetics of coat color in guinea pigs. He reported the following scores for degree of black coloration for the three genotypes at the *c* locus: $c^rc^r = 1.20$, $c^rc^d = 1.06$, $c^dc^d = 0.95$.

- **a.** Calculate *a, d,* and *d/a* for this trait, and place these values, the genotypes, and the raw genotypic values (*G*) on a line graph, using Figures 4.6 and 4.7 as a guide.
- **b.** Calculate the population mean, additive genetic variance, and dominance variance for this trait, assuming the *c* locus is the only locus that affects this trait and that the frequency of the c^r allele is $p = 0.5$.
- **c.** Calculate the population mean, additive genetic variance, and dominance variance for the case when $p = 0.25$. What important principle is illustrated by the difference between these answers and your answers for part (b)?

4.2. Merilä (1997) performed a cross-fostering experiment with collared flycatchers in which unequal numbers of offspring were moved, so that half of the nests had reduced numbers of nestlings and half had enlarged brood sizes. The regression slopes of the offspring tarsus length on their genetic midparent values (not the foster parents that raised them) were 0.48 in the nests with reduced brood sizes and 0.22 in the nests with enlarged numbers of young. What do these slopes represent? What are two possible reasons for the differences in slopes between the two treatments? What role do you think maternal and paternal effects played in determining these slopes?

4.3. The study by Futuyma, Keese, and Scheffer (1993) (see Tables 4.7 and 4.8) also examined genetic variance for adult feeding rate using their nested half-sibling design. The ANOVA table for the amount of leaf consumed by *Ophraella* adults feeding on marsh elder is below. Calculate the sire and dam variance components, V_A and h^2 for these data. Is there evidence for additive variance, dominance variance, or common environmental effects? Compare the results to the larval feeding data for the same species in Table 4.7; what similarities and differences do you see?

	d.f.	*MS*	*F*	*P*
Sire	49	1.07	1.2	>0.25
Dam (sire)	25	0.90	2.0	0.005
Error	150	0.44		

Note: $s = 50$, $d = 1.5$, $k = 3$.

SUGGESTED READINGS

*Falconer, D. S., and T. F. C. Mackay. 1996. *Introduction to Quantitative Genetics*, 4th Edition. Longman, Harlow, UK. Sometimes called the "bible" of quantitative genetics, this has been the standard text and reference for the field for many years.

Janssen, G. M., G. DeJong, E. N. G. Joosse, and W. Scharloo. 1988. A negative maternal effect in springtails. Evolution 42:828–834. This study reports an unusual result: A negative maternal effect.

Lynch, M., and B. Walsh. 1998. *Genetics and Analysis of Quantitative Traits*. Sinauer Associates, Sunderland, MA. Extraordinarily detailed, thorough, and up-to-date, with a very rigorous mathematical treatment of the topic.

Mazer, S. J., and C. T. Schick. 1991. Constancy of population parameters for life-history and floral traits in *Raphanus sativus* L. II. Effects of planting density on phenotype and heritability estimates. Evolution 45:1888–1907. A study using a paternal half-sibling design to measure genetic variation across three environments.

Roff, D. A. 1997. *Evolutionary Quantitative Genetics*. Chapman & Hall, New York. Less mathematical and more focused on studies of natural populations than the other two texts mentioned above.

*Wiggins, D. A. 1989. Heritability of body size in cross-fostered tree swallow broods. Evolution 43:1808–1811. A clear example of the use of offspring–parent regression and cross–fostering to estimate heritability.

**Indicates a reference that is a suggested reading in the field and is also cited in this chapter.*

SUGGESTED READINGS QUESTIONS

The following questions are based on papers from the primary literature that address key concepts covered in this chapter. For the full citation for each paper, please see Suggested Readings.

From Wiggins, D. A. 1989. Heritability of body size in cross-fostered tree swallow broods.

1. Describe how Wiggins performed his experiment. How were the experimental and control families each handled? Was his control a good one? If not, how would you improve it?
2. Describe what animals are involved in the two regressions in Figure 1. What does each point represent in each? Who is being regressed on whom in each?

3. Explain the biological significance of the differences between the slopes of the two regression plots in Figure 1. What is the relationship between the data in these plots and the data in Table 1? Was there any evidence for maternal or paternal effects?
4. Why are measurements of heritabilities of these traits important and interesting?

From Janssen, G. M., G. DeJong, E. N. G. Joosse, and W. Scharloo. 1988. A negative maternal effect in springtails.

1. What is a negative maternal effect?
2. What evidence do Janssen et al. present for a negative maternal effect?
3. Why is this data interpreted as evidence for a negative maternal effect?

From Mazer, S. J., and C. T. Schick. 1991. Constancy of population parameters for life-history and floral traits in *Raphanus sativus* L.

1. The mean phenotypic values of some of the traits differed significantly across density treatments. Which traits were these, and which figure and/or table presents these data? Are the differences in the direction you expect?
2. For which traits is there evidence for significant narrow-sense heritability (additive genetic variance) at each of the three densities individually? What specifically is this evidence, and in which table and/or figure is it presented?

CHAPTER REFERENCES

Agrawal, A. A., J. K. Conner, M. T. J. Johnson, and R. Wallsgrove. 2002. Ecological genetics of an induced plant defense against herbivores: Additive genetic variance and costs of phenotypic plasticity. Evolution 56:2206–2213.

Arnold, S. J. 1981. Behavioral variation in natural populations. I. Phenotypic, genetic and environmental correlations between responses to prey in the garter snake, *Thamnophis elegans*. Evolution 35:489–509.

Bennington, C. C., and J. B. McGraw. 1996. Environment-dependence of quantitative genetic parameters in *Impatiens pallida*. Evolution 50:1083–1097.

Boag, P. T. 1983. The heritability of external morphology in Darwin's ground finches (*Geospiza*) on Isla Daphne Major, Galapagos. Evolution 37:877–894.

Boonstra, R., and P. T. Boag. 1987. A test of the Chitty hypothesis: Inheritance of life-history traits in meadow voles *Microtus pennsylvanicus*. Evolution 41:929–947.

Caldwell, R. L., and J. P. Hegmann. 1969. Heritability of flight duration in the milkweed bug *Lygaeus kalmii*. Nature 223:91–92.

Campbell, D. R. 1996. Evolution of floral traits in a hermaphroditic plant: Field measurements of heritabilities and genetic correlations. Evolution 50:1442–1453.

Conner, J., and S. Via. 1992. Natural selection on body size in *Tribolium*: Possible genetic constraints on adaptive evolution. Heredity 69:73–83.

Conner, J. K., R. Franks, and C. Stewart. 2003. Expression of additive genetic variances and covariances for wild radish floral traits: Comparison between field and greenhouse environments. Evolution 57:487–495.

Conner, J. K., and S. Via. 1993. Patterns of phenotypic and genetic correlations among morphological and life-history traits in wild radish, *Raphanus raphanistrum*. Evolution 47:704–711.

Futuyma, D. J., M. C. Keese, and S. J. Scheffer. 1993. Genetic constraints and the phylogeny of insect-plant associations: Responses of *Ophraella communa* (Coleoptera: Chrysomelidae) to host plants of its congeners. Evolution 47:888–905.

Hoffmann, A. A., and M. Schiffer. 1998. Changes in the heritability of five morphological traits under combined environmental stresses in *Drosophila melanogaster*. Evolution 52:1207–1212.

Houde, A. E. 1992. Sex-linked heritability of a sexually selected character in a natural population of *Poecilia reticulata* (Pisces: Poeciliidae) (guppies). Heredity 69:229–235.

Hunt, J., and L. W. Simmons. 2000. Maternal and paternal effects on offspring phenotype in the dung beetle *Onthophagus taurus*. Evolution 54:936–941.

Kirkpatrick, M., and R. Lande. 1989. The evolution of maternal characters. Evolution 43:485–503.

Kruuk, L. E. B., T. H. Clutton-Brock, J. Slate, J. M. Pemberton, S. Brotherstone, and F. E. Guinness. 2000. Heritability of fitness in a wild mammal population. Proceedings of the National Academy of Sciences of the United States of America 97:698–703.

Littell, R. C., G. A. Milliken, W. W. Stroup, and R. D. Wolfinger. 1996. *SAS System for Mixed Models*. SAS Institute, Inc., Cary, NC.

Merilä, J. 1997. Expression of genetic variation in body size of the collared flycatcher under different environmental conditions. Evolution 51:526–536.

Merilä, J., and B. C. Sheldon. 2000. Lifetime reproductive success and heritability in nature. The American Naturalist 155:301–310.

Mitchell-Olds, T. 1986. Quantitative genetics of survival and growth in *Impatiens capensis*. Evolution 40:107–116.

Mousseau, T., and C. Fox. 1998. *Maternal Effects as Adaptations*. Oxford University Press, New York.

Provine, W. B. 1971. *The Origins of Theoretical Population Genetics*. University of Chicago Press, Chicago.

Simons, A. M., and D. A. Roff. 1994. The effect of environmental variability on the heritabilities of traits of a field cricket. Evolution 48:1637–1649.

Smith, D. A. S. 1980. Heterosis, epistasis and linkage disequilibrium in a wild population of the polymorphic butterfly *Danaus chrysippus* (L). Zool. J. Linn. Soc. 69:87–109.

Westneat, D. F., P. W. Sherman, and M. L. Morton. 1990. The ecology and evolution of extra-pair copulations in birds. Current Ornithology 7:331–369.

Wilcockson, R. W., C. S. Crean, and T. H. Day. 1995. Heritability of a sexually selected character expressed in both sexes. Nature 374:158–159.

5

Quantitative genetics II: Advanced topics

In the preceding chapter we showed how the means and variances of quantitative traits are affected by allele frequencies and gene action, and we described how heritability and additive genetic variance are estimated using offspring–parent regression and sibling analysis. That discussion was based on single environments and single traits. In this chapter we expand on this information to consider the effects of multiple environments and multiple traits, as well as to describe two other techniques for studying quantitative genetic variation.

Here we will first discuss phenotypic plasticity, in which traits are expressed differently in different environments. If different genotypes differ in the level or direction of plasticity they express, then genotype-by-environment interaction exists. We will explain how genetic and phenotypic relationships among multiple traits can arise and how these relationships can be studied using correlations. Artificial selection—selection of certain traits by humans—has produced all our domesticated plants and animals. We will discuss how artificial selection can be used as an alternative to offspring–parent regression and sibling analysis to study genetic variances and correlations. Finally, we will turn to QTL mapping, in which molecular markers are used to find the locations of genes contributing to genetic variation in quantitative traits.

Phenotypic Plasticity and Genotype-by-Environment Interaction

Phenotypic plasticity is exhibited when the same genotype produces different phenotypes in different environments. Figure 5.1 shows different **reaction norms,** which depict the phenotypes produced by different genotypes within a population in two (or more) different environments. Reaction norms are extremely useful for understanding plasticity and genotype-by-environment interaction. The environments in a reaction-norm plot are arrayed along the *x*-axis, and on the *y*-axis are the mean phenotypic values for each genotype or family. Each line connects the genotypic or breeding value in one environment to the genotypic or breeding value in the other environment. Just as in estimating additive variance and heritability, these points represent genotypic values for clonal species, selfed or full-sibling family means for highly selfing species, and breeding values or half-sibling family means for sexual species. In the discussion that follows, we will focus on half-sibling means (sire effects), but in each case clones or full-sibling families could be substituted where appropriate based on the reproductive system of the organism. The spread of points along each vertical axis represents the corresponding variance, for example, additive genetic variance if breeding values are used. For the remainder of this section we will assume that the points represent breeding values from half-sibling families.

The two environments on the *x*-axis represent different "macroenvironments," for example, an open field versus forest, or a pond with fish versus a fishless one. Therefore, this type of environmental variation is different from V_E, because V_E is phenotypic variation within one genotype caused by "microenvironmental" variation—that is, small random differences in the environment within one site (Figure 4.4). V_E is not shown in Figure 5.1 because family means or breeding values are used; the environmental variance is part of the variance within these families (see Chapter 4). The distinction between micro- and macroenvironments is sometimes tricky because they often occur along a spatial continuum. For example, heading up a slope from a wetland to a ridge, the soil becomes progressively drier; the wetland and the top of the ridge are clearly different macroenvironments, but there may or may not be a clear dividing line between them.

The steepness of the slope of the line for each family indicates the amount of plasticity of that family. Figure 5.1A shows a population with no plasticity; the horizontal reaction norms indicate that each family produces the same phenotypic mean in each environment. Note that the families differ phenotypically, indicating additive genetic variance that is equal within each environment. The steep slopes in Figure 5.1B indicate a high amount of plasticity, that is, the phenotypic means are very different between environments. Note that all the families respond to the two environments in exactly the same way—they all decrease by the same amount—which indicates that

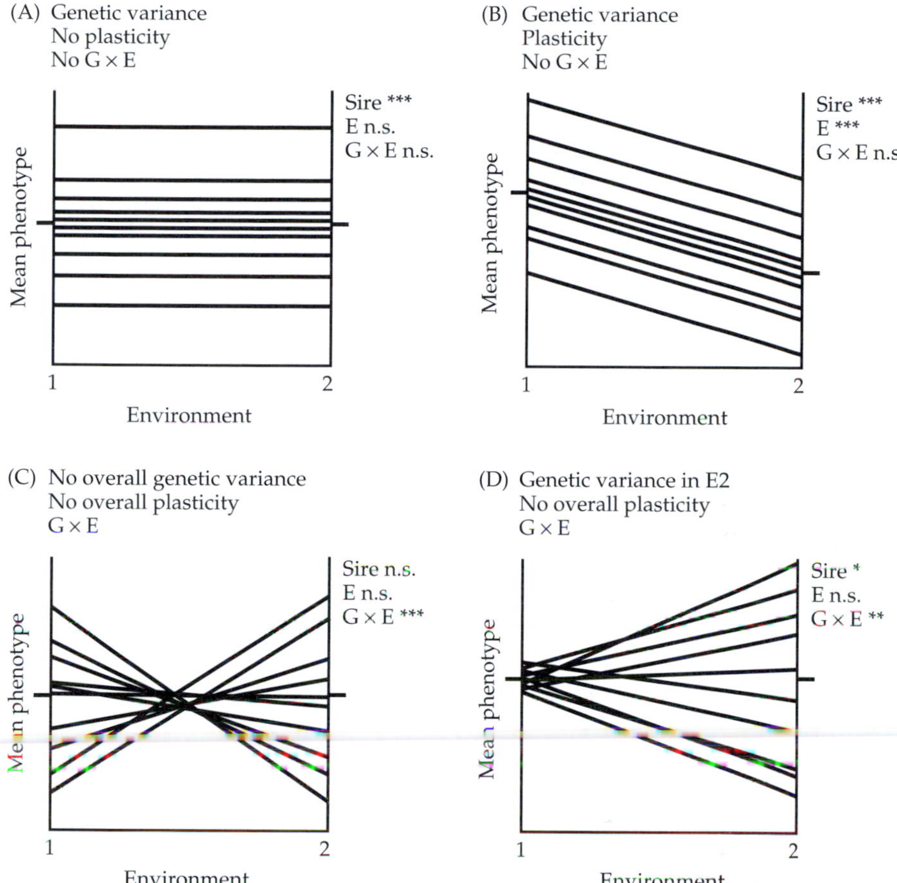

Figure 5.1 Reaction norms for four different hypothetical situations, with the mean phenotype for each clone or family on the *y*-axis and the two environments on the *x*-axis. The tick marks on the outside of the *y*-axes show the overall phenotypic mean within each environment. At the top of each panel is a summary of whether the reaction norms depict overall genetic variance, overall plasticity, or G × E for the trait. To the right of each panel is a summary of what a two-way ANOVA on the data would show; the asterisks denote the *P*-value for each effect, with more asterisks representing a smaller *P*-value, that is, a stronger effect. Factors without a significant difference between levels are denoted with n.s. for not significant. The sire effect would be replaced by clones or full-sibling families in clonal or selfing species, respectively.

there is no genetic variation in plasticity. There is still equal genetic variation within each environment, as in Figure 5.1A, but no differences in plasticity among families.

In Figure 5.1C there is still roughly equal additive genetic variance within each environment, but now there are large differences among families in plasticity, indicated by the different slopes of the reaction norms. The differences among families means that there is additive genetic variation for plasticity. Note that the lines cross, so that the phenotypic rank of the families is different in the two environments, meaning that the families respond differently to these environments. This is called a **genotype-by-environment interaction (G × E)** or genotype-environment interaction (g-e). This type of G × E is sometimes referred to as "crossing" G × E, because the reaction norms cross. Figure 5.1D shows a different kind of G × E, often called "variance" G × E because the amount of genetic variance differs between environments, but the reaction norms rarely cross. In this example, there is significant additive variance in environment 2 but not environment 1. Note that in Figure 5.1C and D many individual families are plastic—that is, their means differ across environments, as reflected in their sloped reaction norms. However, because roughly equal numbers of families increase and decrease their phenotype when raised in environment 2 as compared to environment 1, on average the population is not plastic—the means across all families do not differ between the two environments.

Also in Figure 5.1 are brief summaries of likely results from an analysis of variance for each data set depicted. Shown are simplified results from **two-way ANOVA;** it is called a *two-way* analysis of variance because there are two main factors, sire (or clone or full-sibling family) and environment. These factors are **crossed** (as opposed to nested), which means that individuals at each **level** of each factor are also represented at each level of the other factor (see Box 5.1). Thus, there are members of each half-sibling family (i.e., offspring of each sire) raised in each of the two environments. This is different from the nested ANOVA, in which the levels of the nested factor are not represented across all levels of the factor they are nested within (see Box 5.1 and Chapter 4).

The factors in the two-way ANOVA are called **main effects** to contrast them with the **interaction** between them. In the ANOVA, the *F*-test for the main effects tests whether the phenotypic values at each level of that factor differ significantly when they are averaged over the levels of the other factor(s). For example, a significant sire effect indicates that the sire family means differ from one another when all the sire's offspring are averaged regardless of which environment they were raised in. This average for two equally frequent environments is the midpoint of the reaction norm. Comparing Figure 5.1A and C, we see differences among the midpoints of the reaction norms in Figure 5.1A, and therefore a significant sire effect, but few differences among the midpoints and a nonsignificant sire effect in Figure 5.1C. The nonsignificant sire main effect in part C occurs in spite of the fact that there is significant additive variance within each environment. Since the

BOX 5.1 *Two-Way ANOVA*

Table 5.1 shows hypothetical data for a plant experiment in which four offspring from each of four sires are each grown in each of two environments, sun and shade. The numbers in the table are measurements of leaf length in mm. The $\bar{Y}_{sire}$ rows show the mean leaf lengths for the offspring of each sire in each environment, the bottom row shows the grand mean leaf length of all eight offspring of each sire across both environments, and the right hand column shows the grand mean of all 16 offspring in each environment averaged across sires. Note that in a real experiment of this type there would be many more sires, and dams nested within sires, and therefore more offspring per sire. The accompanying graph (see the next page) for these data has the reaction norm for each sire numbered, and the heavy tick mark on each vertical axis shows the grand mean for that environment. Note the difference from nested ANOVA (see Chapter 4)—here the offspring from all sires are grown in both environments; therefore, the two main effects of sire and environment are crossed with each other rather than one main effect being nested within the other.

TABLE 5.1 *Hypothetical and highly simplified data for two way ANOVA*

Environ.	Sire 1	Sire 2	Sire 3	Sire 4	$\bar{\bar{Y}}_{environ}$
Sun	61	63	66	64	62.75
	64	61	60	67	
	59	66	62	66	
	61	60	62	62	
$\bar{Y}_{Sire}$	61.25	62.5	62.5	64.75	
Shade	66	65	66	71	66.8
	64	68	69	69	
	67	65	65	70	
	62	66	68	68	
$\bar{Y}_{Sire}$	64.75	66	67	69.5	
$\bar{\bar{Y}}_{Sire}$	63	64.25	64.75	67.1	

These data show evidence for differences among sire overall averages (bottom row of table and graph) and a difference of about 4 mm between the leaf lengths in sun and shade (right-hand column and graph), but no evidence for an interaction between the two main effects. The means for each sire differ by approximately 4 mm between the sun and shade environments, suggesting that the offspring of all sires responded similarly to the two environments.

Box 5.1 continued

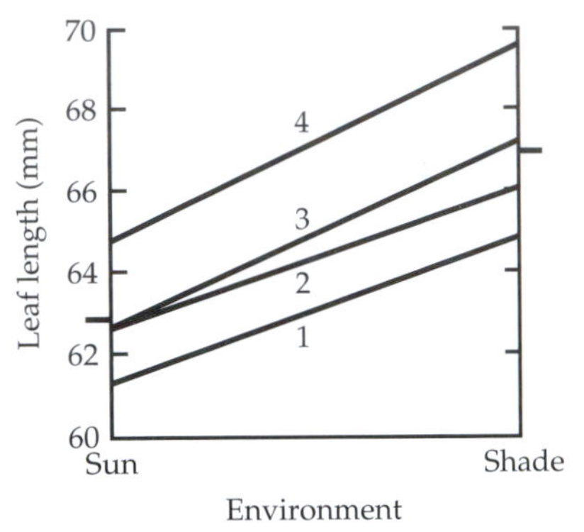

These conclusions are supported by the ANOVA (Table 5.2). For Table 5.2, recall from previous chapters that *d.f.* is degrees of freedom, *SS* stands for sums of squares, *MS* is mean squares (*SS/d.f.*), (which equals the variance for each main effect or interaction), the *F*-ratio is the ratio of two *MS*, and the *P*-value is the probability that each variance (the difference among means) could be greater than zero because of chance alone. Note that the degrees of freedom for an interaction is the product of the *d.f.* for the two main effects in the interaction.

TABLE 5.2 *ANOVA table for the data in Table 5.1*

Source	*d.f.*	*SS*	*MS*	*F* Ratio	*P*
Sire	3	71.6	23.9	5.5	0.005
Environment	1	132.0	132.0	152.7	0.001
Sire-by-environment	3	2.6	0.87	0.2	0.89
Error	24	103.2	4.3		

We conclude from these data that there is additive genetic variation for leaf length as well as plasticity for leaf length (significant main effects of sire and environment, respectively), but no genetic variance for plasticity itself, as is indicated by the absence of genotype (sire)-by-environment interaction. If, for example, sires 2 and 4 showed the reverse pattern, that is, leaves 4 mm longer in sun than in shade, then the results would look more like Figure 5.1C, with no significant main effects of sire (averaged across environments) or environment (averaged across sire), but a strong sire-by-environment interaction.

An actual experiment like this one using sexual species would be more complex than in this simplified example because it would include dams nested within sires as well as more sires. Therefore, it is really a combination of a two-way ANOVA with crossed factors and a nested ANOVA. Additional main effects, nested or not, and interactions can be added by using the same basic principles outlined here, with one caveat—there are no interactions between a main effect and a factor nested within that main effect. The details of more complex ANOVA analyses can be found in statistics books. At the present time these

Box 5.1 continued

analyses are usually done with maximum likelihood methods. Regardless of the statistical details, the basic interpretation remains the same as in our simplified examples in Table 5.2 and Figure 5.1—a significant sire main effect is evidence for overall additive genetic variance, a significant environment main effect is evidence for overall plasticity, and a significant sire-by-environment interaction is evidence for G × E and thus evidence for additive variance for plasticity.

families change in phenotypic rank across environments, there is little variance among sires when the phenotypic values for each sire are averaged across environments, as shown by the midpoints. In Figure 5.1D the sire effect is significant but weaker because there is significant additive variance in environment 2 but not in environment 1, causing the midpoints of the reaction norms to be moderately different. Similarly, the environment main effect tests the overall or average amounts of plasticity. It is based on a comparison of the phenotypic values in the two environments averaged across all sires. In Figure 5.1 there is an average difference only in part B; in the others the average in each environment (marked on the y-axis) is about the same.

The interaction in a two-way ANOVA tests whether different levels of one factor affect the dependent variable (phenotypic trait here) differently depending on the level of the other factor. In the two-environment experiment we have been describing, an interaction occurs when different families respond differently to the two environments, that is, when there is genetic variation in plasticity. In other words, the reaction norms are not parallel. In Figure 5.1A and B, the reaction norms are parallel, so there is not a significant interaction. In contrast, Figure 5.1C and D show marked variation in the slope of the reaction norms and thus a highly significant interaction. This statistical interaction is why it is called genotype-by-environment interaction.

As an example of genotype-by-environment interaction, Mazer and Schick (1991b) grew wild radish plants in a garden at three different densities (environments). The results for one trait, petal area, are shown in Figure 5.2 and Table 5.3. Note that the reaction norms cross extensively, and the sire-by-density interaction (= G × E) is significant. Also note that there is evidence for additive genetic variance averaged over all environments because the sire main effect is also significant; this can be visualized as the broad vertical spread of points in Figure 5.2. Therefore, if there was selection on petal area, or selection for plasticity in petal area, either or both could evolve. There is weak evidence for overall plasticity for petal area (density $P = 0.07$), as the mean petal area is lower at high density.

It is important to note that plasticity may or may not be adaptive. For example, it seems likely that the plasticity in petal area in wild radish is not

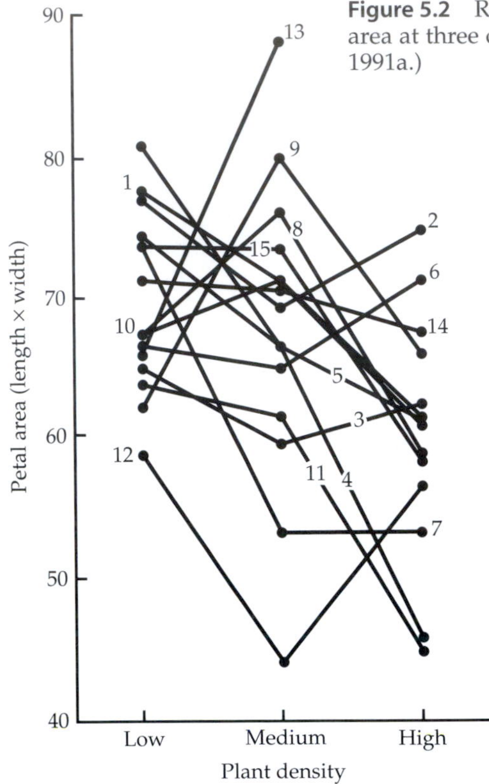

Figure 5.2 Reaction norms for wild radish petal area at three densities. (From Mazer and Schick 1991a.)

Raphanus raphanistrum

adaptive, but rather that the plants are less able to produce large petals under conditions of high intraspecific competition, that is, at high density (Figure 5.2). In addition, the environmental variation among the phenotypes within one macroenvironment, that is, V_E, is generally not thought to be adaptive, and is in fact sometimes called **developmental noise.** This is still plasticity, because different phenotypes are produced by the same genotype due to direct effects of the environment during development (see Figure 4.4).

Clear examples of adaptive plasticity come from induced defenses to predation in invertebrates and plants. A variety of aquatic invertebrates change

TABLE 5.3 *ANOVA table for petal area in wild radish plants grown at 3 densities*

Source	*d.f.*	*SS*	*MS*	*F* ratio	*P*
Sire	14	2.10	0.15	2.16	0.008
Density	2	0.72	0.36	2.92	0.07
Sire-by-density	27	3.24	0.12	1.76	0.01
Error	476	33.32	0.07		

Note: The sire-by-density interaction is the test for G × E.
Source: Data from Mazer and Schick 1991b.

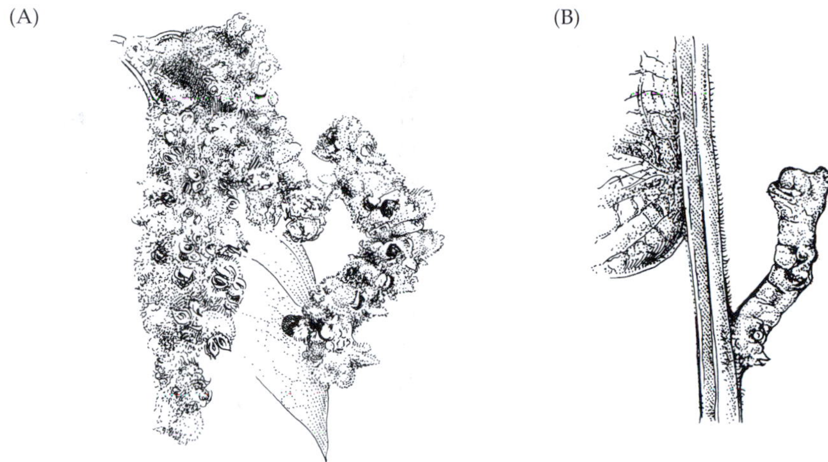

Figure 5.3 Phenotypic plasticity in caterpillar mimetic morphology. (A) A spring catkin morph in its usual vertical hiding position (right side of picture) alongside catkins. (B) An individual expressing the summer twig morph in its usual hiding position. These two individuals were full-siblings raised on different diets. (After Greene 1989.)

the morphology of their exoskeleton in response to the presence of predators (e.g., Harvell 1984), and plants commonly increase the production of defensive chemicals and structures in response to herbivore attack (e.g., Agrawal 1998). Plasticity can also help organisms hide from predators. In one species of inchworm (*Nemoria arizonaria*), caterpillars that hatch in the spring feed on oak catkins (male flowers) and closely mimic the catkins in appearance (Figure 5.3A; Greene 1989). Caterpillars from the same species that hatch in the summer, after the catkins are gone, feed on oak leaves and resemble twigs (Figure 5.3B). The cue is tannins in the diet—leaves are much higher in tannins than are catkins and when caterpillars eat catkins supplemented with tannins, most caterpillars develop the twig morphology.

Multiple Subpopulations—Differentiation vs. Adaptive Plasticity

If different subpopulations inhabit different macroenvironments, then these large environmental differences are likely to cause differences in selection between subpopulations. In Chapter 3 we stated that this situation is likely to lead to local adaptation and genetic differentiation at loci that affect adaptation to these local conditions. Locally adapted populations that differ phenotypically and genetically are sometimes called **ecotypes.** However, we also said that gene flow can be a potent force reducing differentiation, so that the evolution of local adaptation is constrained by high gene flow. This is because alleles that confer high fitness in one environment are moved into the other environment and vice versa. An alternative route to adaptation to variable environments in the face of high gene flow is the evolution of adaptive plasticity. Adaptive plasticity is also favored by temporal variation in the environment, either within or between years.

In many respects temporal variation is analogous to high gene flow between environments, because the environment changes within the same location and population. For adaptive plasticity to evolve, there needs to be additive genetic variation for plasticity, that is, genotype-by-environment interaction. To summarize, spatial differences in environments that are consistent over time combined with low gene flow can lead to local adaptation through genetic differentiation. On the other hand, high gene flow or temporal variation in the environment is more likely to lead to the evolution of adaptive plasticity.

Therefore, if two subpopulations differ in some phenotypic trait, it is important to know whether that difference is due to genetic differentiation or due to direct effects of the environment on the trait—that is, phenotypic plasticity. The simplest method to determine the cause of phenotypic differentiation is a **common garden** experiment, where individuals from both (or several) subpopulations are raised together in the same environment. If the phenotypic differences persist in the common environment, then this is good evidence for genetic differentiation between the subpopulations at loci that affect the trait(s). For example, when grown in a common greenhouse environment, sea-rocket plants (*Cakile edentula*) from a dry site had significantly lower stomatal conductance, higher water-use efficiency, and smaller leaves than plants from a nearby wet site (Dudley 1996). These results suggest genetic differentiation for these traits between the two populations.

Cakile edentula

It is best to intersperse the individuals from the different subpopulations in the garden rather than grow them in adjacent plots to eliminate the possibility of microenvironmental differences causing a phenotypic difference between the subpopulations. Another potential pitfall with a common garden is that differences could be due to nongenetic maternal effects. For example, if one subpopulation of plants was in a higher nitrogen environment and thus had larger individuals than another subpopulation, this size difference could persist in the common garden in the absence of genetic differentiation because the larger individuals also produce larger seeds, which in turn grow into larger plants in the common garden. For this reason researchers sometimes raise the individuals for one or more generations in the common environment before measuring the traits. It is best to do only one or two generations, because selection to adapt to the common environment can occur in the intervening generations and thus reduce the phenotypic differences.

Another drawback to the common garden experiment is that if there is genotype-by-environment interaction for the traits of interest, then the results are only valid for the environment used for the common garden. Because common garden experiments are rarely done in the natural habitat, it is difficult to know if any genetic differences or plasticity found would be expressed in the wild. For these reasons a more difficult technique, called a **reciprocal transplant** experiment, is superior. In a reciprocal transplant, individuals from each subpopulation are experimentally placed in the habitats of all subpopulations (Figure 5.4), creating a common garden at each home site. If there are phenotypic differences between subpopulation means within one or more of the sites, then this is evidence for genetic differentiation because each site is a common garden. If there are differences between the phenotypic means for the same subpopulation when it is planted in different sites, then this is evidence for plasticity. These two outcomes are not mutually exclusive; there could be both differentiation and plasticity simultaneously. Because the reciprocal transplant experiment is really the same as multiple common gardens, the potential problems of microenvironmental variation and maternal effects still apply. However, since the natural habitats of each subpopulation are used, the results are highly applicable to nature.

The classic example of a reciprocal transplant is found in the work of Clausen, Keck, and Hiesey (1940) on California plants. In one of a series of experiments, they reciprocally transplanted sticky cinquefoil (*Potentilla glan-*

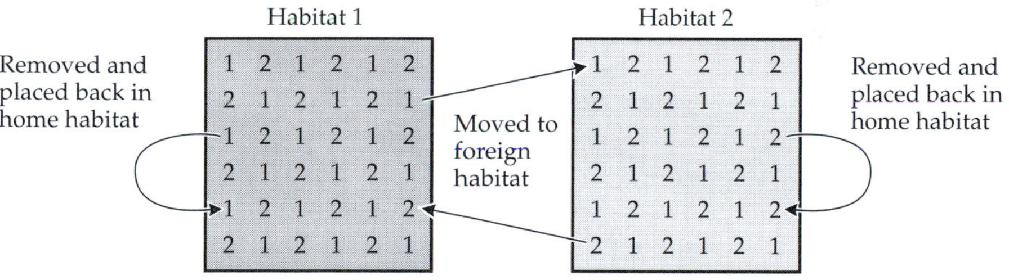

Figure 5.4 Diagram of a reciprocal transplant between two environments (habitats). Individuals from both environments (represented by the numerals 1 and 2) are raised interspersed within each of the habitats, creating a common garden at each site. The arrows depict the movement of individuals by the experimenter (perhaps as seeds) between habitats (straight arrows) and also removed and placed back into the same habitat (curved arrows). It is important to remove and place the individuals to be reared back in their home habitat, rather than just leave them in place, so that all individuals are treated the same.

(A)

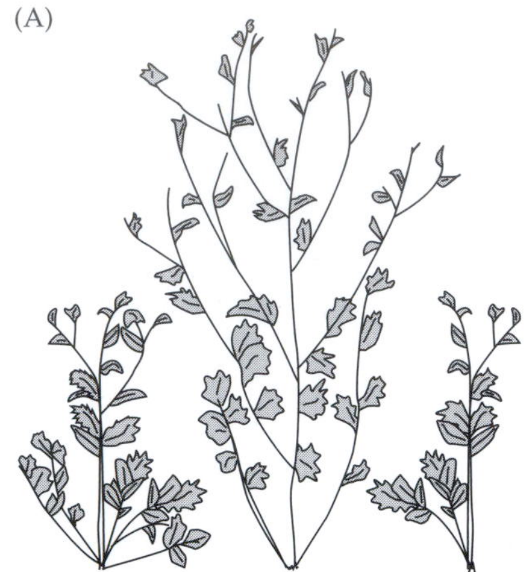

(B)

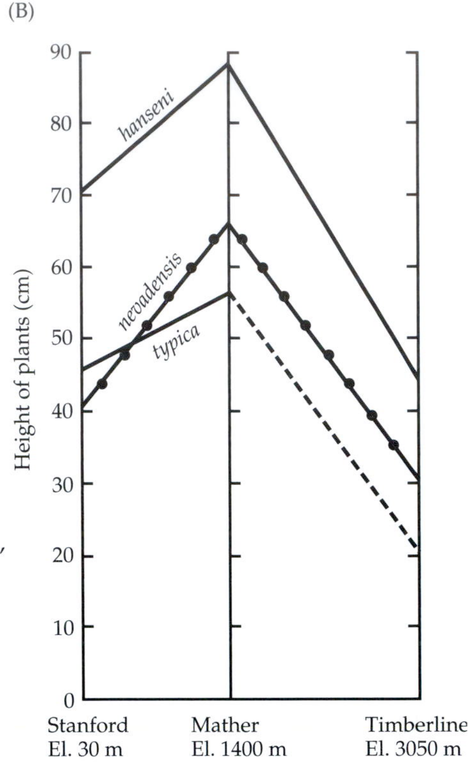

Figure 5.5 Reciprocal transplants in *Potentilla glandulosa*. (A) Drawings of representative plants from low, medium, and high elevation (from left to right) grown at mid-elevation. (B) Results for plant height (on the y-axis) at the three elevations at which the plants were grown (x-axis). The *typica* ecotype is from low-elevation coastal habitats, *hanseni* is from mid-elevation, and *nevadensis* is a high altitude alpine ecotype. The line for *typica* leading to high elevation is dashed because few of these coastal plants survived in the alpine environment. (After Futuyma 1998.)

dulosa) plants between three elevations (Figure 5.5). The results show that variation in plant height is due both to plasticity and to genetic differentiation. Note that plants from each of the three elevations are tallest at 1400 m elevation and shortest at 3050 m, showing plasticity for height. Also note that plants native to the mid-elevation site (*hanseni* ecotype) are always the tallest, regardless of the elevation at which plants were grown, demonstrating genetic differentiation for height.

As you might guess from the names of these techniques, common garden and reciprocal transplant, they were developed and have been used mostly in studies of plants. It is easier to conduct these studies in plants than it is in animals because seeds or plants can be moved around and then stay wherever the experimenter places them. However, these techniques can also be used in animals that rarely move or that can be caged without affecting the relevant environmental variables. This is often true in herbivorous insects, for example. Via (1991) reciprocally transplanted pea aphids (*Acyrthosiphon pisum*) between two different pairs of red clover and alfalfa fields, confining the transplanted individuals on the plants with mesh bags. Fitness of the aphids on their home plant species was four to six times higher than on the alternate host (Figure 5.6), demonstrating strong local adaptation to the host plant. This strong genetic differentiation occurred in spite of a high potential for past gene flow due to the alfalfa and clover fields being interspersed in the landscape.

A newer method for studying adaptive differentiation employs comparisons between quantitative trait differentiation and neutral marker differentiation. As discussed in Chapter 3, neutral marker differentiation is measured using F_{ST}, and by analogy **quantitative trait differentiation** is quantified using a measure called $\boldsymbol{Q_{ST}}$ (Spitze 1993). To estimate Q_{ST}, additive genetic variation (or variation among clones or selfed families where appropriate) is estimated in a common garden design in each of the subpopulations of interest. Then, again by analogy with F_{ST}, the proportion of genetic variation that is due to differentiation among subpopulations is calculated as the estimate of Q_{ST}. A finding of significantly greater Q_{ST} for a given phenotypic trait relative to the average F_{ST} for the neutral markers is evidence that

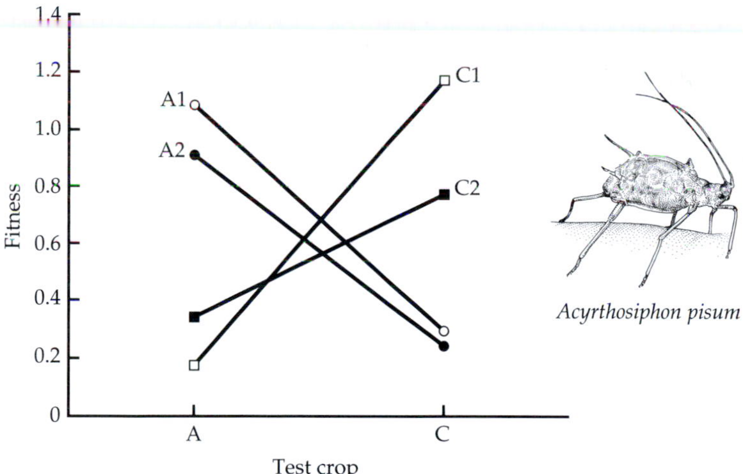

Figure 5.6 Reaction norms for fitness in reciprocally transplanted pea aphids. On the y-axis is fitness, incorporating both longevity and fecundity, and the two crop species on which the aphids were raised ("test crop") are on the x-axis. Each point represents the mean fitness of individuals collected originally from each of two different alfalfa fields (A1 and A2) or two different clover fields (C1 and C2). (After Via 1991.)

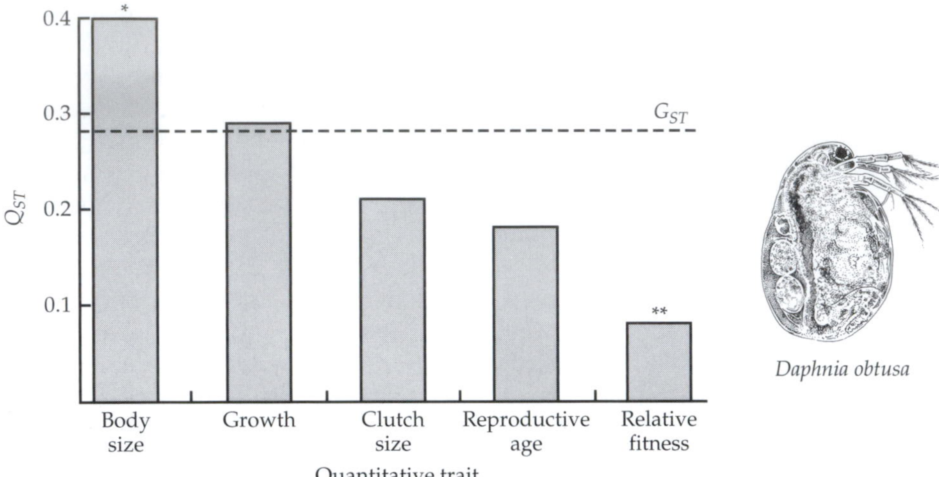

Figure 5.7 Estimates of quantitative trait differentiation (Q_{ST}) among eight populations of *Daphnia obtusa*. The average G_{ST}, which estimates neutral differentiation from seven allozyme loci, is shown with a dashed line. Traits marked with asterisks are those for which Q_{ST} is significantly different from G_{ST}. $^*P < 0.05$; $^{**} P < 0.01$. (Data from Spitze 1993.)

differential selection has acted on the phenotypic trait, because, on average, drift and migration act on all loci and traits equally (Chapter 3).

For example, Spitze (1993) estimated an average G_{ST} (G_{ST} is closely related to F_{ST}; Chapter 3) of 0.28 for seven allozyme loci in eight populations of *Daphnia obtusa* (see Chapter 2). The values of Q_{ST} for five quantitative traits are shown in Figure 5.7. For three of the traits, growth, clutch size, and reproductive age, Q_{ST} was not significantly different than G_{ST} for the neutral markers, suggesting that these quantitative traits varied about as much as expected by chance. The Q_{ST} for body size was significantly greater than average G_{ST}, which is evidence for differential selection among the populations. Conversely, the Q_{ST} for relative fitness was significantly lower than G_{ST}, suggesting that selection acts in the same direction in all populations for this trait. This makes sense, because selection always acts to increase fitness. All these conclusions could be further tested by measuring the direction of selection on these traits in these populations using the techniques discussed in Chapter 6.

Correlations Among Traits

To this point we have dealt almost exclusively with single phenotypic traits. But organisms are not collections of isolated traits. Individual organisms are made up of genetically, functionally, developmentally, and physiologically interconnected traits. In addition, selection and the ecological forces causing selection act on whole organisms, not single traits (see Chapter 6). Therefore, we need a method to understand the genetics and evolution of groups of traits simultaneously; for this purpose biologists use phenotypic, genetic, and environmental correlations among traits. As for most concepts in ecological genetics, the correlation has no meaning in an individual, but rather it is a property of the population.

The **phenotypic correlation (r_P)** measures the degree to which two traits covary among individuals in the population. If two traits covary, it means that variance in one trait is related to variance in the other trait. Figure 5.8 shows examples of phenotypic correlations of various magnitudes; each point in the figures represents the phenotypic values for two traits measured in the same individual, and all the points together are a random sample of

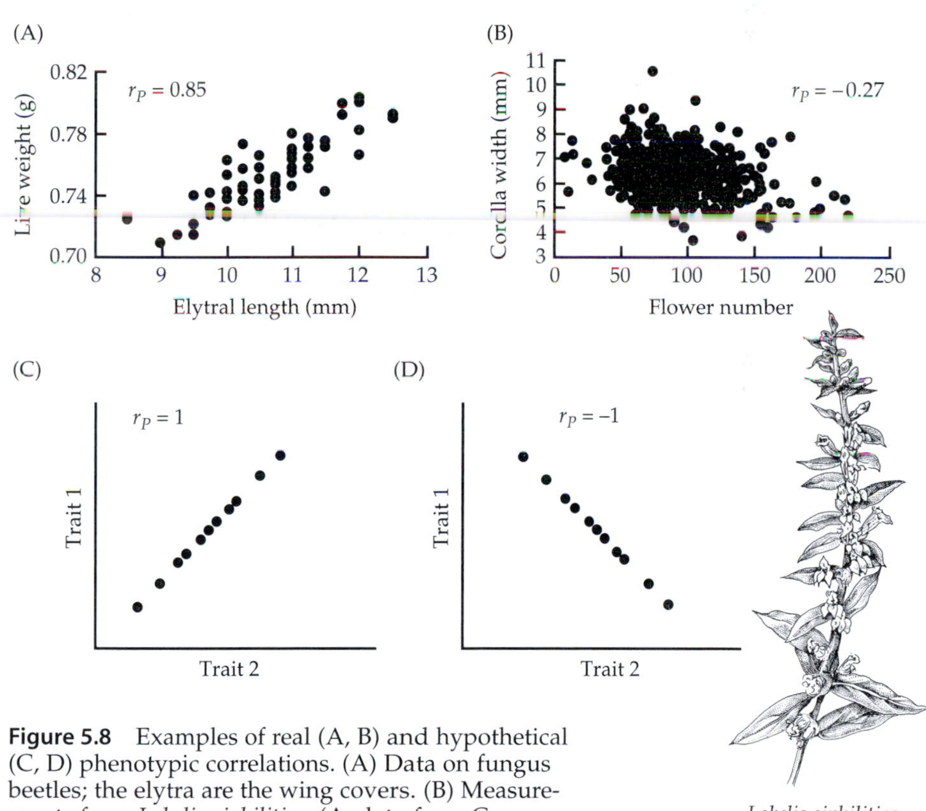

Figure 5.8 Examples of real (A, B) and hypothetical (C, D) phenotypic correlations. (A) Data on fungus beetles; the elytra are the wing covers. (B) Measurements from *Lobelia siphilitica*. (A, data from Conner 1988; B, data from Caruso, in press.)

the population. Figure 5.8A shows a strong positive correlation ($r_P = 0.85$), so that beetles with long elytra also are heavier, whereas beetles with short elytra are lighter. Although there is some scatter in this relationship, generally one can predict the weight of a beetle fairly well from its elytral length, and vice versa. Therefore, little of the variation in these two traits is independent. Note that a single individual has no correlation by itself, but rather the correlation emerges from the collection of individuals in the population. Figure 5.8B shows a weak negative correlation ($r_P = -0.27$); there is a significant negative relationship between flower size (as measured by corolla width) and flower number, meaning that plants with more flowers tend to have smaller flowers. However, the greater scatter around this relationship compared to Figure 5.8A means that there is more independent variation in the two traits. Figure 5.8C and D show hypothetical "perfect" correlations of 1 and –1, in which none of the variation in the two traits is independent of variation in the other trait.

Correlations are standardized versions of the more fundamental **covariance;** because the former is a standardized version of the latter, they are conceptually (but not mathematically) interchangeable in most cases. When two traits covary, the situation is exactly as it sounds—they vary together. Thus, Figure 5.8A shows a strong positive covariance, because the variance in weight is tightly related to the variance in elytral length, and Figure 5.8B shows a weaker negative covariance. Covariances are closely related to variances, as is shown by the formula for estimating the covariance (Cov):

$$\mathrm{Cov}_{xy} = \frac{\sum_{i=1}^{n}\left(X_i - \overline{X}\right)\left(Y_i - \overline{Y}\right)}{n-1} \qquad 5.1$$

The only difference between this formula and that for a variance (see Equation 4.3) is that, instead of summing the squared deviations from the mean for one variable as is done for a variance, for a covariance the **cross-product** is summed across individuals. The covariance therefore depends upon the products of the deviations from the means for each of two variables. This means that, unlike variances, covariances can be positive or negative. High values for covariances (and thus correlations) occur when individuals with values that deviate strongly from the mean for one trait also deviate strongly from the mean in the other trait. If they tend to deviate in the same direction, the result is a positive covariance (Figure 5.8A). If the deviations from the mean for one trait are less related to the deviations from the mean for the other trait, then lower covariances result. If positive deviations for one trait tend to occur in individuals with negative deviations for the other trait and vice versa, then negative covariances and correlations result (Figure 5.8B). Correlation coefficients are calculated by dividing the covariance by the square root of the product of the variances for each trait:

$$r_{x,y} = \frac{\text{Cov}_{x,y}}{\sqrt{V_X V_Y}} \qquad \textbf{5.2}$$

The advantage of a correlation compared to covariance is that it varies from −1 to 1, facilitating comparisons among traits measured on different scales. Covariances depend on the scale of measurement, and are therefore not easily compared across traits; however, covariance is the more fundamental measure, and as such is more useful in equations for evolutionary change (Chapter 6).

The phenotypic correlation is made up of two components: the genetic correlation and the environmental correlation. The **genetic correlation** estimates the degree to which two traits are affected by the same genes (pleiotropy) or pairs of genes (linkage disequilibrium; discussed later in the chapter). The **environmental correlation (r_E)** estimates the degree to which two traits respond to variation in the same environmental factors. The additive **genetic correlation (r_A)** is the correlation of breeding values, and it is most important in evolutionary change for the same reason as the additive genetic variance: It measures the degree of transmission from parent to offspring. As we will see in the next section on artificial selection and in the next chapter, selection on one trait will cause an evolutionary change in a second trait if there is an additive genetic correlation between the traits.

The relative importance of r_A and r_E in determining the phenotypic correlation between two traits X and Y depends on the heritability of the two traits:

$$r_P = h_X h_Y r_A + e_X e_Y r_E \qquad \textbf{5.3}$$

where h is the square root of the narrow-sense heritability, and $e^2 = 1 - h^2$. In other words, e^2 is the complement of heritability, or the proportion of phenotypic variance caused by environmental and nonadditive genetic variance. Equation 5.3 shows that if heritability is high, then the phenotypic correlation is determined mainly by r_A, whereas when heritability is low, the environmental correlation is more important.

Note here that r_E also includes correlation due to nonadditive genetic effects and thus lumps both true environmental correlation and nonadditive genetic correlation together. However, as we discussed in the last chapter, in studies of natural populations it is often difficult to estimate additive variance and covariance, so full-sibling families are sometimes used. This estimates the broad-sense genetic correlation, r_G, which includes dominance and the effects of common rearing environment (Chapter 4).

Table 5.4 shows examples of phenotypic, genetic, and environmental correlations among a variety of traits in a variety of organisms. Note that the

TABLE 5.4 *Examples of phenotypic (r_P), genetic (r_A or r_G), and environmental (r_E) correlations among traits in natural populations.*

Species	Trait 1	Trait 2	r_P	r_A or r_G	r_E	References
Garter snake	Color pattern	Reversal behavior	–0.18	–0.75	0.29	Brodie 1989; Brodie 1992
Darwin's finch	Wing length	Bill length	0.60	0.95	–0.68	Boag 1983
House mouse	Brain size	Body size	0.22	–0.23	0.34	Leamy 1990
Spring peeper	Larval duration	Final larval size	0.31	0.53	NA	Woodward et al. 1988
Water strider	Abdomen length	Femur width	0.33	0.71	0.10	Preziosi and Roff 1998
Daphnia (crustacean)	Size at maturity	Clutch size	0.35	0.11	0.49	Pfrender and Lynch 2000
Phlox (plant)	Height	Petal width	0.17	0.42	0.13	Schwaegerle and Levin 1991

Note: NA = not available.

most common pattern, especially for morphological traits, is for all three correlations to be positive, which means that large organisms tend to be larger for all traits, due to the effects of both genetic and environmental variation. This makes intuitive sense, because alleles that increase the number or size of cells or increase rates of nutrient acquisition will tend to increase the sizes of all traits, whereas other alleles at these loci will tend to decrease the sizes of all traits. Similarly, organisms that find themselves in an environment favorable for growth, (i.e., with optimal temperatures and abundant resources) will tend to be large in all traits compared to conspecifics in less favorable environments.

In some cases the genetic and environmental correlations are opposite in sign, for example, the house mouse brain:body-size correlation in Table 5.4. The negative genetic correlation could reflect a **trade-off** between the two traits in the allocation of the fixed amount of resources available to an individual during development; in other words, alleles that cause larger resource allocation to body size cause fewer resources to be available to brain development, and vice versa. The positive environmental correlation reflects differences in the overall resources available to an organism. Here the phenotypic correlation is positive, due to the environmental correlation being larger and to fairly low heritabilities of the two traits ($h^2 < 0.5$; see Equation 5.3). In contrast, the negative genetic correlation between color pattern and reversal behavior in garter snakes is likely to be an adaptation rather than a trade-off; this example is discussed further in Chapter 6.

To estimate genetic correlations and covariances, one uses exactly the same mating designs as are used for estimating genetic variance and heritability, that is, offspring–parent and sibling analyses. The only differences are that more than one trait is measured and the analyses are somewhat different. For offspring–parent analysis, instead of regressing the offspring values on the parental values for the same trait, two times the covariance of trait X in the offspring and trait Y in the midparents is taken as an estimate of the additive genetic covariance between the two traits. The opposite covariance is also calculated, that is, the covariance of trait Y in the offspring and trait X in the parents. The additive genetic correlation is calculated by standardizing these additive genetic covariances by the square root of the product of the two offspring–parent covariances for each trait:

$$\begin{aligned} r_A(1) &= \frac{\mathrm{Cov}(X,y)}{\sqrt{[\mathrm{Cov}(X,x)][\mathrm{Cov}(Y,y)]}} \\ r_A(2) &= \frac{\mathrm{Cov}(x,Y)}{\sqrt{[\mathrm{Cov}(X,x)][\mathrm{Cov}(Y,y)]}} \end{aligned} \qquad \textbf{5.4}$$

(where uppercase letters refer to the midparent values and lowercase letters to the average of the offspring). These equations provide two estimates of the genetic correlation, which can then be averaged.

When using full-sibling or half-sibling data, ANOVA can estimate covariance components as well as variance components, and the correlations can be constructed from these using Equation 5.2. A simpler method is to calculate the correlation of breeding values, especially when these are estimated using BLUP (see Chapter 4). Both these methods have advantages and disadvantages, and neither is clearly superior to the other. In the past either full-sibling or half-sibling **family mean** correlations were used as estimates of the genetic correlation; this is very similar to the breeding value correlation, but is inferior to using the BLUP estimates.

Campbell (1996) used both paternal half-sibling family mean correlation and father–offspring covariance to estimate genetic correlations among floral traits in a field study of scarlet gilia, *Ipomopsis aggregata*. The two methods agreed well for the correlation between corolla length and anther position (r_A = 0.89 and 0.95 for the two methods, respectively) but less well for the correlation between corolla length and width (r_A = 0.44 and 0.13).

Sources of genetic covariance among traits

Mendel's law of independent assortment states that different traits are inherited independently. One of Mendel's experiments demonstrated that whether

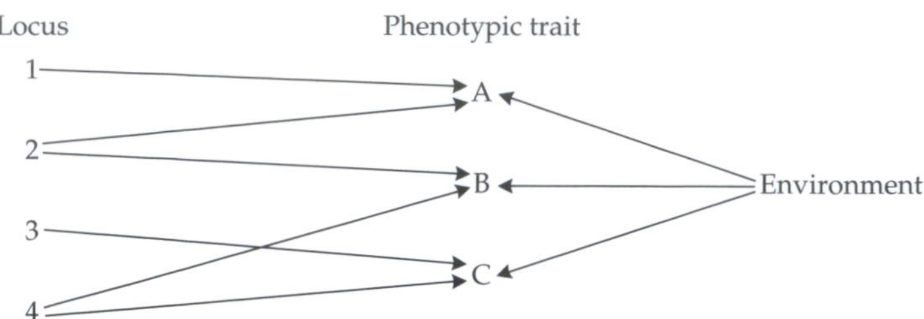

Figure 5.9 Simplified diagram of pleiotropic gene action. All three phenotypic traits are oligogenic (affected by two loci), and loci 2 and 4 are pleiotropic, affecting two traits each.

a pea seed was round or wrinkled was inherited independently from whether the seed was yellow or green. A genetic correlation between two traits is a violation of the law of independent assortment.

The two causes of genetic covariance (and thus correlation) are pleiotropy and linkage disequilibrium. **Pleiotropy** occurs when one locus affects more than one trait (Figure 5.9). For example, genes affecting digestive enzyme activity or photosynthetic rates are likely to affect the sizes of many morphological traits. This hypothetical example emphasizes that the degree of pleiotropy depends on the level of the trait in the phenotypic hierarchy (see Chapter 1); higher level traits (like morphology) will be affected more by pleiotropy than lower level traits (like physiology). At higher phenotypic levels, there may be multiple pleiotropic loci affecting the same pair of traits, and some of these may be positively pleiotropic (that is, causing a positive genetic correlation) and some may be negatively pleiotropic.

For example, the *W* locus in morning glories (*Ipomoea purpurea)* determines the amount of floral pigment, with *WW* plants having dark flowers, *Ww* light flowers, and *ww* white flowers. The floral pigments are related to compounds that affect levels of herbivory. Simms and Bucher (1996) randomly placed larvae of the herbivorous beetle *Charidotella* on morning glories with each of the three genotypes at the *W* locus. When placed on *WW* and *Ww* morning glories, 50% of the larvae died, while 61% of the larvae died when placed on *ww* plants ($P = 0.04$). Therefore, the *W* locus has pleiotropic effects on both floral color and plant toxicity to herbivores.

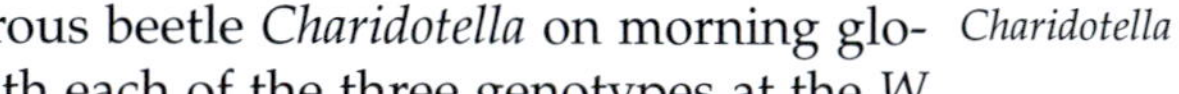

Charidotella

Ipomoea purpurea

In addition to pleiotropy, genetic correlations can also be caused by nonrandom relationships between the alleles present at two or more loci. The situation can be made clear by first considering the opposite case of random association between alleles. Suppose p_1 and p_2 are the allele frequencies of A_1 and A_2, respectively, for one gene locus, and q_1 and q_2 are the allele frequencies of B_1 and B_2, respectively, for another locus and that these are the only alleles present at these loci. When the alleles of the two loci are randomly associated with each other, it means that the frequencies of the four possible types of gametes are as follows:

$$f(A_1B_1) = p_1q_1$$
$$f(A_1B_2) = p_1q_2$$
$$f(A_2B_1) = p_2q_1$$
$$f(A_2B_2) = p_2q_2$$

In other words, the frequencies of the gametes are those expected at random, determined by the products of the frequencies of the alleles at each locus. This condition is called **linkage equilibrium.** When this condition is not met, that is, when a nonrandom relationship between the alleles present at two or more loci exists, it is called **linkage disequilibrium.** Linkage disequilibrium is the second cause of genetic correlation, in addition to pleiotropy. However, linkage between loci is not necessary for disequilibrium to occur, so a more descriptive term, less often used in the literature, is **gametic phase disequilibrium.** The term *linkage disequilibrium* is derived from the fact that genetic linkage among loci helps maintain the nonrandom association between alleles against the randomizing effects of recombination.

Recombination can be defined as the mixing of maternal and paternal genes into new combinations in the gametes during meiosis. The evolutionary significance of recombination is that it produces a set of genetically variable, unique progeny from a single pair of parents. Although recombination can seem simple, it is a source of confusion, partly because there are three generations involved—we are concerned with whether genes came from the mother or father of an individual who is producing gametes that will be used to form the offspring of that individual. There are two sources of recombination (Figure 5.10). The first is during early meiosis (prophase I) when the homologous chromosomes pair up. During this pairing **crossing over** can occur, in which sections of the homologous chromosomes from each parent are exchanged. Thus, crossing over is the cause of recombination between loci on the *same* chromosome.

The second source of recombination, between loci on *different* chromosomes, occurs because there is a random probability that the maternal or paternal chromosome of each homologous pair will line up on either side of the metaphase plate (**random alignment;** Figure 5.10). Therefore, if two loci A and B are **unlinked** (on different chromosomes), the probability of recombination between them is 0.5 due to the 50% chance of the maternal A allele

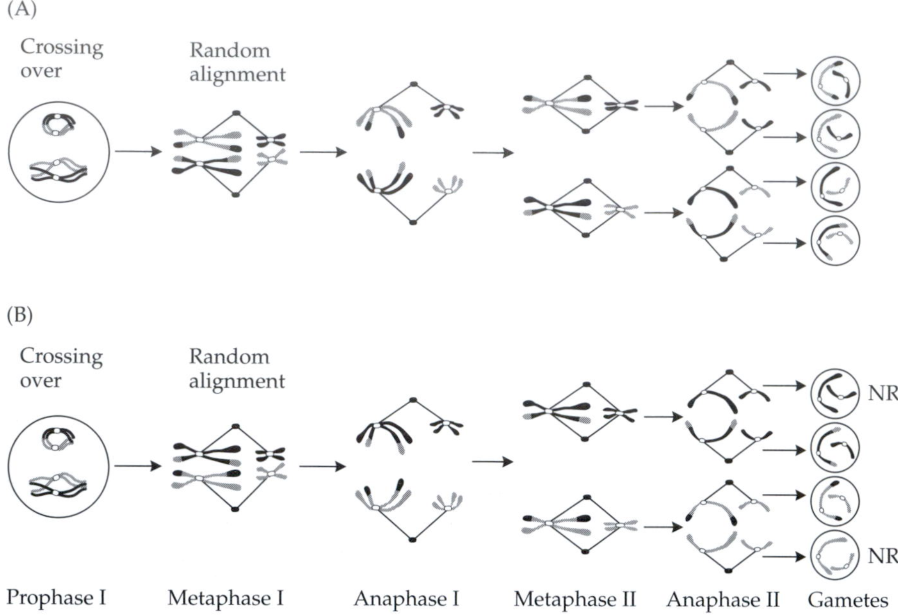

Figure 5.10 Simplified diagram of meiosis showing an organism with $N = 2$ pairs of chromosomes (one long and one short). The chromosomes that came from this organism's father are shown in black and the maternal chromosomes in gray. The two times that recombination can occur are labeled *crossing over* and *random alignment*. A and B show two separate meioses. In (A) recombination occurs between most of the maternal and paternal loci on the two nonhomologous (short and long) chromosomes, because they aligned with an opposite orientation during metaphase I. Because of crossing over, in some cases there was recombination between loci on the same chromosome (denoted by gray or black sections in the chromosomes of the opposite shade). In (B) the only recombination occurred due to crossing over within chromosomes, because the maternal and paternal origin chromosomes lined up on the same side of the metaphase plate. Therefore, two of the gametes lack recombination entirely—these are labeled NR for nonrecombinant.

lining up on the same side of the metaphase plate as the paternal *B* allele. If *A* and *B* are sufficiently close together on the same chromosome, that is, they are **linked,** the probability of recombination is proportional to the physical distance between them. This proportionality occurs because there is a random chance of crossing over at any site on the chromosome, so there is an increased probability of crossing over with increased distance. **Tight linkage** means that the loci are very close together. Loci far apart on the same chromosome can appear to be unlinked if crossovers between them are very common.

For these reasons, the probability of recombination or the **recombination rate, *c*** (sometimes referred to as *r*) between two loci varies between a minimum of 0—there is no crossing over because they are the same locus—and a maximum of 0.5 meaning that the loci are either far apart in one chromosome or on different chromosomes. The recombination rate determines how fast a population in linkage disequilibrium goes to equilibrium. If $c = 0.5$, then disequilibrium is halved each generation (Figure 5.11). Therefore, even if the loci are unlinked, there is only an asymptotic approach to equilibrium. Why doesn't the population return to linkage equilibrium for unlinked loci after a single generation, as is the case for HWE? It is because, by chance, half of the chromosomes from each parent line up on the same side of the metaphase plate, so half of the allele combinations across loci remain the same each generation (e.g., Figure 5.10B). Figure 5.11 shows that linkage is not necessary for linkage disequilibrium, but it does help main-

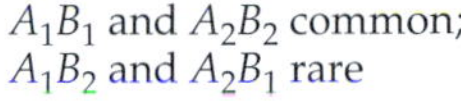

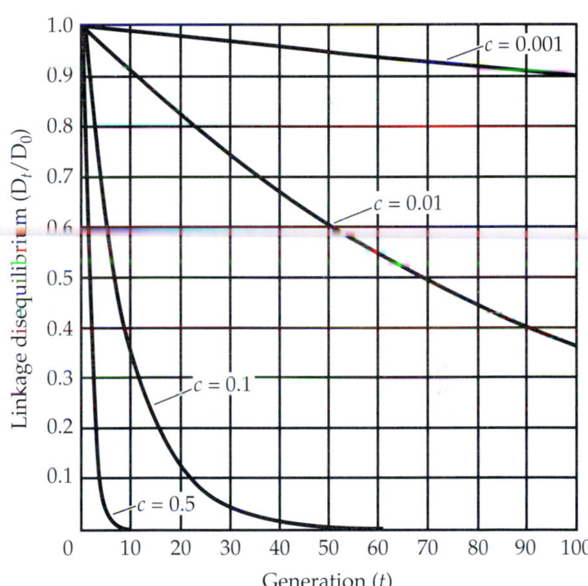

Random frequencies of
A_1B_1, A_2B_2, A_1B_2, and A_2B_1

Figure 5.11 Theoretical plots of the return to linkage equilibrium ($D_t/D_0 = 0$) under different degrees of physical linkage. D_t/D_0 is the degree of disequilibrium at generation t relative to the initial level at generation 0, and c is the per-generation recombination frequency between the loci. The curve for $c = 0.5$ is that for unlinked loci, such as loci on different chromosomes, whereas the other three curves represent pairs of loci that are located on the same chromosome. The progressively lower recombination frequencies, and thus lower rates of decline of linkage disequilibrium, are due to decreasing physical distance between the pairs of loci. The equation for these curves is $D_t/D_0 = (1 - c)^t$.

tain disequilibrium in the face of constant recombination—tighter linkage (lower *c*) slows the return to linkage equilibrium because recombination is less likely. Figure 5.11 illustrates why the state is called a *disequilibrium*; after the nonrandom association is created, recombination tends to return the population to linkage equilibrium, that is, to the combination of alleles at the two loci that is expected to occur at random, based on the allele frequencies.

What can cause linkage disequilibrium?

The most common causes of linkage disequilibrium are (1) selection for particular combinations of alleles (sometimes called **epistatic selection,** because fitness depends upon interactions of alleles at different loci), (2) a mixture of two populations with different allele frequencies, and (3) random genetic drift. The classic example of selection creating linkage disequilibrium is heterostyly in the English primrose (*Primula vulgaris*). The "pin" morph has a long style and the anthers are placed well below the stigma in the middle of the corolla tube. The "thrum" morph has a short style, placing the stigma at mid-tube, and anthers placed up high at the opening of tube (Figure 5.12). The two "homostyle" morphs, with the stigma and anthers at the same height, are extremely rare. Therefore, the genes for style length and anther placement are in linkage disequilibrium, with the pin and thrum combinations common and the homostyle combinations rare. This disequilibrium is thought to be caused by selection for increased outcrossing, because the close proximity of the anthers and stigma in the homostyle morphs makes selfing more likely than in the pin and thrum flowers.

Mechanisms of genetic correlations in nature

Note that the above examples of traits affected by pleiotropy and linkage disequilibrium involve simple polymorphisms rather than pairs of quantitative traits. This is because genetic correlations among quantitative traits, while very important to evolutionary change, are difficult to study at a mechanistic level in natural populations. It is important to understand the underlying mechanisms, because genetic correlations will cause evolutionary constraints (see Chapter 6) only if they are caused by pleiotropy; correlations caused by linkage disequilibrium can be changed quickly by recombination and selection. Also, a finding of a correlation caused by linkage disequilibrium suggests the possibility that selection has acted to increase functional integration among the correlated traits (see Chapter 6). For oligogenic or polygenic traits, pleiotropy and linkage disequilibrium are not mutually exclusive. A pair of correlated traits might have some shared pleiotropic loci and other nonpleiotropic loci that are in linkage disequilib-

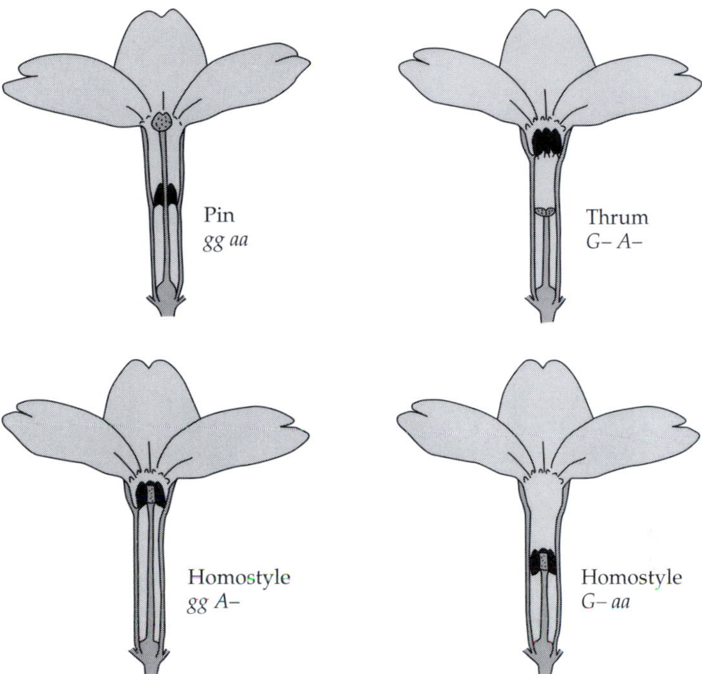

Figure 5.12 Cross sections of four types of primrose flowers, with their names and genotypes shown. The anthers are shown in black and the stigmas are stippled. The *G* locus determines style length (short style dominant) and the *A* locus determines anther position (higher anthers dominant). The pin and thrum phenotypes are common and the homostyle very rare due to linkage disequilibrium between these loci. In both cases the thrum phenotype is produced by the dominant allele. The dashes in the genotypes signify that either allele may be present with the dominant allele to produce the phenotype. (After Ford 1975.)

rium. Theoretical work indicates that tight linkage, inbreeding, extremely strong epistatic selection, or some combination of the three is necessary to maintain linkage disequilibrium. How often these conditions hold in nature is unknown.

For these reasons, we know very little about the mechanisms underlying genetic correlations in nature. Some insight is now being gained from QTL mapping (see the section entitled QTL Mapping later in this chapter). Conner (2002) used the decline in linkage disequilibrium through recombination (Figure 5.11) to distinguish this mechanism from pleiotropy. Two large ($N_e \approx 600$, to minimize drift) replicate populations of wild radish were randomly mated for nine generations in the greenhouse. Each plant contributed equal numbers of offspring to the next generation, minimizing

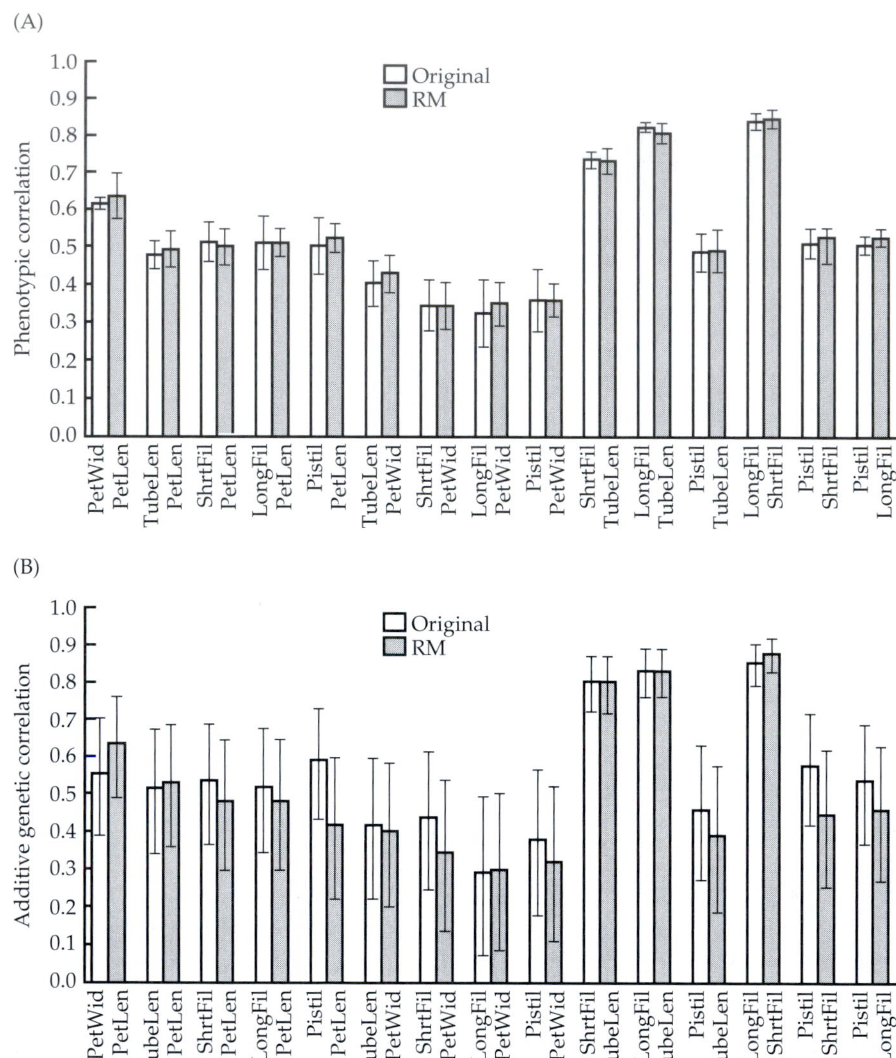

Figure 5.13 Correlations between pairs of wild radish floral traits before and after nine generations of random mating. Each pair of bars represents correlations between the two traits on the *x*-axis; the lighter bar is the correlation in the original population, and the darker bar is the mean of the two populations that were random-mated for nine generations. The error bars are statistical 95% confidence intervals; the fact that they broadly overlap means the correlations did not change significantly due to the random mating. (A) Phenotypic correlations. (B) Additive genetic correlations. PetLen and PetWid are petal length and width, TubeLen is the length of the corolla tube, ShrtFil and LongFil are the lengths of short and long filaments, and Pistil is pistil length. (From Conner 2002.)

selection. Under these conditions, linkage disequilibrium should be greatly reduced unless linkage is very tight. Phenotypic and genetic correlations among six floral traits were then estimated simultaneously in both replicate populations and in an equal-sized sample of the original natural population grown from stored seeds. The results (Figure 5.13) show no significant change in the correlations, indicating that the correlations are caused by pleiotropy or linkage disequilibrium with extremely tight linkage.

Artificial Selection

Artificial selection refers to the process of selective breeding of plants and animals by humans to produce populations with more desirable traits. All of our domesticated species, whether crops, farm animals, or pets, were modified from wild species by humans practicing artificial selection. For example, many of our crop plants, such as wheat and corn, have seed heads that do not shatter and scatter the seeds, unlike their wild relatives. Most pets have been selected to be more docile and thus better companions, and farm animals have been selected to produce greater quantity and quality of meat and milk. We are interested in artificial selection in this book because it causes human-directed evolution, and it is a very useful tool for measuring heritabilities and genetic correlations. Artificial selection is analogous to natural selection, and it was for this reason that Darwin devoted most of the first chapter of the *Origin of Species* to artificial selection.

To perform artificial selection, the phenotypic trait to be selected is measured in a population, and members of the population at one or the other phenotypic extremes are chosen for breeding (Figure 5.14). The rest of the individuals do not contribute to the next generation. This selection procedure is called *truncation selection* because there is a discrete phenotypic value (the truncation point) above which the organisms have high fitness and below which they have zero fitness. In nature, selection usually does not work this way because fitnesses are continuously distributed (see Chapter 6). The strength of truncation selection is measured by the **selection differential, *S*,** which is the difference in the mean of the selected group and the mean of the entire population before selection (Figure 5.14). Note that this is not the same as the selection coefficient (see Chapter 3), which is denoted by a lowercase letter *s*. The selection coefficient measures selection on genotypes, whereas the selection differential measures selection on phenotypes.

Sometimes the selection differential is referred to as the change in the mean *within* a generation, that is, before versus after the selective event (the choosing of organisms to be bred in artificial selection). The breeder and the evolutionary biologist are usually more interested in the change in the mean *across* generations, in other words, between the parental and offspring generations. This is short-term phenotypic evolution, and is called the **response**

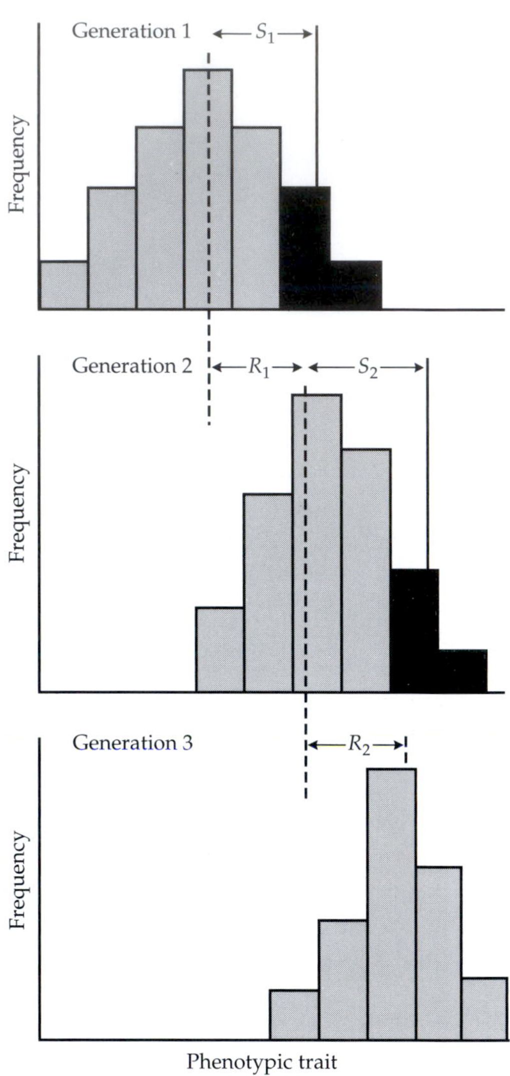

Figure 5.14 Two rounds of artificial selection with associated responses. The histograms represent the distribution of the phenotypic trait in the population, with the black bars representing the individuals selected for breeding to form the next generation. Dashed lines show the means of entire population each generation, solid lines are the means of the group of selected individuals, and the selection differential (S) depicts the difference between the two in each of the generations with selection. The response to selection (R) indicates the differences across generations in the population means (the dashed lines). Note that each of the responses to selection is only slightly smaller than the selection differential that gave rise to it (i.e., those with the same subscripts), meaning that this is a highly heritable trait. In addition to the change in mean across generations (R) caused by this selection, note that the selection also reduced the variance in the population.

to selection, *R*, and is calculated as the mean of all offspring minus the mean of all parents (Figure 5.14).

The magnitude of R for a given strength of selection S depends on the narrow-sense heritability. Recall from the last chapter that narrow-sense heritability is the proportion of the phenotypic variance that is caused by additive genetic variance, and that the latter is what is transmitted directly from parents to offspring. Selection within a generation (measured as S) acts on the phenotypic variance, and the change across generations R depends on the product of S and the fraction of this variance that is additive genetic (that is, the heritability):

$$R = h^2S \tag{5.5}$$

This is often called the *breeder's equation*, and is a fundamental equation for phenotypic evolution. It clearly illustrates the crucial point that R increases when either the strength of selection (S) or the heritability (h^2) increases.

Palmer and Dingle (1986) selected for both increased and decreased wing length over nine generations in a natural population of milkweed bugs (*Oncopeltus fasciatus*). Wing length is associated with differences in migratory behavior in this insect. Figure 5.15 shows two common ways of presenting artificial selection results. In Figure 5.15A and B, the change in mean wing length is plotted against generation number for each of two replicate lines selected for long wings, short wings, and random-mated controls. Note the strong and consistent response to selection over time. Figure 5.15C and D shows the response as a deviation from the control means, and presents plots of this response against the cumulative selection differential rather than against time. After nine generations, the long lines gained about 2 mm of wing length from a total selection differential of 4 mm; from a rearrangement of Equation 5.5 we can calculate a **realized heritability** of about 0.5. The realized heritability is usually estimated as the slope of a linear regression of cumulative response across generations on the cumulative selection differential, in other words, the slopes in Figure 5.15C and D. Note that this estimate works because Equation 5.5 is in the form of the equation for a line. The short lines lost about 1.2 mm of wing length from the 4 mm total selection differential, giving a realized heritability of about 0.3.

The disparity in response between the two directions of selection is called an **asymmetrical response,** and it is common in artificial selection studies. There are many possible reasons for asymmetry (Falconer and Mackay 1996, pp. 211–215), and it is often hard to determine the cause in any particular experiment. For example, inbreeding depression is one possible cause of asymmetrical response. Artificial selection lines are often small, and thus inbreeding depression might cause a decline in the size of the selected trait. This in turn might cause the lines selected for larger size to exhibit a smaller response than the lines selected for small size.

If there are genetic correlations between the selected trait and other traits, then a **correlated response** to selection can occur, that is, an evolutionary response in one or more other unselected traits. Correlated responses occur because selection on one trait changes allele frequencies at pleiotropic loci or loci in linkage disequilibrium, thus changing the phenotypes of the correlated traits. Just as heritability determines the magnitude of the direct response to a given strength of selection on the same trait, the genetic corre-

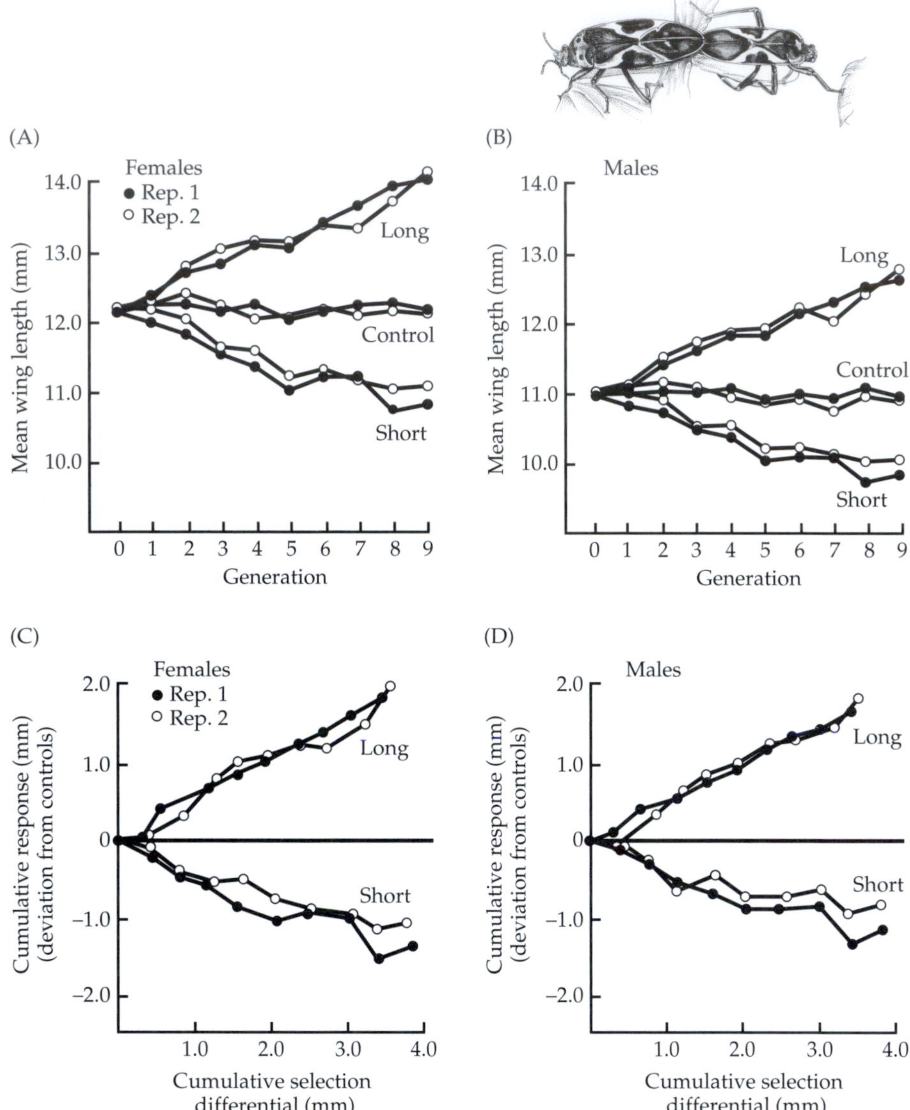

Figure 5.15 Responses to artificial selection on wing length in milkweed bugs (*Oncopeltus fasciatus*). There were two replicates each of lines selected for longer wings, lines selected for shorter wings, and random-mated controls. Because females are larger than males, results for the two sexes are separated. (A,B) Responses plotted as wing length versus generation number. (C,D) The same responses plotted as a cumulative deviation from controls versus cumulative selection differential. (The latter is the sum of the individual selection differentials up to that generation, and is a measure of total selection imposed to that point.) (After Palmer and Dingle 1986.)

lation determines the magnitude and direction of correlated response to selection on a correlated trait.

Correlated responses are common, indicating that genetic correlations are common. In their selected lines, Palmer and Dingle found significant differences from controls in body length, head and thorax width, development time, and fecundity. In all cases these were positive correlated responses, indicating positive genetic correlations. In other words, the lines selected for long wings were larger in many dimensions, had higher fecundity (often correlated with body size), and took longer to develop compared to controls, whereas lines selected for short wings were smaller, less fecund, but developed faster. Correlated responses to natural selection will be discussed in the next chapter.

In the vast majority of artificial selection experiments, a response to selection occurs; this means that additive genetic variation exists for most quantitative traits in natural populations. In fact, in selection experiments that have continued for many generations, the population mean in the selected lines is often moved far beyond the range of phenotypes that were present in the original population. For example, originating from a single presumed wild ancestor, the breeds of domestic dogs now vary over almost two orders of magnitude in weight (Stearns and Hoekstra 2000). An experiment started in 1896 has selected continuously on oil content in corn, moving the population mean far outside the original phenotypic range (Figure 5.16).

At first glance it is hard to understand how selection could move the mean beyond the existing range, because selection can act only on available variation in the population, and only on the additive genetic portion of this phenotypic variation. The first answer to this apparent paradox is new mutation, which provides additional variation to select upon. New mutations may have been very important in animals and plants that have been domesticated and selected upon over millennia. However, even long-term experiments, such as the corn-oil study, do not encompass enough generations for mutation to be a sufficient explanation for the large response.

The more important reason that long-term experiments can move the mean beyond the phenotypic range of the original generation is that without strong directional selection, no individual will be homozygous for alleles that increase or decrease the trait at all loci affecting the trait. After a few dozen generations of strong artificial selection, however, many or most of the individuals in the population may be homozygous at most or all relevant loci, moving the population outside the original range. When this happens, additive variance for the trait will be depleted, the population will stop responding to selection, and will reach a **selection plateau** or **limit to selection.**

Plateaus are common in selection experiments, but often they are not caused by a depletion of additive variance. One line of evidence for this is that often when selection is **relaxed** by stopping the artificial selection for sev-

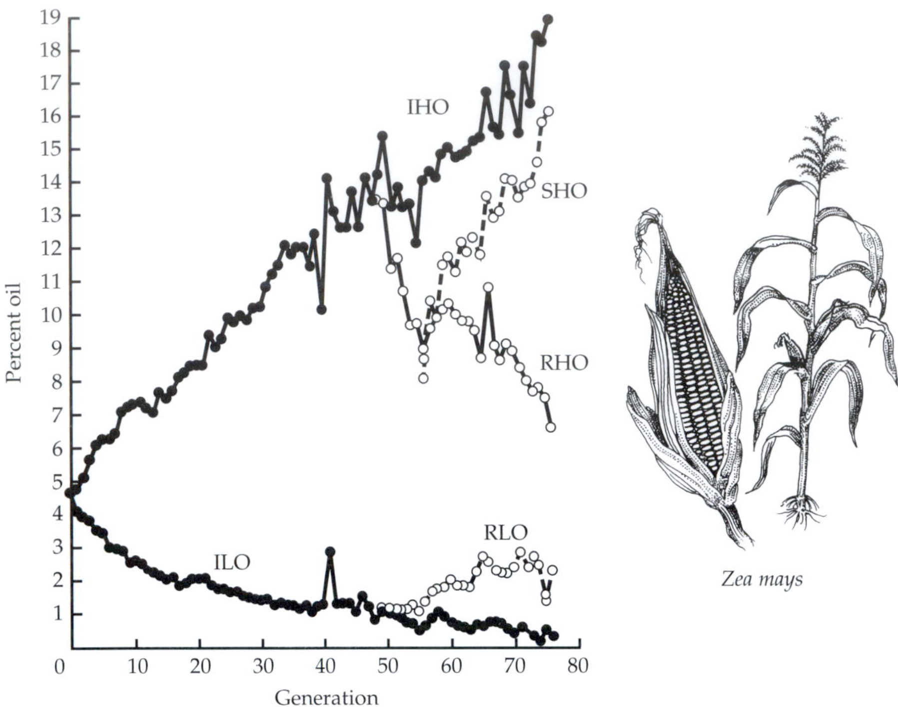

Figure 5.16 Results from the world's longest continuous artificial selection experiment, selecting for increased and decreased oil content in corn kernels at the University of Illinois. IHO and ILO are the original high-and low-oil selection lines, respectively. RHO and RLO are reverse selection lines, originated from the high and low lines but then selected in the opposite direction (i.e., for low oil content in the RHO line). SHO is a secondary high-oil selected line started from the RHO line; the strong response in this line is further evidence that additive genetic variation for oil content has not been depleted. (After Dudley 1977.)

eral generations, the traits evolve back towards their original mean. The fact that the traits can still evolve clearly shows that additive variance has not been entirely depleted. Because the lines spontaneously evolve in the opposite direction to the direction of artificial selection, this suggests that some form of natural selection in the laboratory caused the selection plateau. In other words, the artificial selection was opposed by an equal amount of natural selection in the opposite direction, halting any further response. Selection is sometimes **reversed** in long-term artificial selection lines by imposing selection in the opposite direction to the original artificial selection (Figure 5.16); a response to reversed selection is also excellent evidence for remaining additive genetic variance. On the other hand, some artificial selection lines have been observed to remain at the selected mean for several generations even after reverse selection is imposed, suggesting that additive variance has been depleted in these cases (Falconer and Mackay 1996). Another cause of selection plateaus is called an **intrinsic limit;** a good example is the low oil lines of maize (see Figure 5.16), in which response is diminishing only because the mean phenotype is pushing toward the intrinsic limit of no oil.

Advantages and disadvantages of artificial selection

Artificial selection has several advantages for determining genetic variation and covariation over the alternative methods of offspring–parent regression and sibling analysis (hereafter, referred to as *single-generation methods*). First and foremost, artificial selection answers directly whether the trait can evolve in response to selection, whereas single-generation methods are indirect. Artificial selection is simpler conceptually and practically than sibling analysis, and it has greater statistical power for the same number of individuals measured. This is primarily because artificial selection tests differences between line means, whereas sibling analysis relies on variance and covariance components (Falconer and Mackay 1996). This greater statistical power means that artificial selection is a good choice for traits that are difficult to measure (e.g., physiology, behavior), because generally fewer individuals need to be measured at one time compared to single-generation methods.

Artificial selection also has some disadvantages compared to single-generation methods. Each artificial selection experiment can measure additive variance only for the selected trait and can estimate genetic covariances only between this trait and other measured traits. Therefore, a single artificial selection experiment cannot be used to measure the entire matrix of genetic variances and covariances among a group of traits (***G***); a separate artificial selection experiment for each trait is necessary. In other words, artificial selection provides no information on the genetic covariance or correlation between two unselected traits, and does not provide a quantitative estimate of genetic variance or heritability for these traits. This information is necessary for making quantitative predictions of the speed of evolutionary change for several traits (Campbell 1996; see Chapter 6).

Artificial selection experiments are not practical for many organisms because they typically require maintaining the organisms in the laboratory or greenhouse, sometimes for long periods, and controlled matings need to be performed. In some organisms single-generation experiments can be done relatively easily in the field. In monogamous birds, for example, offspring–parent regression has often been carried out with natural matings in natural populations (Schluter and Smith 1986; reviewed in Boag and van Noordwijk 1987; see Chapter 4). The magnitude of genetic variance and covariance can be strongly affected by the environment (genotype-by-environment interaction), so that confining experiments to unnatural environments can become a serious problem if quantitative estimates of variance are of interest. A final practical problem is that artificial selection is most efficient if the individuals that are measured can then be mated, so that traits that cannot be measured on live individuals pose more difficulty. This difficulty can be overcome by measuring some individuals and using their full-siblings for the matings, but this will slow progress because the measured and mated individuals only share half of their genes.

QTL Mapping

The techniques of statistical quantitative genetics described in this chapter and in Chapter 4 provide great insight into the evolution of quantitative traits, particularly when combined with measures of the strength of natural selection on these traits (see Chapter 6). However, these techniques use statistical abstractions such as means, variances, and correlations to represent the genome. Because statistical quantitative genetics treats the genome as a "black box," it cannot directly address a number of important questions concerning the **genetic architecture** of phenotypic traits, such as how many genes affect a given trait, where those genes are located in the genome, how much variation in the phenotype is determined by individual gene loci, what is the mode of action of these genes (additive, dominant, or overdominant), what other traits are affected by the genes (pleiotropy), how the genes interact with each other to produce the phenotype (epistasis), and what genetic mechanisms underlie genotype-by-environment interaction. Recent advances in molecular genetics combined with theoretical and statistical innovations have given rise to a new technique that is a first step in breaking open the black box of the genetics of quantitative traits. This technique, called **quantitative trait locus (QTL) mapping,** locates regions of the genome containing genes affecting quantitative traits (reviewed in Lynch and Walsh 1998; Tanksley 1993). Once these regions have been identified, further work can determine what the genes actually code for, but this functional knowledge is still difficult to obtain.

The first step in QTL mapping is to create a *genetic map*, preferably covering the entire genome. This is the step that has been greatly facilitated by the proliferation of DNA-based genetic markers (Chapter 2), because earlier genetic maps were limited to small numbers of visible polymorphisms and allozymes. The goal is to have markers evenly and relatively closely spaced throughout the genome. To make a map, a **mapping population** is often created by crossing genetically divergent populations, most often inbred lines or closely related species. A single individual from one population is mated to a single individual from another population to create the F_1 generation. The F_1 tends to be highly heterozygous because the divergent parental populations are fixed for different alleles at many loci. It also has high linkage disequilibrium because the chromosomes from the parental populations have not had the opportunity to recombine. An F_1 individual is then either self-fertilized (if possible) or mated to another F_1 individual to create an F_2 mapping population, or alternatively, an F_1 individual can be crossed with an individual from one of the parental populations to create a **backcross population.** In this second generation of sexual reproduction, recombination produces unique combinations of the genomes from the two parental populations in each F_2 individual. Note that most of these designs involve inbreeding in the F_1 mating, which can be a problem if inbreeding depression affects the traits of interest.

It is the frequency of recombination between markers from the two parents that is used to construct the map (Figure 5.17) because the frequency of crossing over increases with increasing distance on the chromosome. The **map distance** between two markers is usually measured in centimorgans (cM), named after the pioneering geneticist T. H. Morgan. One cM is equivalent to a 1% recombination rate (i.e., $c = 0.01$), a relationship that holds only for short mapping distances. In spite of the fact that recombination frequency increases with increasing physical distance on the chromosome, the map distance is not always a good measure of the **physical distance,** in terms of the number of nucleotides in the DNA. Physical distance is usually measured in units of thousands of base pairs or **kilobases (kb).** Discrepancies between map and physical distances can arise because the rate of recombination per kb may differ among different regions of one chromosome, among different chromosomes, between the sexes, or among species.

The result of this analysis is a genetic map (Figure 5.18) showing the positions of markers in various linkage groups, often with the distance in cM between markers shown. The linkage groups are groups of markers that are on the same chromosome. In species that are not genetic model organisms, it is often not known to which chromosome each linkage group corresponds. Depending on the density of the map (that is, how many markers and how close together they are), it is possible that two linkage groups are in fact on the same chromosome. This seeming paradox can happen because pairs of markers that are far apart on the same chromosome can have recombination rates of 50%, and so they appear to be on different chromosomes. It is also possible with lower-density maps that some chromosomes may not be represented by any of the markers examined, and thus these chromosomes would not appear on the map.

Once a linkage map has been constructed, the next step is to test for associations between variability in DNA markers and variation in the phenotypic trait (Figure 5.19). If certain marker bands occur with certain values of the phenotypic trait at a higher probability than that expected by random chance, then this is evidence that a QTL affecting this trait is linked to the

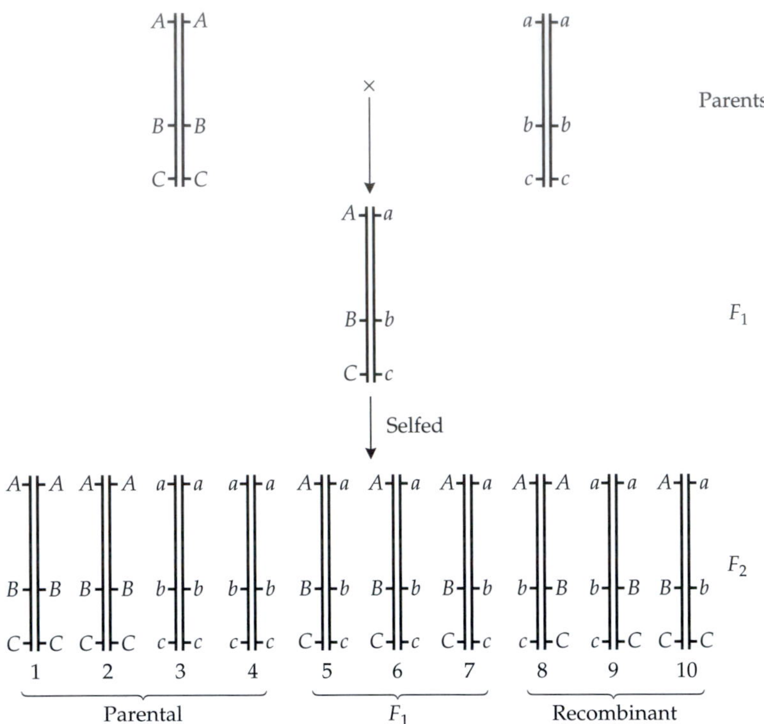

Figure 5.17 Diagram illustrating recombination among three marker loci on the same chromosome in an F_2 mapping population. Each individual is represented by one pair of homologous chromosomes containing the three marker loci. Note that the parents are homozygous for opposite alleles at all three loci, the ideal situation for mapping often achieved by using divergent inbred lines or different species for the parents. The markers are linked, and there is no recombination in 70% of the F_2 chromosomes—they represent either the parental or the F_1 genotypes (numbers 1–7). Two of the F_2 individuals (numbers 8 and 9) show recombination between the *A* and *B* marker loci on one of their chromosomes, whereas only individual 10 shows recombination between the *B* and *C* loci (one chromosome). If these numbers were representative of a large sample of F_2 individuals, then the recombination fraction *r* would be 0.1 between loci *A* and *B* (2 out of the 20 chromosomes in the F_2 showing recombination) and 0.05 between loci *B* and *C* (1 out of 20). Loci *A* and *B* are therefore about 20 cM apart and *B* and *C* about 10 cM apart; this is why in the diagram the former pair of loci are shown twice as far apart as the latter.

marker. Testing for this association between the phenotype and markers has been facilitated by a variety of new statistical techniques. The most common statistical test for QTL starts with a **log-odds ratio** or **LOD score:**

$$LOD = log_{10} (L_1/L_0) \qquad \textbf{5.6}$$

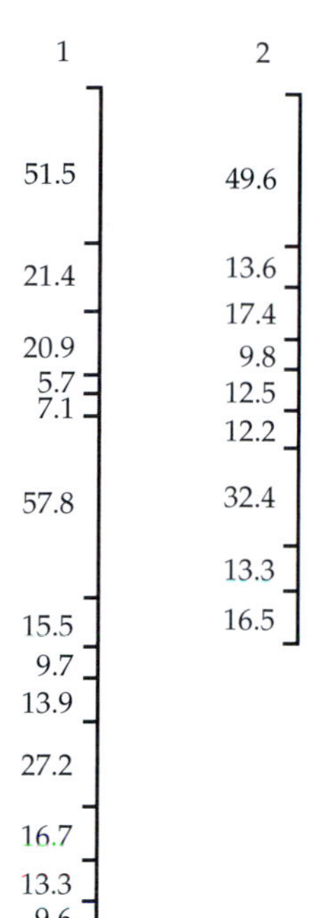

Figure 5.18 Linkage map for four representative linkage groups (chromosomes or portions of chromosomes) from a backcross mapping population resulting from a cross between two monkey flower species (*Mimulus guttatus* and *M. platycalyx*). The position of each RAPD marker locus is depicted as a tick mark on the vertical line that represents the linkage group. The numbers indicate the map distances between each pair of adjacent markers in cM. (From Lin and Ritland 1997.)

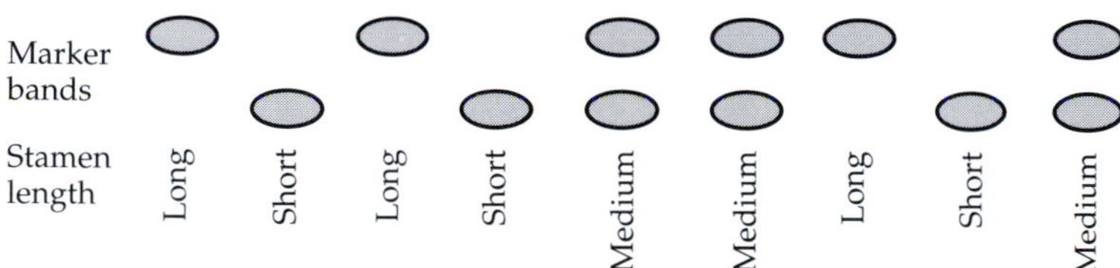

Figure 5.19 Hypothetical raw data for QTL mapping. In this simplified example, genotypes at a codominant molecular marker locus (the ovals represent bands on an electrophoretic gel) are correlated with the size of a quantitative trait (stamen length, for example). There are nine individuals (nine lanes on the gel). Individuals that are homozygous for the upper band always have long stamens, homozygotes for the lower band always have short stamens, and heterozygotes have intermediate stamen length.

where L_1 is the likelihood that there is a QTL linked to a particular marker, given the data, and L_0 is the probability of no QTL near that marker, for whatever phenotypic trait is under analysis. The results are commonly presented as LOD plots for a given linkage group (Figure 5.20). Each horizontal line represents a statistical **threshold.** A LOD score above the threshold is taken as evidence that there is a QTL at that location in the linkage group, whereas a LOD score below the threshold is judged to be attributable to chance alone. There are a number of methods for setting the threshold, and clearly a LOD score that falls just below the threshold should not be ignored

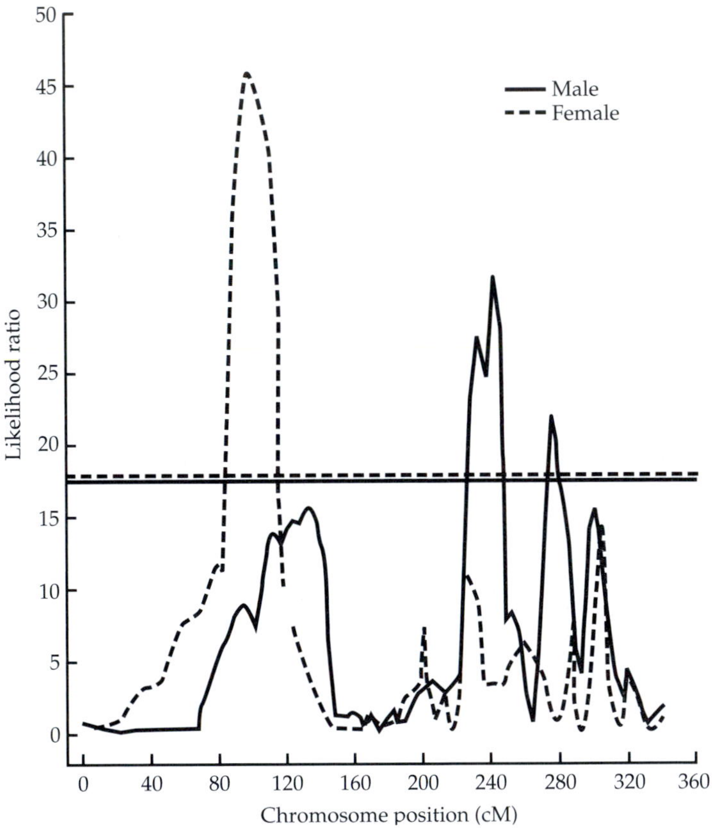

Figure 5.20 LOD plot showing QTLs for lifespan in *Drosophila* chromosome 3 in males and females separately. The x-axis shows the map position along the chromosome in cM, the y-axis shows the likelihood ratio, and the horizontal lines show the significance thresholds. Note that some peaks lie just below the threshold (e.g., at about 140 and 300 cM.) (After Leips and Mackay 2000.)

entirely, because the statistical support for this is only a little worse than the support for a LOD score just above the threshold (see Figure 5.20).

When there is an association between phenotypic variance and a given marker, then this is evidence that a locus that affects the trait (that is, a QTL) is linked to that marker. These QTL are then placed on the genetic map (Figure 5.21). On this particular map, the positions of the QTLs on the genetic map are indicated by the horizontal lines to the right of the vertical lines, and the length of each rectangle connected to these lines represents a **confidence interval** that indicates the statistical uncertainty in this position. The maps in Figures 5.18 and 5.21 are from a study that mapped floral differences in a cross between *Mimulus guttatus*, a plant that has a high rate of

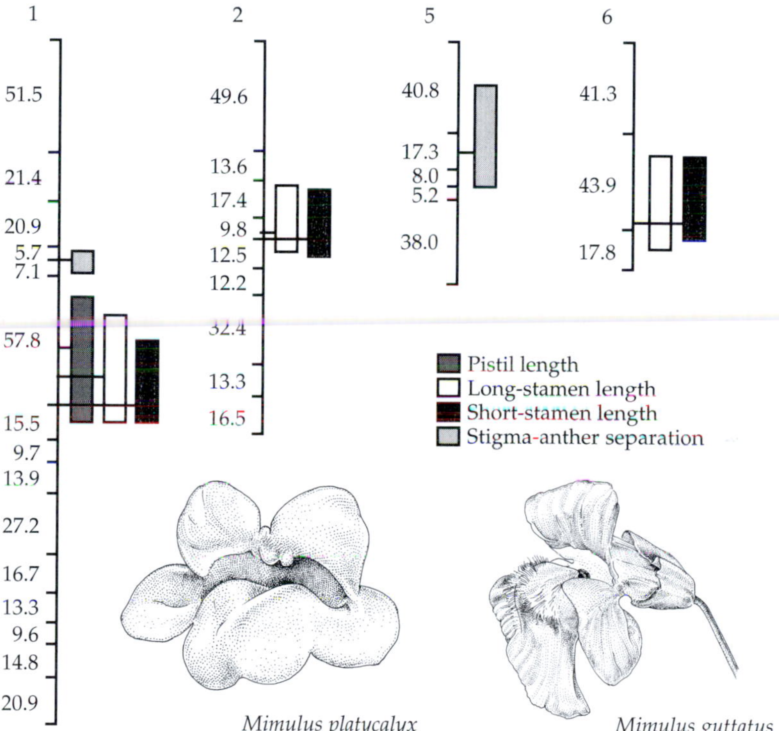

Figure 5.21 QTL positions for several floral traits in monkey flowers. These are the same linkage maps as in Figure 5.18 with the QTL added. The lengths of the rectangles for each QTL represent confidence intervals, meaning that there is strong statistical support for the QTL occurring somewhere in that interval. Recall that the numbers to the left of the vertical lines represent the map distances between adjacent markers, and the positions of the markers are indicated by tick marks. (After Lin and Ritland 1997.)

outcrossing, and *Mimulus platycalyx*, which has a high rate of selfing (Lin and Ritland 1997). Lin and Ritland found three gene loci that affect the lengths of the stamens in three different linkage groups, two loci that affect stigma-anther separation (a trait that affects selfing rate), and one locus that affects pistil length. The loci affecting the long and short stamens in linkage groups 2 and 6 are so close together that it likely indicates a pleiotropic locus affecting both traits in each of these linkage groups. The QTLs affecting the two stamen lengths and pistil length in linkage group 1 are farther apart, but still within each other's confidence intervals, so that this could also be one pleiotropic locus, or it could be two or three linked loci.

The percentage of phenotypic variance explained by each of these QTLs, along with its LOD score, is shown in Table 5.5. Note that the effect of each locus on the trait variance is moderate to fairly strong, ranging from 7.6% to 28.6% of the phenotypic variance explained. Also note that the statistical support for each QTL (that is, the LOD score) generally increases with increasing effect of the locus on the phenotype.

Lin and Ritland (1997) had a fairly small mapping population, consisting of 96 individuals, whereas theory suggests mapping populations of about 500 are necessary to increase power to detect QTLs and reduce an inherent upward bias in the estimates of their effects (Beavis 1998). A study of floral QTLs in a cross between two different species of *Mimulus* with 465 individuals in the mapping population was undertaken by Bradshaw et al. (1998). *Mimulus cardinalis*, a species with tubular red flowers typical of hummingbird pollinated plants, was crossed with *Mimulus lewisii*, whose flowers are pale with flat petals and are likely adapted to its bumblebee pollinators. This

TABLE 5.5 *The percentage of phenotypic variance explained and LOD score for each of the QTL in Figure 5.21*

Trait	Linkage group	% phenotypic variance explained	LOD score
Long stamen length	1	17.7	3.43
	2	7.6	2.04
	6	13.0	3.35
Short stamen length	1	12.9	3.40
	2	8.3	2.25
	6	12.4	2.97
Stigma-anther separation	1	10.7	3.20
	5	10.5	2.47
Pistil length	1	28.6	4.68

Source: Data from Lin and Ritland (1997).

larger study found twice as many QTLs per trait, with an average of almost four per trait compared to less than two in Lin and Ritland's study, and a wider range of percent variance explained, from 3.3% to 84.3%. This comparison illustrates that with larger sample sizes QTLs of smaller phenotypic effect can be detected.

Bradshaw et al. (1998) also reported that more than two-thirds of their QTLs exhibited evidence for dominance, with the rest being additive. Leips and Mackay (2000) provide more detailed information on mode of action in their study of QTLs affecting lifespan in *Drosophila melanogaster* at two larval densities (Table 5.6). Recall from the previous chapter that a measures the magnitude of the additive effect of the locus on the trait; therefore, a provides another measure (in addition to percentage of variance explained) of the magnitude of effect of a locus. Also recall that d/a measures the degree of dominance. The additive effects vary fourfold, from 1.5 to 6.2, while the degree of dominance varies from partial dominance ($d/a < 1$) to strong overdominance for QTL 1 in males at low density ($d/a = 3.7$). More interesting is the fact that both measures of mode of action differ between the sexes for the same QTL. This is seen most strongly for QTL 1, which has weak additive effects but is strongly overdominant in males at low density; it has almost three times the additive effect but shows only partial dominance in females at low density. At high density there is not merely a difference in mode of action of the same QTL; rather, the QTLs affecting lifespan are completely different between males and females at high density (i.e., QTL 2 and 3 in males versus QTL 4 in females).

TABLE 5.6 *Mode of action and candidate genes for several QTLs for lifespan in* Drosophila

Sex	Density	QTL	*a*	*d*	*d/a*	Candidate genes
Male	Low	1	1.5	5.5	3.7	*Adh*
		5	2.9	3.4	1.2	
		6	4.8	3.3	0.7	*Pgm, Cat*
Male	High	2	6.2	4.6	0.7	*EF1α*
		3	3.4	5.8	1.7	
Female	Low	1	4.2	3.2	0.8	*Adh*
		4	3.9	3.0	0.8	*Sod*
		5/6	3.8	4.2	1.1	*Pgm, Cat, Ide, Hsp70*
Female	High	4	5.2	5.5	1.1	*Sod*

Note: The additive effect of each QTL is given by a and the degree of dominance by d/a.
Source: Data from Leips and Mackay 2000.

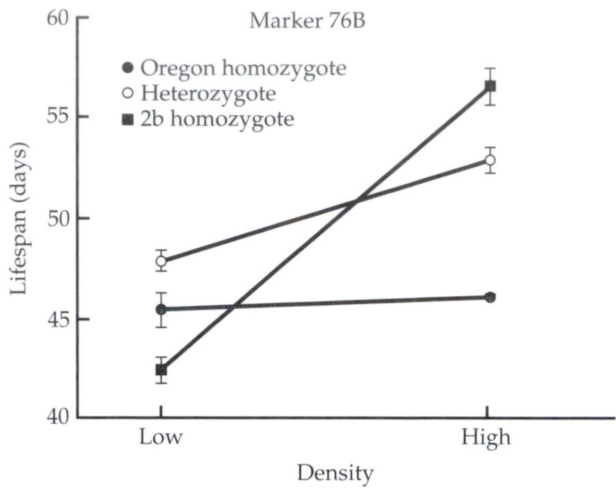

Figure 5.22 Genotype-by-environment interaction at a single QTL affecting lifespan in *Drosophila*. Note that the 2b homozygote has the shortest lifespan of the three genotypes at low density and the longest at high density. Oregon and 2b are the parental strains that were used to create the mapping population. One error bar is smaller than the marker symbol. (After Leips and Mackay 2000.)

Leips and Mackay also present evidence for genotype-by-environment interaction at single QTLs; this interaction has similarities to the genotype-by-sex interaction (discussed in the previous paragraph). Different QTLs affect lifespan in males across the two densities (QTL 1, 5, and 6 at low density and QTL 2 and 3 at low density). Other loci affect lifespan at both densities, but the mode of action changes (see Figure 5.22). Note that at low larval density in Figure 5.22, the heterozygote for the QTL near marker 76B has the longest lifespan, indicating that there is heterozygote advantage at this QTL, and the homozygous genotype from the 2b parental line has the shortest lifespan. At high density, however, the 2b homozygote has the longest lifespan, and this QTL shows only weak partial dominance for lifespan because the heterozygote is nearly intermediate between the two homozygotes. Therefore, the mode of action of this QTL for lifespan differs between environments. These observations provide a first step in understanding the mechanisms underlying genotype-by-environment interactions.

Epistasis can also be detected with QTL analyses, although this is often difficult statistically due to the large number of possible interactions between different QTLs. We can again turn to Leips and Mackay for an example (Figure 5.23). Note that when the QTL at marker 76B is homozygous for the *B* allele, the 2b homozygote at the 50D QTL is associated with

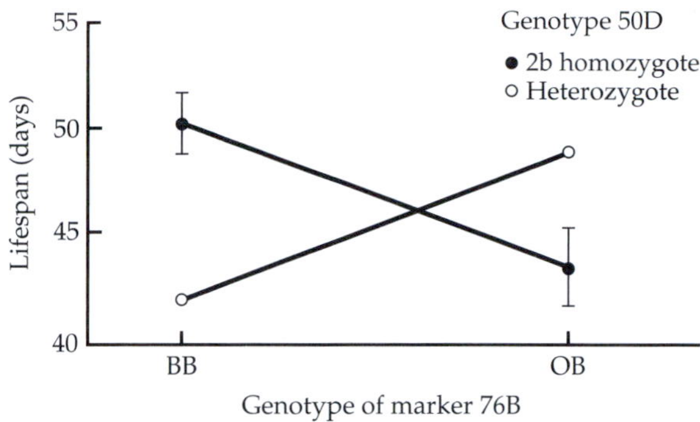

Figure 5.23 Epistasis between two QTLs. The relative lifespans of the two genotypes at the 50D QTL is reversed depending on the genotype at the 76B QTL. The error bars for the heterozygote are smaller than the marker symbol. (After Leips and Mackay 2000.)

an eight-day increase in lifespan compared to the heterozygote. However, when 76B has the *OB* heterozygous genotype, the 2b homozygote decreased lifespan by about five days relative to the heterozygote.

An important thing to remember about all current QTL studies, especially those on natural populations, is that the resolution is still quite low—most studies have markers approximately every 10 cM or more. There can be many loci in 10 cM. For example, in the genetic model plant *Arabidopsis*, 10 cM corresponds to 2130 kb and about 500 loci (Mauricio 2001). Thus, the term QTL is a bit of a misnomer, because individual loci are not being identified, but rather gene regions that affect the trait of interest. One ultimate goal is to identify and determine the function of genes affecting the phenotype. There are several techniques that researchers are taking to achieve this goal, but to date we know little about individual gene loci affecting quantitative traits. One of these techniques is the **candidate gene** approach. This technique is used in genetic model organisms, in which the molecular functions of genes have been identified using mutants, knockouts, and other genetic techniques. If one of these known genes maps to the same region as a QTL for a given quantitative trait, then this is a candidate gene for the quantitative trait. Some possible candidate genes for lifespan in *Drosophila* identified in the Leips and Mackay study are listed in Table 5.6. These include genes involved in basic metabolism (*Adh, Pgm*), breakdown of insulin (*Ide*), removal of toxic metabolic byproducts (*Sod, Cat*), protein synthesis (*EF1α*), and response to stress (*Hsp70*). Although each of these functions could easily be related to lifespan and it is not hard to envision how their effects on lifespan could vary across environments, much more work will be needed to determine whether and how each actually contributes to this phenotypic trait.

For correct inferences to be drawn from any QTL study it is very important to keep in mind how the mapping population was created. For natural populations to date, the most common mapping population is an F_2 or backcross from a cross between species (e.g., Bradshaw et al. 1998; Lin and Ritland 1997). The QTLs identified in this kind of study are those that differ between the two species. These QTLs are very important if much of adaptive divergence is associated with the creation of new species. These studies suggest that loci of large effect (i.e., explaining more than 25% of the phenotypic variance) commonly are responsible for differences between species, but that some trait differences are caused entirely by loci with smaller effects (e.g., Fishman, Kelly, and Willis 2002). The Leips and Mackay study discussed earlier was based on a cross between two laboratory strains that originated from different natural populations. What we do not know yet is whether the QTLs identified in these studies (or in most QTL studies to date) are the loci responsible for segregating variation in natural populations and thus are those that underlie the response to natural selection. It seems likely that often they will be, especially in the cases of traits that are clearly adaptations, but there are undoubtedly other important loci contributing to natural genetic variation that remain to be identified. A few studies have mapped QTLs from divergent artificial selection lines derived from a single natural population (e.g., Long et al. 1995). These kinds of studies identify the loci that are responsible for the response to selection and correlated responses, and therefore they represent at least some of the loci with segregating variation in the original population. A greater understanding of the genetic mechanisms underlying variation within natural populations is one of the greatest challenges in ecological genetics today.

PROBLEMS

5.1. On the facing page are reaction norm plots and summaries of ANOVA results for four traits of the weed *Polygonum persicaria* grown under three different light intensities (Sultan and Bazzaz 1993). Each line in the reaction norms represents means for replicates of one clone. For each trait, describe whether or not these data provide evidence for:

a. Overall genetic variance for the trait.

b. Overall plasticity for the trait. If so, is this plasticity likely to be adaptive? Why or why not?

c. Genotype by environment interaction. Could adaptive plasticity evolve?

Defend your answers with specifics from the data.

	Total leaf area	Mean leaf size	Total fruit mass	Mean fruit weight
Genotype	n.s.	***	n.s.	***
Environment	***	***	**	***
G × E	**	**	n.s.	n.s.

Note: n.s., not significant; ** $P < 0.01$; *** $P < 0.001$.

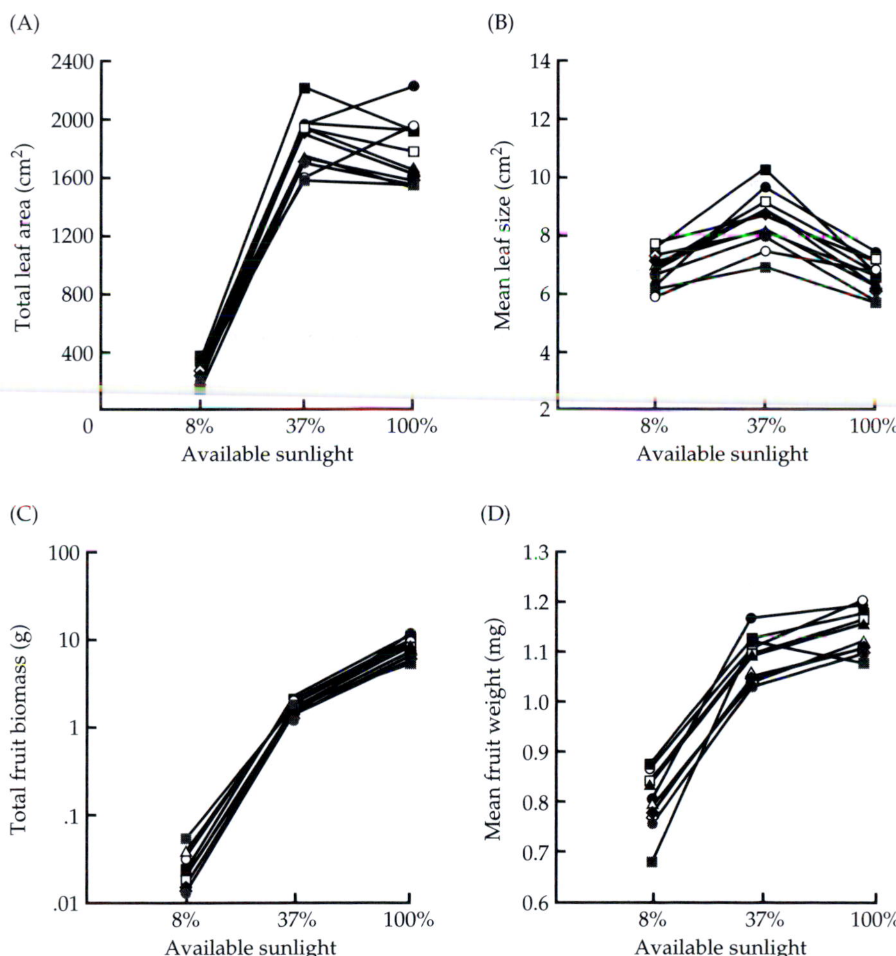

5.2. The accompanying table shows heritabilities and genetic and environmental correlations for pairs of traits from three of the same studies in Table 5.4. Different traits from those in Table 5.4 are shown for snakes and finches, while the same traits measured at a different age are given for mice. Calculate the phenotypic correlation in each case, and describe how each of the parameters in the table affected each of the correlations you calculated.

Species	Trait 1	Trait 2	h^2_1	h^2_2	r_A or r_G	r_E	References
Garter snake	Color pattern	Sprint speed	0.71	0.53	0.28	–0.23	Brodie 1989; Brodie 1992
Darwin's finch	Weight	Bill depth	0.91	0.79	0.87	0.19	Boag 1983
House mouse	Brain size	Body size	0.16	0.20	–0.43	0.31	Leamy 1990

5.3. Endler et al. (2001) artificially selected guppies (*Poecilia reticulata*) for increased sensitivity to red light in two replicate selection lines (R1 and R2) for seven generations. The results, plotted as the cumulative response (R) versus the negative of the cumulative selection differential ($-S$), are shown below.

a. Is there evidence for a selection plateau? If so, describe this evidence and briefly outline two experiments that could be done to examine possible causes of the plateau. Be sure to explain how results of these experiments could be interpreted.

b. Endler and colleagues calculated the realized heritability of sensitivity to red light from regressions on these data as 0.39 and 0.30 in R1 and R2, respectively. Estimate the realized heritability using the breeder's equation, but leaving out the last three generations in the figure (that is, use only the data up to the generation marked with the arrow). Give a possible reason for any discrepancy between your estimate and that of the authors.

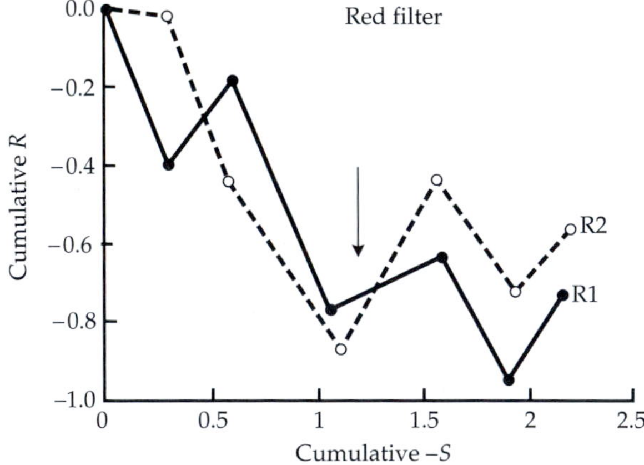

5.4. The accompanying figure shows a portion of the QTL map from the cross between *Mimulus lewisii* and *M. cardinalis* discussed in the text (Bradshaw et al. 1998). Shown are the QTL positions for four floral traits on three of the linkage groups.

a. How many gene loci for each of the traits did they find in these three linkage groups?

b. Is there evidence for pleiotropic effects of any of these loci? If so, describe this evidence.

c. What are the approximate map distance and recombination rate between the loci affecting aperture height and pistil length in linkage group A? What is the recombination rate between the two loci that affect aperture height?

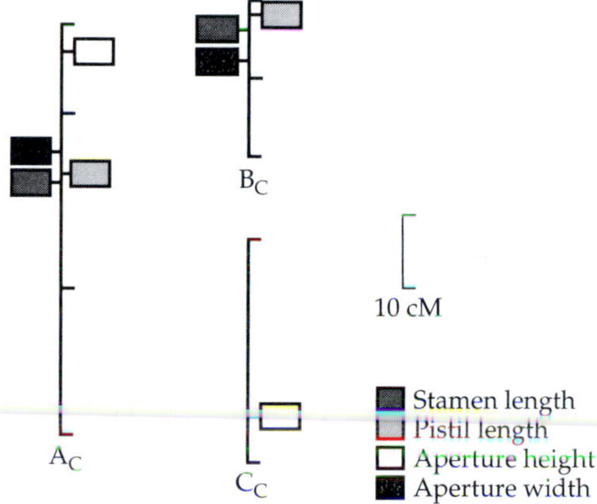

SUGGESTED READINGS

Agrawal, A. A. 2003. Special Feature: Selection studies in ecology: Concepts, methods, and directions. Ecology 84:1649–1712. A group of papers that discusses applications of artificial selection, QTL mapping, and other techniques to ecological questions.

*Falconer, D. S., and T. F. C. Mackay. 1996. *Introduction to Quantitative Genetics*. Longman, Harlow, UK. Sometimes called the "bible" of quantitative genetics, this has been the standard text and reference for the field for many years.

Gromko, M. H., A. Briot, S. C. Jensen, and H. H. Fukui. 1991. Selection on copulation duration in *Drosophila melanogaster*: Predictability of direct response versus unpredictability of correlated response. Evolution 45:69–81. A study of artificial selection on a behavioral trait, showing clear genetic variation but less consistent effect of genetic correlation.

*Lynch, M., and B. Walsh. 1998. *Genetics and Analysis of Quantitative Traits*. Sinauer Associates, Sunderland, MA. Extraordinarily detailed, thorough, and up-to-date, with a very rigorous mathematical treatment of the topic.

*Mauricio, R. 2001. Mapping quantitative trait loci in plants: Uses and caveats for evolutionary biology. Nature Trends Genetics 2:370–381. A readable and up-to-date review of methods and research.

*Schluter, D., and J. N. M. Smith. 1986. Genetic and phenotypic correlations in a natural population of song sparrows. Biol. J. Linn. Soc. 29:23–36. An example of genetic correlations estimated using offspring-parent regression.

*Tanksley, S. D. 1993. Mapping polygenes. Annual Review of Genetics 27:205–233. An authoritative overview by one of the pioneers of QTL mapping.

Via, S. 1984. The quantitative genetics of polyphagy in an insect herbivore. I. Genotype–environment interaction in larval performance on different host plant species. Evolution 38:881–895. The first application of sibling analysis to the study of genotype by environment interaction in a natural population

Via, S. 1994. The evolution of phenotypic plasticity: What do we really know? Pp. 35–57 *in* L. A. Real, ed. *Ecological Genetics*. Princeton University Press, Princeton, NJ. A review of quantitative genetic approaches to phenotypic plasticity and genotype-by-environment interaction by a leader in the field.

**Indicates a reference that is a suggested reading in the field, and is also cited in this chapter.*

SUGGESTED READINGS QUESTIONS

The following questions are based on papers from the primary literature that address key concepts covered in this chapter. For full citations for each paper please see Suggested Readings.

Bradshaw, H. D., Jr., K. G. Otto, B. E. Frewen, J. K. McKay, and D. W. Schemske. 1995. Genetic mapping of floral traits associated with reproductive isolation in monkeyflowers.

1. What is a QTL? What is the general technique used by Bradshaw et al. to find QTLs?
2. What phenotypic traits did they measure, and how do each of these affect pollination by bees and hummingbirds? Which phenotype for each trait is favored by bees? Which by hummingbirds?
3. Do their results provide evidence for pleiotropy? If so, what is the evidence, and for which trait pairs are there likely to be pleiotropic loci?
4. What is their main conclusion regarding the genetic basis of floral differences between the two species? Does their evidence support this conclusion? How does this conclusion affect our understanding of speciation?

Gromko, M. H., A. Briot, S. C. Jensen, and H. H. Fukui. 1991. Selection on copulation duration in *Drosophila melanogaster*.

1. What trait did they artificially select? In what traits did they predict a correlated response would occur? Did they predict that the unselected traits would respond in the same or opposite direction as the selected trait?
2. Focusing on generation 10 of the Chain Cross population, are the correlated responses in the predicted directions? What table or figure contains these data?

Schluter, D., and J. N. M. Smith 1986. Genetic and phenotypic correlations in a natural population of song sparrows.

1. Are there any possible sources of bias in the measurements of heritability that Schluter and Smith did not consider? If so, what specific effects would these biases have on the heritability estimates?
2. What *basic* method did they use to estimate genetic correlations?
3. How similar to each other are the phenotypic and the genetic correlations for the same pairs of traits? For what pairs of traits are there major differences between the phenotypic and genetic correlations? Use the corrected values throughout.

Via, S. 1984. The quantitative genetics of polyphagy in an insect herbivore.

1. Why is genotype-by-environment interaction (g-e, same as G × E) important in agriculture? Why is it important in evolution?
2. Describe the experimental design used by Via. Pay particular attention to where the flies were collected from, how they were mated, and what host plants the offspring from each female were reared on.
3. From what data was the additive genetic variation (V_A) calculated?
4. For which traits was there significant additive genetic variance? Significant G × E?
5. Why could G × E across the two host plants lead to genetic differentiation among the populations on the two host species ("host races")? What is the evidence that host races have not formed in *Liriomyza*? What is one possible reason why this has not happened?

Via, S. 1994. The evolution of phenotypic plasticity: What do we really know?

1. Define phenotypic plasticity and reaction norm.
2. Distinguish between noisy plasticity and adaptive reaction norms. What environmental conditions give rise to each, and what is the phenotypic response to these environments in the two types of plasticity?
3. Explain the meanings of each of the terms *main effects, interaction,* and *error term* in the two-way ANOVA analyzing plasticity.
4. Sketch reaction norms for two environments depicting the following conditions:
    a. No plasticity
    b. Plasticity but no variance in plasticity (i.e., no G × E)
    c. Variance in plasticity
5. Which of the first three terms in the ANOVA (the two main effects and the interaction) would be statistically significant in each of your sketches?
6. Which of the cases in question 4 allow for the evolution of a new average reaction norm (new form of plasticity) in response to new environments?

CHAPTER REFERENCES

Agrawal, A. A. 1998. Induced responses to herbivory and increased plant performance. Science 279:1201–1202.

Beavis, W. D. 1998. QTL analyses: Power, precision, and accuracy. Pp. 145–162 *in* A. H. Paterson, ed. *Molecular Dissection of Complex Traits*. CRC Press, Boca Raton, FL.

Boag, P. T. 1983. The heritability of external morphology in Darwin's ground finches (*Geospiza*) on Isla Daphne Major, Galapagos. Evolution 37:877–894.

Boag, P. T., and A. J. van Noordwijk. 1987. Quantitative genetics. Pp. 45–78 *in* F. Cooke and P. A. Buckley, eds. *Avian Genetics*. Academic Press, London.

Bradshaw, H. D., Jr., S. M. Wilbert, K. G. Otto, and D. W. Schemske. 1995. Genetic mapping of floral traits associated with reproductive isolation in monkeyflowers (*Mimulus*). Nature 376:762–765.

Bradshaw, H. D., Jr., K. G. Otto, B. E. Frewen, J. K. McKay, and D. W. Schemske. 1998. Quantitative trait loci affecting differences in floral morphology between two species of monkeyflower (*Mimulus*). Genetics 149:367–382.

Brodie, E. D., III. 1989. Genetic correlations between morphology and antipredator behaviour in natural populations of the garter snake *Thamnophis ordinoides*. Nature 342:542–543.

Brodie, E. D., III. 1992. Correlational selection for color pattern and antipredator behavior in the garter snake *Thamnophis ordinoides*. Evolution 46:1284–1298.

Campbell, D. R. 1996. Evolution of floral traits in a hermaphroditic plant: Field measurements of heritabilities and genetic correlations. Evolution 50:1442–1453.

Caruso, C. M. The quantitative genetics of floral trait variation in *Lobelia*: potential constraints on adaptive evolution. *Evolution*, in press.

Clausen, J., D. D. Keck, and W. M. Hiesey. 1940. *Experimental Studies on the Nature of Species. I. Effect of Varied Environments on Western Northern American Plants*. Carnegie Institution of Washington Publication No. 520, Washington, D.C.

Conner, J. 1988. Field measurements of natural and sexual selection in the fungus beetle, *Bolitotherus cornutus*. Evolution 42:736–749.

Conner, J. K. 2002. Genetic mechanisms of floral trait correlations in a natural population. Nature 420:407–410.

Dudley, J. W. 1977. 76 generations of selection for oil and protein percentage in maize. Pp. 459–473 *in* E. Pollak, O. Kempthorne and T. B. J. Bailey, eds. Proceedings of the International Conference on Quantitative Genetics. Iowa State University Press, Ames, Iowa.

Dudley, S. A. 1996. The response to differing selection on plant physiological traits: Evidence for local adaptation. Evolution 50:103–110.

Endler, J. A., A. Basolo, S. Glowacki, and J. Zerr. 2001. Variation in response to artificial selection for light sensitivity in guppies (*Poecilia reticulata*). The American Naturalist 158:36–48.

Fishman, L., A. J. Kelly, and J. H. Willis. 2002. Minor quantitative trait loci underlie floral traits associated with mating system divergence in *Mimulus*. Evolution 56:2138–2155.

Ford, E. B. 1975. *Ecological Genetics*. Chapman & Hall, London.

Futuyma, D. J. 1998. *Evolutionary Biology*, 3rd Edition. Sinauer Associates, Sunderland, MA.

Greene, E. 1989. A diet-induced developmental polymorphism in a caterpillar. Science 243:643–646.

Harvell, C. D. 1984. Predator-induced defense in a marine bryozoan. Science 224:1357–1359.

Leamy, L. 1990. The evolution of brain and body size: Genetic and maternal influences. Pp. 144–159 *in* M. E. Hahn, J. K. Hewitt, N. D. Henderson, and R. H. Benno, eds. *Developmental Behavior Genetics: Neural, Biometrical, and Evolutionary Approaches*. Oxford University. Press, New York.

Leips, J., and T. F. C. Mackay. 2000. Quantitative trait loci for lifespan in *Drosophila melanogaster:* Interactions with genetic background and density. Genetics 155:1773–1788.

Lin, J.-Z., and K. Ritland. 1997. Quantitative trait loci differentiating the outbreeding *Mimulus guttatus* from the inbreeding *M. platycalyx*. Genetics 146:1115–1121.

Long, A. D., S. L. Mullaney, L. A. Reid, J. D. Fry, C. H. Langley, and T. F. C. Mackay. 1995. High resolution mapping of genetic factors affecting abdominal bristle number in *Drosophila melanogaster*. Genetics 139:1273–1291.

Mazer, S. J., and C. T. Schick. 1991a. Constancy of population parameters for life history and floral traits in *Raphanus sativus* L. I. Norms of reaction and the nature of genotype by environment interactions. Heredity 67:143–156.

Mazer, S. J., and C. T. Schick. 1991b. Constancy of population parameters for life-history and floral traits in *Raphanus sativus* L. II. Effects of planting density on phenotype and heritability estimates. Evolution 45:1888–1907.

Palmer, J. O., and H. Dingle. 1986. Direct and correlated responses to selection among life–history traits in milkweed bugs (*Oncopeltus fasciatus*). Evolution 40:767–777.

Pfrender, M. E., and M. Lynch. 2000. Quantitative genetic variation in *Daphnia*: Temporal changes in genetic architecture. Evolution 54:1502–1509.

Preziosi, R. F., and D. A. Roff. 1998. Evidence of genetic isolation between sexually monomorphic and sexually dimorphic traits in the water strider *Aquarius remigis*. Heredity 81:92–99.

Schwaegerle, K. E., and D. A. Levin. 1991. Quantitative genetics of fitness traits in a wild population of *Phlox*. Evolution 45:169–177.

Simms, E. L., and M. A. Bucher. 1996. Pleiotropic effects of flower-color intensity on herbivore performance on *Ipomoea purpurea*. Evolution 50:957–963.

Spitze, K. 1993. Population structure in *Daphnia obtusa*: Quantitative genetic and allozymic variation. Genetics 135:367–374.

Stearns, S. C., and R. F. Hoekstra. 2000. *Evolution: An Introduction*. Oxford University Press, New York.

Sultan, S. E., and F. A. Bazzaz. 1993. Phenotypic plasticity in *Polygonum persicaria*. I. Diversity and uniformity in genotypic norms of reaction to light. Evolution 47:1009–1031.

Via, S. 1991. The genetic structure of host plant adaptation in a spatial patchwork: Demographic variability among reciprocally transplanted pea aphid clones. Evolution 45:827–852.

Woodward, B. D., J. Travis, and S. Mitchell. 1988. The effects of the mating system on progeny performance in *Hyla crucifer* (Anura:Hylidae). Evolution 42:784–794.

6

Natural selection on phenotypes

Natural selection and adaptation have been recurring themes throughout this book. We discussed selection on genotypes (and discrete phenotypes) and how selection affects allele frequencies at individual loci (see Chapter 3). Now that we have an understanding of the genetics of continuously distributed traits, we turn to selection on these common and ecologically important phenotypes. We describe the general and widely used regression-based approaches to measuring selection, and go on to examine ways to identify the phenotypic traits that are the targets of natural selection, as well as ways to determine the environmental agents that cause selection. Identifying these selective agents and targets provides us with a powerful approach to understanding adaptation. Finally, we integrate this material with the concepts covered in Chapters 4 and 5 to show how short-term phenotypic evolution can be modeled and predicted, and how this undertaking sheds light on constraints on adaptive evolution. Throughout this chapter the effects of genetic and phenotypic correlations among traits (see Chapter 5) are highlighted, to emphasize that selection acts on whole organisms rather than individual traits in isolation.

Evolution by natural selection has three parts (Figure 6.1; Endler 1986):

1. There is phenotypic variation for the trait of interest.
2. There is some consistent relationship between this phenotypic variation and variation in fitness.

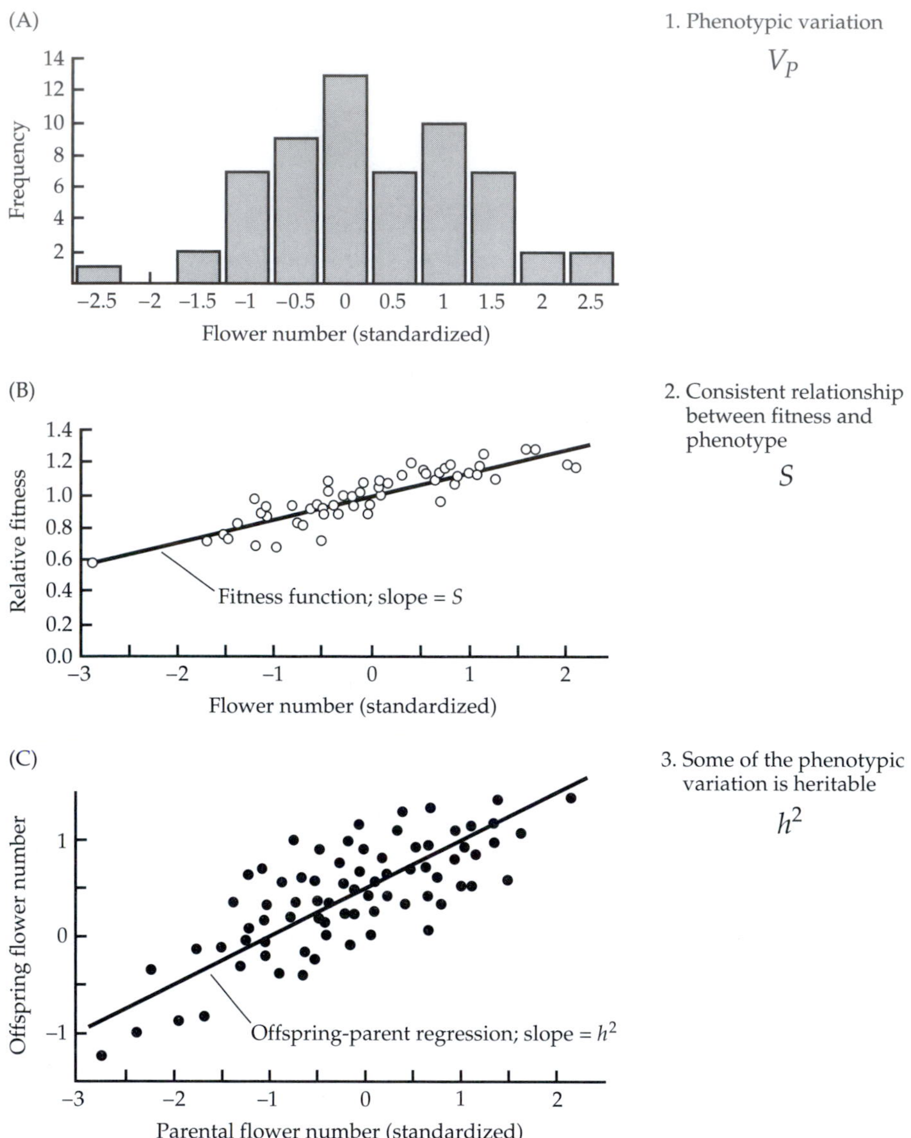

Figure 6.1 Graphical depiction of the three elements of phenotypic evolution by natural selection. (A) Histogram showing the frequency distribution for flower number in wild radish. Flower number is standardized to have a mean of zero and a standard deviation of 1. (B) The relationship between flower number and fitness. The spread of points along the x-axis shows exactly the same phenotypic variation as in part A. The fitted regression line is the estimated fitness function; the slope of this line is the selection differential S. (C) Hypothetical offspring–parent regression for flower number, showing that the trait is heritable. (A,B data from Conner, Rush, and Jennetten 1996.)

3. A significant proportion of the phenotypic variation is caused by additive genetic variance; in other words, the trait is heritable.

Numbers 1 and 2 represent selection on phenotypes, which occurs within a single generation and can be quantified using the selection differential (S) just as with artificial selection. Number 3 describes heritability, which can be thought of as determining the amount of phenotypic change due to selection that can be passed on to the next generation. As with artificial selection (see Chapter 5), the amount of evolutionary change, R, is determined by the product h^2S. Increases in either h^2 or S produce a greater evolutionary change across generations.

The Chicago School Approach to Phenotypic Evolution

An extremely fruitful approach to studying phenotypic evolution in nature is based on the scheme depicted in Figure 6.1. This is a quantitative genetic approach, sometimes referred to as the "Chicago School," because it was developed at the University of Chicago in the late 1970s and early 1980s (Arnold and Wade 1984a,b; Lande 1979; Lande and Arnold 1983). In this scheme, the bivariate plot shown in Figure 6.1B is the fundamental depiction of natural selection. On the x-axis the phenotype is shown; it can be any quantitative trait in the phenotypic hierarchy (see Chapter 1), including physiology, morphology, behavior, or life-history traits. For measurements of selection it is very useful to standardize the phenotypic values by taking each individual value and subtracting the population mean and dividing by the population standard deviation. This produces measures of selection that have useful statistical properties and are comparable across traits and organisms (Kingsolver et al. 2001). In these models the phenotypic traits are often given the symbol z.

The y-axis depicts fitness, which is a fundamental but difficult concept. In general, **fitness** refers to the ability of an organism to survive, reproduce, and thus have descendents in future generations. Unfortunately, fitness is difficult to define more specifically so that it can be measured and understood more clearly. Therefore, a number of different definitions of fitness have been proposed, each with strengths and weaknesses (Dawkins 1982, see Chapter 10; Endler 1986, see Chapter 2; deJong 1994). In addition to the conceptual difficulty, in practice, fitness is very difficult to measure in natural populations. For the Chicago School methods, one of the best measures is lifetime number of offspring produced, and this is the working definition we will use in this chapter. This definition is certainly an excellent practical measure of fitness for organisms with nonoverlapping generations and a stable population size. Most field studies to date have been based not on total lifetime offspring production, but rather a component of this fitness measure such as survival, mating success, or offspring production over a defined interval.

Many definitions of fitness measure the number of offspring surviving to a certain age (e.g., independence from the parents or reproductive maturity) rather than measure just the number of offspring produced. The problem with these definitions is that they include traits of the offspring, which are newly created individuals with unique phenotypes and genotypes (except in clonal organisms). Therefore, it is difficult to separate selection on parental phenotypes from selection on offspring phenotypes with a fitness definition that spans more than one generation, and these types of fitness measures can sometimes be misleading (Wolf and Wade 2001).

For any actual analysis, relative fitness values are used; these are calculated by dividing the fitness of each individual by the average fitness of the population. This is different from how relative fitness was calculated for genotypes (see Chapter 3), but the symbol (w) and the rationale are the same—selection operates through the fitnesses of individuals relative to other individuals in the population, not through absolute numbers of offspring produced.

The Chicago School methods estimate the strength of natural selection by regressing fitness on the phenotype. This regression provides a statistical estimate of the **fitness function,** which describes the relationship between fitness and the phenotype and determines the strength and form of natural selection. There are three basic types of phenotypic selection defined according to the shape of the fitness function. **Directional selection** is characterized by a linear fitness function (a straight line), which can be positive or negative depending on whether fitness increases or decreases with increasing phenotypic value (see Figure 6.1B and Figure 6.2A, respectively). Linear regression is used to fit a line through the data, and the slope of this line measures the strength of selection. If standardized data are used, then this slope equals the standardized selection differential (S; the standardized selection differential is also sometimes called the intensity of selection, i). This is the same measure as used for artificial selection (see Chapter 5), but is calculated in a different way because fitnesses in natural populations are continuously distributed. Recall that artificial selection is usually truncation selection (which is rare or nonexistent in nature), and the selection differential is the difference in mean between the entire population and the subset that is selected. The most fundamental definition of the selection differential is that it is the covariance between fitness and the trait ($\text{Cov}(w,z)$); (Price et

Figure 6.2 Hypothetical fitness functions (A, C, E) and the effect of each kind of selection on the frequency distribution of the population for that trait (B, D, F). (A) and (B) show negative directional selection, (C) and (D) stabilizing selection, and (E) and (F) disruptive selection. In (B, D) and (F), the gray curves are the population distribution before selection, and the black curves are the population distribution after selection. ▶

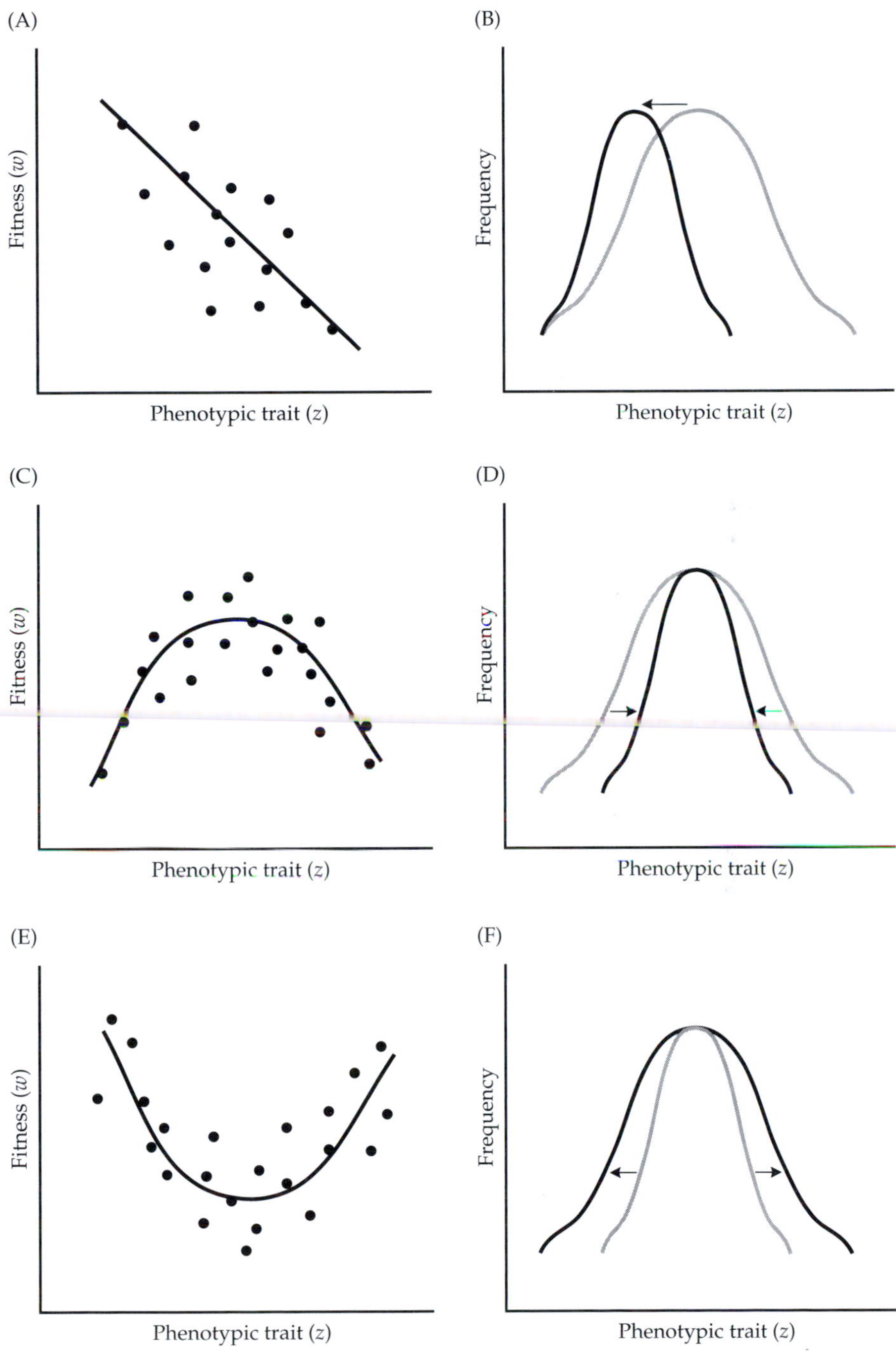
(A)
Fitness (w)
Phenotypic trait (z)
(B)
Frequency
Phenotypic trait (z)
(C)
Fitness (w)
Phenotypic trait (z)
(D)
Frequency
Phenotypic trait (z)
(E)
Fitness (w)
Phenotypic trait (z)
(F)
Frequency
Phenotypic trait (z)

al. 1970); with standardized data, the regression slope equals the covariance (see Appendix). Also remember that the selection differential refers to phenotypic selection and is not the same as the selection coefficient (lowercase s) that is used to measure genotypic selection (see Chapter 3). The results of directional selection are similar to truncation selection; if the trait is heritable, it will change the mean and decrease the variance (Figure 6.2B).

Fitness functions are not always straight lines, so when the fitness function has curvature, quadratic regression is used to estimate the strength of selection:

$$w = \alpha + \beta z + \frac{\gamma}{2} z^2 \qquad \textbf{6.1}$$

In this equation, α is the y-intercept, β is the slope of the fitness function (equivalent to S with standardized data), and γ measures the rate of change of the slope with increasing z. In other words, γ estimates the amount of curvature in the fitness function, and is variously called the **nonlinear, variance,** or **stabilizing/disruptive selection gradient.** (Nonlinear is preferable, and stabilizing/disruptive is cumbersome, sometimes misleading, and should be avoided.) If $\beta = 0$ and γ is negative, then there is no overall increasing or decreasing trend in the data but the slope is constantly decreasing (Figure 6.2C); this case is called **stabilizing selection.** The key characteristic of stabilizing selection is that there is an intermediate optimum for fitness; in other words, fitness is highest at some intermediate phenotype and is lower at the phenotypic extremes. Stabilizing selection by itself (i.e., $\beta = 0$) does not change the trait mean but does decrease trait variance (Figure 6.2D).

A particularly clear pattern of stabilizing selection occurs on the size of galls formed on goldenrod plants by flies in the genus *Eurosta* (Weis and Gorman 1990; Figure 6.3). Female flies lay their eggs inside goldenrod stems, and this causes the plant cells in the area to divide profusely, producing a spherical gall inside which the larva feeds and develops. The size of the gall is determined by the fly and thus can be considered a fly trait. A parasitoid wasp often attacks these larvae, but only in galls that are small enough in diameter that the ovipositor of the wasp can reach the fly larva. The fly larvae are also eaten by birds, who are more attracted visually to the larger galls. Therefore, the two extreme gall sizes are each attacked at a higher frequency by different predators, reducing survivorship relative to the intermediate phenotypes.

If $\beta = 0$ and γ is positive, then the slope is steadily increasing with z (Figure 6.2E); this type of selection is called **disruptive selection.** The key characteristic of disruptive selection is opposite to that for stabilizing selection—there is an intermediate minimum for fitness. In other words, fitness is lowest at some intermediate phenotype and is higher at the phenotypic extremes. Like stabilizing selection, disruptive selection by itself does not change the population mean, but it can increase the variance, at least in the

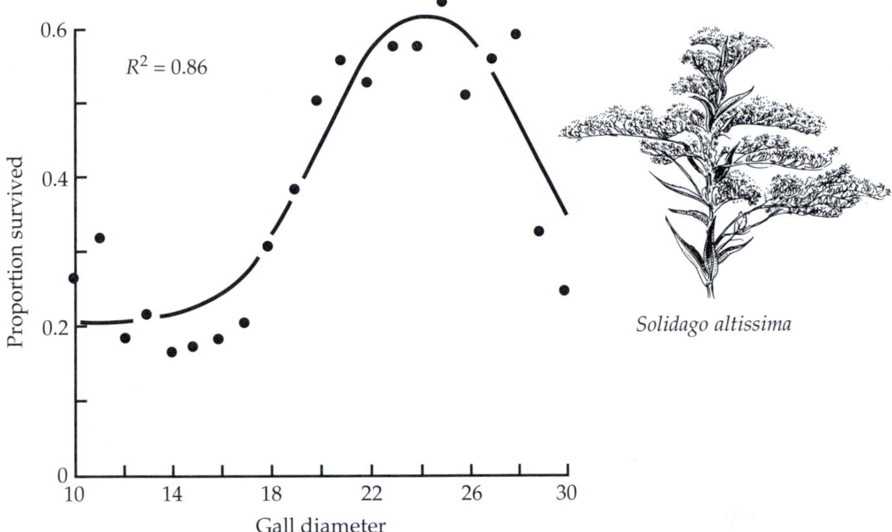

Figure 6.3 Stabilizing selection on gall size of *Eurosta* flies on goldenrod. Survival of the fly larvae was the measure of fitness, and the phenotypic trait was gall diameter measured to the nearest mm. Over 3500 galls were measured and scored for survival to adulthood; each point in the graph represents the proportion of individuals in one population that survived at each gall size. (After Weis and Gorman 1990.)

short term (Figure 6.2F), opposite to the effect of stabilizing selection. These effects on variance are why γ is sometimes called the *variance selection gradient*, and the sign of γ indicates the effect on variance—negative is stabilizing selection, which decreases variance; and positive is disruptive, which increases variance.

Disruptive selection is potentially important because it can maintain phenotypic and genetic variation in the short term, and it could result in adaptive differentiation and even speciation if the two phenotypic extremes became reproductively isolated. However, there is little strong evidence for disruptive selection in natural populations. Some of the best evidence comes from bill size in finches, in which disruptive selection can occur through feeding on seeds from different species of plants. In Darwin's finches from the Galapagos as well as in African finches, large-billed birds have high fitness through their ability to crack abundant hard seeds, whereas small-billed birds survive well on smaller, softer seeds. Birds with intermediate bill sizes have the lowest fitness because they cannot use either resource as efficiently as birds at the phenotypic extremes (Schluter, Price, and Grant 1985; Smith 1990).

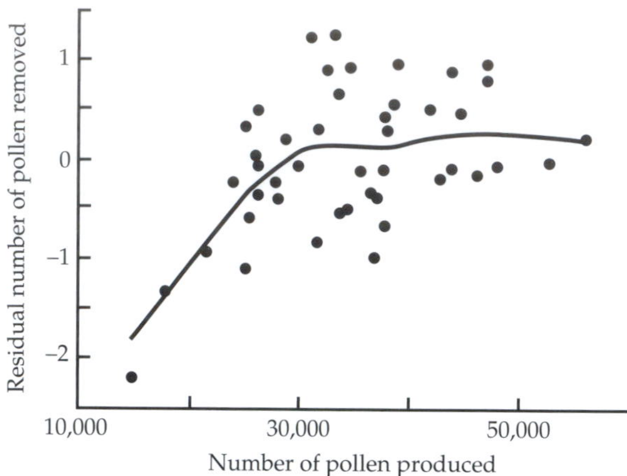

Figure 6.4 Example of a curved fitness function without an intermediate maximum or minimum in fitness. The fitness measure is the number of pollen grains removed from a flower during a visit by small bees (a component of male fitness), and the trait is the number of pollen grains produced by the flower. The y-axis is the residual variation after removing the effects of other variables. The curve was fit using locally weighted least squares; both the linear and quadratic terms in the regression analysis were significant. (After Conner, Davis, and Rush 1995.)

Often more complicated forms of selection can occur that do not exactly fit the simple definitions of directional, stabilizing, or disruptive selection. Sometimes the overall trend of the fitness function is increasing or decreasing without an intermediate maximum or minimum as in directional selection, but the fitness function curves rather than being linear (Figure 6.4). The curve in Figure 6.4 is sometimes called a *saturating fitness function,* because after a certain point (about 30,000 grains produced in this case) further increases in trait value no longer result in increased in fitness. In this case both β and γ can be significant, so a significant γ by itself is not good evidence for stabilizing or disruptive selection, only evidence that the fitness function is not linear. This is why the term stabilizing/disruptive selection gradient for γ can be misleading. Standard regression techniques can also be misleading when there is curvature in the fitness function, so related curve-fitting techniques such as cubic splines and locally weighted least-squares (**lowess** or **loess**) can be very useful in determining the shape of the fitness function (Schluter 1988).

Selective Agents and Targets

The regression-based and related techniques described above provide measures of the strength and pattern of selection, but by themselves they tell us little about the causes of the selection. The fitness differences along the y-axis in a fitness-function plot (see Figures 6.1–6.4) are caused by biotic and abiotic factors in the environment called **selective agents.** We can define a selective agent as an environmental cause of fitness differences among organisms with different phenotypes. Examples of selective agents are the seeds eaten by Darwin's finches or African finches, the bird predators and wasp parasitoids of gall flies (see Figure 6.3), and a winter storm that killed sparrows in a classic example of natural selection reported by Bumpus (1899). However, fitness differences do not always lead to selection, because these differences may be random with respect to a given phenotypic trait. For example, the chance of encountering a mate, getting caught in a storm, or coming in contact with a disease may be random with respect to most or all phenotypic traits, even though these environmental factors may have strong effects on fitness. Even if there is a consistent relationship between fitness and some traits, there may be many other traits that do not affect fitness in a given generation. This is one reason why it is critical to focus on specific phenotypic traits, called **targets of selection,** when thinking about selection. We define targets of selection as phenotypic traits that selection acts directly upon. In the examples above, bill size in finches and gall size in flies were the targets of selection. Much confusion results from thinking about the strength of selection on species or populations in general, without reference to a specific trait.

Therefore, when attempting to understand the causes of selection, we need to simultaneously consider both the *x* and *y*-axes in the fitness function, that is, specific phenotypic traits and the causes of fitness differences. It is very useful to regard selection as an interaction between the phenotype and the environment (i.e., between the targets of selection and the selective agents); it is this interaction that determines the shape of the fitness function. The target of selection is a trait that helps the organism deal with the selective agent, that is, a challenge in the environment. In the examples of natural selection above, the fitness functions for finch bill size are determined by the interaction of the beak dimensions found in a given population and the size and hardness of seeds in the environment. Similarly, the fitness function for *Eurosta* gall size depends on the distribution of gall sizes and the predators and parasitoids in the population's environment. If there is additive genetic variation for the target of selection, it will evolve as an adaptation to the selective agent.

It is extremely important to realize that changes in either the target of selection or the selective agent can change the fitness function. Figure 6.5 shows hypothetical fitness functions for a large range of phenotypic values, greater than the range spanned by any one population ("overall" fitness functions). The pairs of vertical solid, dotted, and dashed lines denote the phenotypic ranges for three different populations. Note that the fitness function within each population (and therefore the pattern of selection) depends on the phenotypic mean and variance of the population relative to the overall fitness function. In Figure 6.5A populations 1 and 2 experience directional

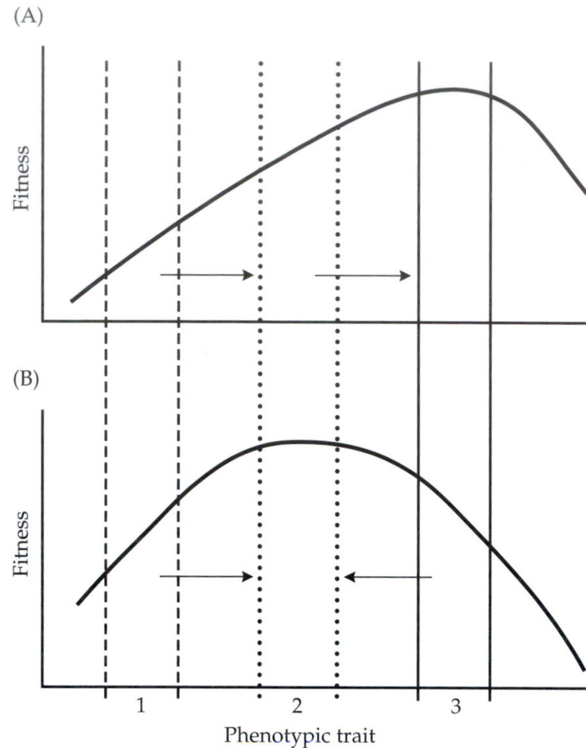

Figure 6.5 Two hypothetical fitness functions, with the phenotypic ranges of three populations mapped onto each with pairs of dashed, dotted, or solid lines. As shown, all have similar phenotypic variances but the means are 1 < 2 < 3. Arrows show the evolutionary changes expected if the trait is heritable. (After Endler 1986.)

selection for an increase in the trait, whereas population 3 is undergoing weak stabilizing selection. Therefore, changes in the mean and variance of the target of selection (the phenotypic trait) can change selection on the trait. If the trait is heritable, then the trait means in populations 1 and 2 will increase across generations (indicated by the arrows in the figure) until their distributions are similar to population 3. At this point the trait mean is at the optimum and the populations will be at an equilibrium for this trait, because unless the trait distribution or the fitness function changes, the trait mean will no longer evolve. This equilibrium is stable, because if something perturbs the trait mean away from the optimum, selection will return the mean to the optimum.

Figure 6.5B shows the effects of changes in the selective agent. When the environment changes, the position or shape of overall fitness function can change, which can also change the selection in each population without a change in the mean or variance of the population. In this case, without any change in the trait distribution compared to Figure 6.5A, population 1 still experiences positive directional selection, population 2 comes under weak stabilizing selection, and population 3 experiences negative directional selection. If the trait is heritable, both populations 1 and 3 will evolve in the direction of population 2.

Populations 1–3 in Figure 6.5 could represent subpopulations that are differentiated for the phenotypic trait of interest, or they could represent one population at different points in time. Similarly, the two overall fitness functions in parts (A) and (B) could result from different local environments, or environmental differences from season to season or from year to year. This discussion emphasizes that fitness functions and the position of populations on the overall fitness function can vary spatially and temporally. For an example of temporally shifting fitness functions, we can turn again to one of the well-studied Darwin's finches, *Geospiza fortis* (reviewed in Grant 1986; Grant and Grant 2002). In drought years caused by La Niña events, large seeds predominate and there is directional selection for increased bill size. During wet El Niño years, small seeds are in greater abundance and there is directional selection for smaller bills. Therefore, the size distribution of seeds in the environment determines the shape and position of the fitness function, as shown in Figure 6.5.

Note that the overall fitness functions in Figure 6.5 both have an intermediate optimum. This seems like a reasonable assumption for most phenotypic traits, because it means that there is some trait value that is too small or too large to function optimally. It is not true for fitness, for which there is always positive directional selection, and it may not be true for fitness components like fecundity and lifespan.

We have been discussing the fitness function as an interaction between the selective agent and the target of selection, but estimating the fitness function by itself does not generally give specific information about either the agents or the targets. This is because one can measure fitness without knowing the reasons for fitness differences among individuals with different phenotypes, and the phenotypic trait measured may not be the true target of selection but rather a trait correlated with the true target. The next several sections cover techniques and difficulties involved in identifying selective agents and targets of selection.

Multiple traits; direct and indirect selection

Direct selection takes place when there is a causal relationship between a phenotypic trait and fitness. The trait under direct selection is the target of selection and is the trait that helps the organism cope with the selective

Figure 6.6 Illustration of how phenotypic correlations cause indirect selection. (A) and (B) scatterplots show linear fitness functions fit by simple regression for horn and elytra (wing cover) length. Standardized selection differentials (S) are shown, which are the slopes of these lines after standardization. The scatterplot in (C) shows the strong phenotypic correlation between horn and elytra length. As examples, the data points for three individual beetles are denoted on all three plots. (Data from Conner 1988.)

agent. Only direct selection leads to adaptive evolution, and evidence for direct selection on a trait indicates that the trait is a present-day adaptation. **Indirect selection** is a covariance between a trait and fitness within a generation caused by a phenotypic correlation between that trait and another trait that experiences direct selection. In general, indirect selection does not contribute to adaptation. The way that correlations cause indirect selection is illustrated in Figure 6.6, which shows selection on horn and elytra (wing cover) length in male fungus beetles (Conner 1988). Figure 6.6A and B show the fitness functions for each of these traits considered separately. These plots show almost identical fitness functions for the two traits, and therefore the standardized selection differentials (S) are very similar.

Why are the selection differentials for horn and elytra length so similar? There are two interconnected reasons. First, there is a strong phenotypic correlation between horn and elytra length (Figure 6.6C), which means that, in general, a male with long horns also has long elytra. This correlation means that the individuals are arranged in roughly the same order along the x-axis in both fitness functions. (Note the numbered points in Figure 6.6A and B.) Second, each individual beetle has only one measure of fitness (lifetime number of females inseminated in this case), which means that they are arranged in *exactly* the same order along the y-axis in the two fitness function plots. Therefore, the two plots are not independent, but are measuring almost the same relationship in two different ways.

The univariate selection differentials, considering each trait separately, measure the total selection acting on the trait, including both direct and indirect selection, but they are not very informative with regard to the target of selection. Selection differentials are uninformative because they do not separate the direct from the indirect selection, so we have no way of knowing what trait causes the higher fitnesses. How could one determine if it is long horns or long elytra that causes higher insemination success? One approach is described in the next section.

Selection gradients

To obtain more information about the target(s) of selection, one can estimate **selection gradients** (β), which measure direct selection on each trait after

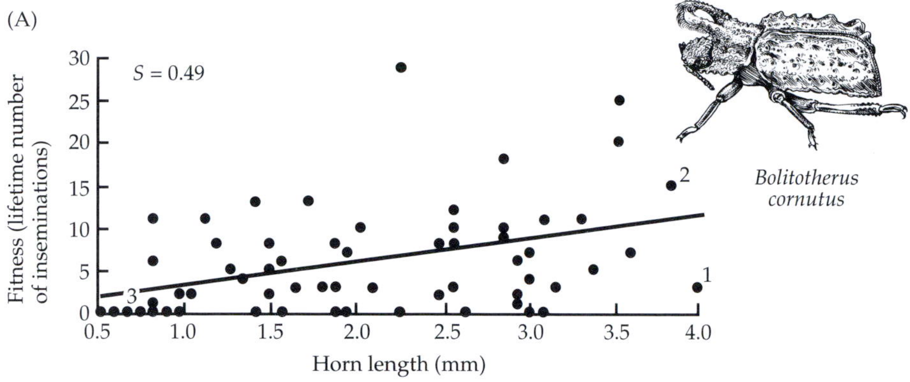

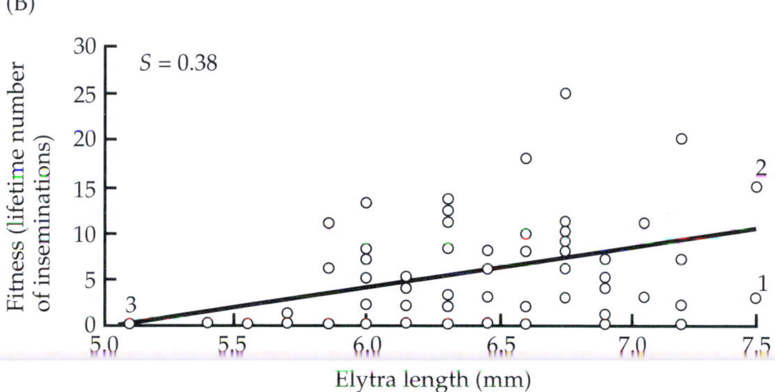

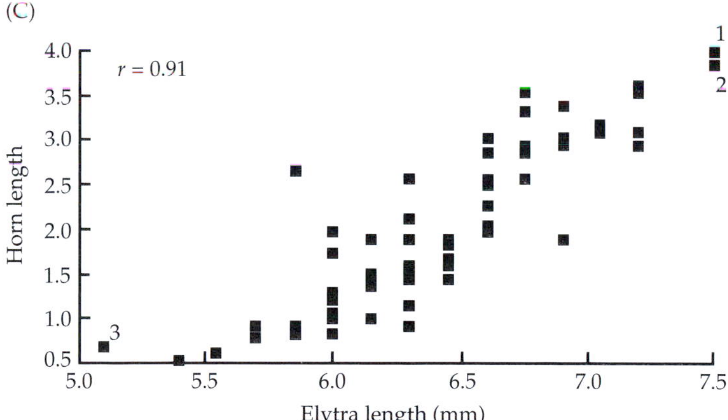

removing indirect selection from all other traits that are in the analysis. Instead of using simple regression, which estimates total selection as the selection differential (S), we use selection gradients that are estimated using multiple regression. In multiple regression, a single dependent variable is regressed on multiple independent variables simultaneously, and the effects of correlations among the independent variables are controlled for statistically. The standard equation for multiple regression with n independent variables is:

$$y = a + b_1x_1 + b_2x_2 + b_3x_3 \ldots b_nx_n \quad \textbf{6.2}$$

where a is the intercept and b represents the partial regression slopes of the relationships between each x variable and y, removing the effects of correlations among the x-variables. Applying this same equation to selection gives:

$$\text{Fitness} = \text{intercept} + \beta_1 \text{ trait } 1 + \beta_2 \text{ trait } 2 + \beta_3 \text{ trait } 3 \ldots \beta_k \text{ trait } n \quad \textbf{6.3}$$

where β_k is the selection gradient, estimated as the partial regression slope. Therefore, each selection gradient measures the slope of relationship between fitness and that trait, just as with a selection differential. However, with gradients, indirect selection is removed by controlling for phenotypic correlations with the other measured traits.

We can gain a better conceptual understanding of how this analysis is done by taking another look at the scatterplot depicting the correlation between horn and elytral lengths (Figure 6.7). Here we have drawn representative vertical ellipses around three groups of beetles; within each group the beetles have the same elytra length but vary in horn length. Additional ellipses like these could be drawn. In multiple regression, the variance in

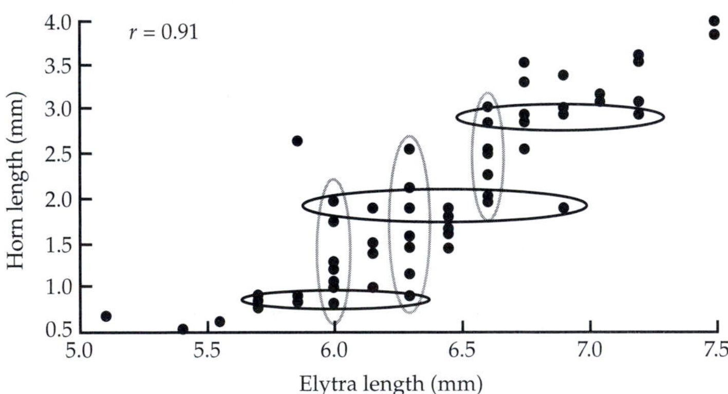

Figure 6.7 Scatterplot of the correlation between horn and elytra length in fungus beetles (from Figure 6.6C), with added ellipses showing groups of beetles with variation in only one of the two traits. Such variation is used in multiple regression to estimate selection gradients, that is, direct selection on each trait corrected for phenotypic correlations among traits.

horn length (within each ellipse) that is independent of variance in elytra length can be used to determine the slope of the relationship between fitness and horn length, thereby correcting for the strong correlation between horn and elytra length. Similarly, the horizontal ellipses show groups of beetles in which there is variation in elytra length but little to no variation in horn length; this variation can be used to estimate the selection gradient for elytra length.

Estimates of direct and indirect selection from fungus beetles and Darwin's finches are given in Table 6.1. In each species, live weight and two or three other linear dimensions were measured. All these traits are highly positively correlated with each other (r = 0.69 to 0.91). Just as we saw with two of the fungus beetle traits in Figure 6.6, all of the selection differentials measuring total selection on the traits are significantly positive. The total selection includes strong indirect selection because these traits are highly positively correlated with each other. Using multiple regression to estimate direct selection, we see that only one or two traits are under significant positive direct selection, namely, horn length in beetles and weight and beak depth in finches. In fact, the direct selection on beak width is negative, and there may also be negative direct selection on elytra length (Conner 1988).

The traits under direct selection are likely to be adaptations to specific selective agents. The horns in male fungus beetles are used in fights over females, and larger-horned males win most fights, so this behavior probably explains the direct selection for increased horn length (discussed in a later section). In the finches, there was direct selection for heavier birds with

TABLE 6.1 *Estimates of total and direct selection in fungus beetles and Darwin's finches*

	Total selection (S)	Direct selection (β)
Fungus beetles		
Elytra	0.38**	–0.33
Horn	0.49***	0.94**
Weight	0.39**	–0.16
Darwin's finches		
Weight	0.62*	0.51*
Beak length	0.49*	0.17
Beak depth	0.60*	0.79*
Beak width	0.49*	–0.47*

Note: *$P < 0.05$, ** $P < 0.01$, *** $P < 0.001$.
Source: Fungus beetle data from Conner (1988); Darwin's finch data from Price et al. (1984).

deeper, narrower bills. The selection was measured during a drought caused by La Niña, in which woody fruits from the plant *Tribulus cistoides* became an important food item. The finches crack the seeds by placing the side of their beaks across a corner of the fruit and twisting; deep narrow beaks are more efficient at performing this task.

Therefore, the significant positive total selection on elytra length and weight in beetles, and on beak length and width in finches, was not caused by direct, adaptive selection. Instead the total selection was caused by strong positive indirect selection from the positively correlated traits that are under positive direct selection (horn length in beetles and weight and beak depth in finches). The traits that are not under direct selection are not targets of selection, and are therefore unlikely to be adaptations or to be undergoing current adaptive evolution. This example illustrates why selection differentials are not informative for determining targets of selection or for inferring adaptiveness of traits, in contrast to selection gradients, which are useful in these contexts.

The contributions of direct and indirect selection to total selection are summarized in the following equation:

$$S_1 = \beta_1 + \beta_2 r_{12} + \beta_3 r_{13} + \ldots \qquad \textbf{6.4}$$

The total selection on trait 1 (as measured by the standardized selection differential) is the sum of the direct selection on that trait (measured as the standardized selection gradient β_1) and the direct selection on all correlated traits weighted by the correlations between these traits and trait 1. Substituting in the numbers for elytra length in fungus beetles, where elytra length is trait 1, horn length is trait 2, and weight is trait 3, we obtain:

		Direct		Indirect		
Total		Elytra		Horn		Weight
0.38	=	−0.33	+	0.94 × 0.91	+	−0.16 × 0.85
	=	−0.33	+	0.85	+	−0.14

Note that the phenotypic correlation between horn and elytra lengths is 0.91 and between weight and elytra is 0.85. Here the negative contributions of both the direct selection on elytra length (–0.33) and the indirect selection through weight (–0.16 × 0.85 = –0.14) are overmatched by the much larger positive contribution made by indirect selection through horn length (0.94 × 0.91 = 0.85), resulting in positive total selection on elytra length.

Indirect selection is different from the correlated responses across generations discussed in Chapter 5. Indirect selection occurs within a generation and is caused by phenotypic correlations, whereas correlated responses occur across generations and are caused by genetic correlations. Correlated responses are more important than indirect selection because indirect selection does not directly affect evolutionary change. However, indirect selection can indicate possible correlated responses if the phenotypic correlations reflect the underlying genetic correlations. The most important reason to understand and identify indirect selection is to separate it from direct selection, so that the targets of selection can be determined.

Selection gradients help uncover the traits that direct selection is acting upon and thereby help identify the targets of selection. However, selection gradients can never prove which trait is the target of selection because of the existence of unmeasured traits. Selection gradients are a strictly *observational*, rather than an *experimental*, approach. By observational approach we mean that only natural variation in the traits and environments are used. The phenotypic traits and fitnesses are measured on the organisms without any experimental manipulation. Selection gradient analyses can only remove indirect selection due to traits that are in the analysis. Because it is impossible to measure or even identify all phenotypic traits, there may be other traits that are actually the targets of selection. Suppose, for example, that in the fungus beetle study, horn length was not measured, and the selection gradients were calculated with just elytra and weight. The analysis may have identified elytra length as the target of selection, when the full analysis with three traits shows that horn length is the target. On the other hand, it could be that horn length itself is not the true target, but it is correlated with an unmeasured fourth trait that is the actual target of selection. This possibility seems unlikely in this case, based on our understanding of the function of horns in gaining access to females, but it cannot be ruled out. Knowledge of the biology of the organism is critical in deciding which traits should be measured and included in any selection gradient analysis.

Experimental manipulation

Although selection gradient analyses cannot prove which traits are the targets of direct selection, they are an excellent place to start because they suggest which traits to focus on for further study. In particular, a trait that does not have a significant selection gradient is not likely to be a true target of selection and thus may not warrant further study. To prove which are the targets, experimental manipulation of the phenotypic traits is necessary. For example, the beetle horns could be cut off or artificially extended and the fitness consequences of this manipulation measured in the natural population. If beetles are placed randomly into different groups for manipulation, then this eliminates the problem of unmeasured correlated traits. With random assignment of beetles to the different treatment groups (reduced vs. extended horns), the groups should not differ in any other traits. Therefore, any differences in fitness between the groups could reliably be attributed to the manipulated differences in horn length. If the extended horn group has higher fitness, then this proves that horns are a target of selection.

Experimental manipulation was used to study the effects of tail length on mating success in male long-tailed widowbirds, *Euplectes progne* (Andersson 1982). As their name implies, males of this species have tails about half a meter long, and these tails are used in displays above their territories. The number of females nesting on the territories of 36 males were recorded. (This was the measure of fitness used because it probably is correlated with the number of offspring sired by the male.) The males were then assigned randomly to one of four experimental treatments. The tails of one group were shortened to about 17 cm by cutting the tails and then gluing a short section back on. The rest of the cut tails were glued on to the tails of another group of males, who thereby had their tails elongated by 25 cm on average. There were also two controls; one group was merely caught and banded for identification as all birds were, and the other had their tails cut and then the entire cut tail glued back on, to control for effects of cutting and gluing.

The results were dramatic (Figure 6.8)—the four groups had similar numbers of nests on their territories before the manipulation, but afterward the elongated males had far more than the shortened males, with controls intermediate. This is strong evidence that tail length is a target of selection, with females acting as the selective agent.

An important consideration with experimental manipulation is how much variation to create; the answer depends on the question being asked. If the goal is to determine if the trait is a target of current selection in a specific natural population, then the manipulation should not go beyond the trait variation present in that population. The reason for this was shown in Figure 6.5; the phenotypic range expressed in any environment can determine how selection is acting on the trait. If, however, the question is more generally about how the trait interacts with the environment to affect fitness (that is, whether or not it is an adaptation), then manipulations beyond the range of any given population are appropriate. These larger manipulations help to determine the overall fitness function as depicted in Figure 6.5. The widowbird study is an example of this approach.

Although experimental manipulation is a powerful method to determine causation, it does have some disadvantages. Unlike selection gradient studies, only one trait can be studied at a time. This can be a serious drawback, because measuring fitness is usually difficult. It can also be difficult to design a proper control for the manipulation. In the case of the widowbirds, cutting and then re-gluing the same feathers to maintain the same tail length was an excellent control, because all the birds had their feathers cut and glued in the same way. However, due to the complexity and integration of traits in living organisms, it is difficult to ensure that some other trait is not also being affected, and that this other trait actually is the target of selection.

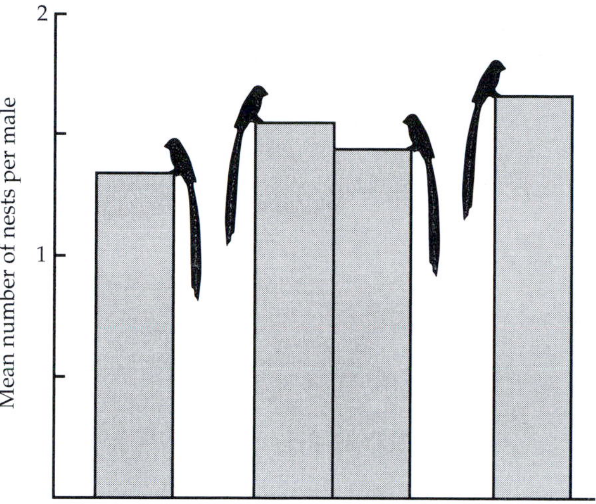

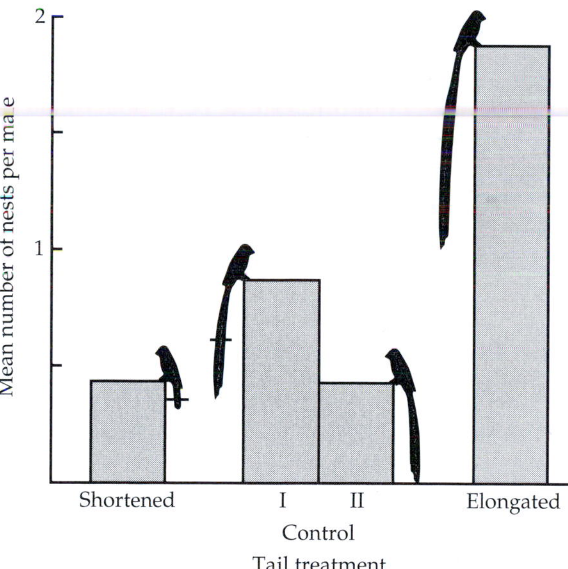

Figure 6.8 Results from experimental manipulation of widowbird tail lengths. The measure of fitness is the number of female nests per male. The top panel (A) shows the numbers before the tails were manipulated and the bottom (B) is after manipulation. Control I birds had their tails cut and glued back on, whereas control II birds were merely captured and banded. (After Andersson 1982.)

For example, in the widowbirds, it is possible that changing the length of the tail changed flight patterns or some other aspect of the male's behavior that the females were actually responding to. This seems unlikely in this particular case (Andersson showed that tail-shortened males actually displayed more often), but it is possible. Once again, detailed knowledge of the biology of the organism (e.g., display behaviors in widowbirds) is crucial.

A closely related experimental problem is that the manipulation may not produce biomechanically equivalent structures. For example, there is some evidence that fungus beetle horns have nerve cells and sensory hairs, so that cutting and re-gluing them as was done with the widowbird tails may fundamentally alter their function. Finally, there are many traits for which experimental manipulation is extraordinarily difficult or perhaps impossible. A good example may be the finch beak and weight traits in Table 6.1; it is hard to see how they could be manipulated without injuring the birds or altering many other traits.

Therefore, the observational selection gradient and experimental manipulation approaches are complementary. Only selection gradient analysis can produce a quantitative estimate of the strength of natural selection in a population for use in predicting speed and direction of evolution. Only experimental manipulation can demonstrate causality. In general, experiments are a very powerful way of discovering what could be happening in nature and the underlying causes, but not necessarily a way of proving what is happening in nature. Observational studies are often better suited for determining what is happening in nature, but not as good at determining causality. An excellent approach, rarely used, is to employ a selection gradient approach first to measure selection on a number of traits identified from knowledge of the biology of the organism as potential adaptations, and then to experimentally manipulate those traits with significant selection gradients to test for a causal relationship between phenotypic variation and fitness.

Correlational selection

From the earlier discussion of direct and indirect selection, and the section on correlations among traits in the previous chapter, it is clear that single traits cannot be considered in isolation. The selection estimates for finches and fungus beetles in Table 6.1 show the effects of correlations on selection, but it is also possible to turn the question around and study the effects of selection on correlations. **Correlational selection** occurs when two traits interact to determine fitness, or in other words, when certain combinations of trait values have higher fitness than other combinations. If both traits are heritable, correlational selection can cause evolutionary change in the correlation between the two traits.

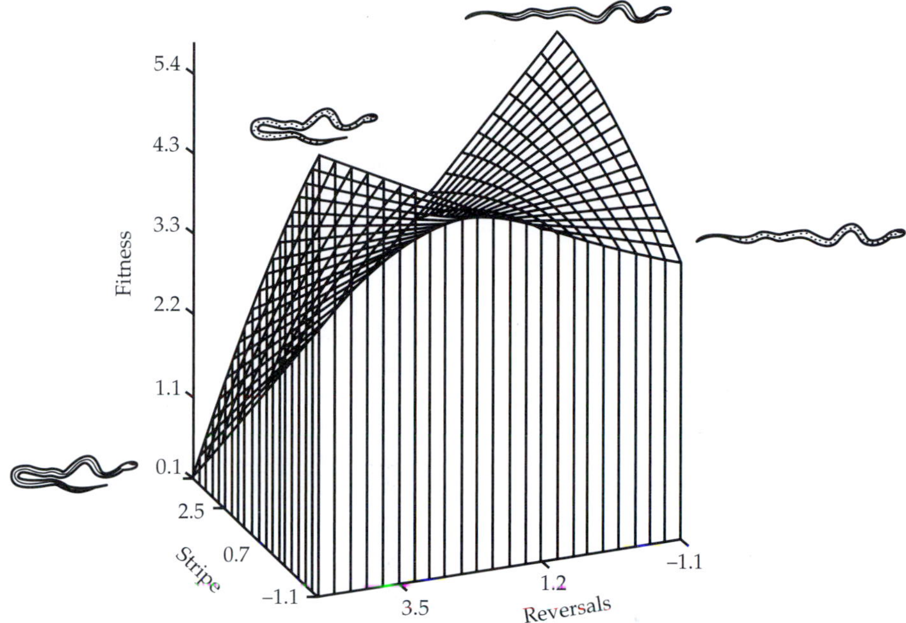

Figure 6.9 Bivariate fitness surface for two antipredator traits in garter snakes. Positive scores for stripe indicate highly striped snakes, whereas negative scores indicate that the snakes are blotchy. Positive scores for reversals indicate snakes that reverse frequently. Fitness is shown on the vertical axis. Note that fitness depends on the combination of the two traits—high values of both lead to low fitness, as do low values of both. The two adaptive peaks are snakes that were highly striped and rarely reversed (back right of figure) and blotched snakes that often reversed (left foreground). (From Brodie 1992.)

So far the fitness functions we have been examining have been univariate (one trait only; see Figures 6.1–6.5), but to understand correlational selection we need a three-dimensional phenotypic **fitness surface** (Figure 6.9). As in the fitness function, individual fitness is on the vertical axis, but with a surface there are two horizontal axes representing two phenotypic traits instead of a single trait in a fitness function. Our example of a fitness surface comes from a study of selection on juvenile garter snakes (Brodie 1992). Brodie's measure of fitness was juvenile survival, measured using a mark-recapture study (see Chapter 3). Brodie was mainly interested in two antipredator traits, the degree of striping on the body and the tendency for the snake to perform reversals—a rapid doubling-back when being pursued by a predator. There were no significant linear or quadratic selection gradients for the traits individually, but there was a significant correlational gradient. The correlational gradient is also estimated using multiple regression.

Instead of squaring the value for one of the traits as is done to estimate the quadratic gradient, the correlational selection gradient for a pair of traits is the regression coefficient for the product of the two traits; this is called a cross-product.

The correlational selection gradient for the cross-product of stripes and reversals was negative, which means there is selection for a negative correlation. Snakes that were highly striped (a positive stripe score in Figure 6.9) had high fitness if they also had low (negative) scores for reversals (the fitness peak in the back right of Figure 6.9), but low fitness if they were prone to reversals (the very low fitness valley at the rear left of the figure). Conversely, unstriped snakes had highest fitness if they also had a high tendency to reverse (front left peak in figure), but had low fitness if they did not reverse often (valley at the front right). Therefore, there was no consistent pattern of selection on each trait individually; the fitness associated with a certain value of stripes depends on the value for reversals and vice versa. This is a key point that causes a great deal of confusion. Finally, note the "saddle" in the figure, where both traits are at intermediate values; fitness is intermediate also.

Therefore, very similar to the case with epistasis, the fitness of an individual with a given phenotypic value for one trait depends on its value for the other trait. Selection in this case does not act on traits individually, but on combinations of traits that are functionally integrated—that is, those that work together to affect organism function and fitness. In the garter snakes, the longitudinal stripes make speed of a straight-moving snake appear to be slower than it actually is, so that predators misplace their strikes. The reversals are often combined with a pause in movement, and the unstriped snakes are harder to see when motionless. Therefore, striped snakes that reverse and unstriped snakes that move in a straight line are the easiest for predators to catch and have lowest survival.

Correlational selection is very closely related to the epistatic selection that is at the heart of the shifting balance theory (see Chapter 3). The term *correlational selection* is best used when discussing selection on phenotypes, whereas *epistatic selection* refers to selection on the underlying genotypes. Correlational selection can give rise to epistatic selection if there are separate gene loci affecting the two phenotypic traits, and the epistatic selection can create linkage disequilibrium between the loci (see Chapter 5).

There is also a close relationship between the fitness surface and the **phenotypic adaptive landscape.** The key difference is that the fitness surface represents the relationship between individual phenotypes and individual fitness, as does the fitness function for single traits. In contrast, the phenotypic adaptive landscape represents the relationship between population *means* for the traits and population *mean* fitness. Because they are based on means, the adaptive landscapes are smoother, with less well-defined peaks and valleys, compared to fitness surfaces based on individual values (Schluter 2000). The phenotypic adaptive landscape is analogous to Wright's adaptive landscape. In both cases, the vertical (z) axis is population mean fitness. In the version of Wright's adaptive landscape that we discussed in Chapter 3, the horizontal (x and y) axes are allele frequencies at two different loci, while in a phenotypic adaptive landscape they are the population mean values of two different phenotypic traits. Therefore, either the allele frequencies or the mean and variance of the traits, respectively, determine a population's position on the two types of landscapes. The overall shape of Wright's landscape is determined by the environment and epistatic interactions between the two loci; similarly, the shape of the phenotypic adaptive landscape is determined by the environment and functional interactions between the two phenotypic traits. In both landscapes, selection can usually only increase population mean fitness by changing allele frequencies or mean phenotypes in ways that move the population up the vertical axis of the landscape. Random genetic drift can move populations to lower mean fitness (see Chapter 3).

How do we study selective agents?

We have defined fitness as the lifetime number of offspring produced, and selection gradients based on this fitness measure are likely to provide a good estimate of the strength of selection. However, because so many different environmental factors affect lifetime offspring production, it is difficult to identify the selective agents from the selection gradients. A good first step in determining which selective agents are causing selection is to divide total lifetime fitness up into biologically meaningful **fitness components.**

Darwin was probably the first to separate total fitness into fitness components when he introduced the idea of **sexual selection.** Darwin recognized that many traits, such as the showy plumage of many male birds, should decrease survivorship by making the animals more vulnerable to predators. How then could these traits evolve by natural selection? Darwin also recognized that these types of traits were often **sexually dimorphic,** that is, much more pronounced in one sex, usually the males. In Darwin's process of sexual selection, these traits evolve through differences in mating success, which can lead to differential offspring production. Sexual selection is the process that resulted in the long tails of widowbirds. Therefore, in Darwin's scheme, there are two fitness components, survivorship and mating success.

Sexual selection is best regarded as a subset of natural selection, because differential mating success is only one of several ways that fitness differences can occur. Some authors use natural selection in a narrower sense, to mean everything but sexual selection, whereas others use the term *nonsexual selection*. Nonsexual selection is often broken into mortality selection, caused by differences in lifespan among individuals with different phenotypic values, and fecundity selection, caused by differences in rates of offspring production. This complexity can cause confusion, and often it is difficult to decide where to draw the line between sexual and nonsexual (especially fecundity) selection.

An excellent way to avoid the confusing terminology and to focus on the real goal of identifying the selective agents is to define **multiplicative fitness components** that are measurable and biologically meaningful for the organism under study. By *multiplicative* we mean they are defined so that the numerator of each component is the denominator of the next component (Table 6.2). For the fungus beetle example in Table 6.2, lifespan is the number of days between the first and last sightings of each marked beetle, attendance is the number of nights the male was seen on the surface of the fungi where mating occurs, courtships is the number of females courted in the male's life, and similarly copulation attempts and inseminations are the lifetime number of copulations attempted and successful inseminations over the male's lifetime. Therefore, the attendance/lifespan component is the proportion of days a male was alive that he was in the mating area, courtships/attendance is the average number of females courted per day that the male was in attendance, and so on. The mean values are shown below each component; for example, males lived about 56 days on average, were in the mating area two-thirds of their lives, courted one female per night in attendance, and so on.

Defining fitness components in this way has two key advantages. First, it means that when the components are multiplied together they equal total fitness, because all but the last numerator (which is total fitness) "cancel out." Second, it means that they are likely to be largely uncorrelated and independent of each other, so that one can examine selection and selective agents for each component without being confounded by earlier components. For example, in fungus beetles, the total number of females a male courts is determined by the male's success in fights with other males, as well as by his lifespan and attendance at the mating area. If we measure selection using total lifetime number of females courted as the fitness measure, we would not know if the selection was due to lifespan, attendance, or success in male

TABLE 6.2 *Multiplicative fitness components for fungus beetles*

	Lifespan ×	$\frac{\text{Attendance}}{\text{Lifespan}}$ ×	$\frac{\text{Courtships}}{\text{Attendance}}$ ×	$\frac{\text{Copulation attempts}}{\text{Courtships}}$ ×	$\frac{\text{Inseminations}}{\text{Attempts}}$ =	Lifetime inseminations
Mean	56.2	.65	1.01	0.60	0.24	5.31
β	0.26	0.03	0.19	0.30*	0.26	0.94**

Note: * $P < 0.05$, ** $P < 0.01$.
Source: Data from Conner 1988.

competition. By dividing courtships by the total number of nights a male spent in the mating area in his lifetime, however, the fitness component is number of female courted per night the male was in attendance, which is largely or entirely independent of lifespan or attendance. Therefore, we can isolate selection due to each component and more easily identify the selective agents.

Table 6.2 also shows the horn size selection gradients (β) for each fitness component and for total fitness (lifetime inseminations). We discussed the latter gradient, 0.94, earlier in this chapter (see Table 6.1). There we said the gradient was probably due to male competition for mates, because that is how the horns are used, but we did not offer direct evidence. The selection gradients for the multiplicative fitness components provide this evidence, because the largest and only statistically significant gradient occurs in the copulation attempts per courtship component. Males often attempt to break up other male's courtships before the courting male can attempt to copulate, but this is usually only successful if the attacker has longer horns than the courting male. Therefore, the selection gradients fit with knowledge of the beetle's biology, and they strongly suggest that the selective agent for horn size evolution is other conspecific males, and that horns are an adaptation to increase access to mates.

Measuring selection on components of fitness is an excellent first step in identifying selective agents, but just as selection gradients cannot prove targets of selection, additional evidence is usually needed to provide truly convincing evidence for selective agents. One way to provide additional evidence without altering the natural populations is to measure selection gradients in different locations in which the putative selective agent varies (Wade and Kalisz 1990). If the selection gradient varies across locations in a way that is correlated with some environmental variable, then this is evidence that the environmental variable is the selective agent. This approach is illustrated in Figure 6.10. Here selection is measured in eleven hypothetical populations that differ for some environmental variable. The points on the graph show that the strength of selection increases as the value of the environmental variable increases; in other words, there is a positive relationship between the selection gradient and the environmental variable, which is evidence that this particular environmental variable is a selective agent.

In a rare example of this approach, Stewart and Schoen (1987) measured selection on a number of vegetative size and phenology (timing) traits in pale jewelweed, *Impatiens pallida*, at 24 locations within one population. They found that the strength of selection on some of these traits increased in intensity with increases in light, soil nutrients, and soil moisture.

The main limitation of this approach to determining selective agents is analogous to that of using selection differentials in determining targets of

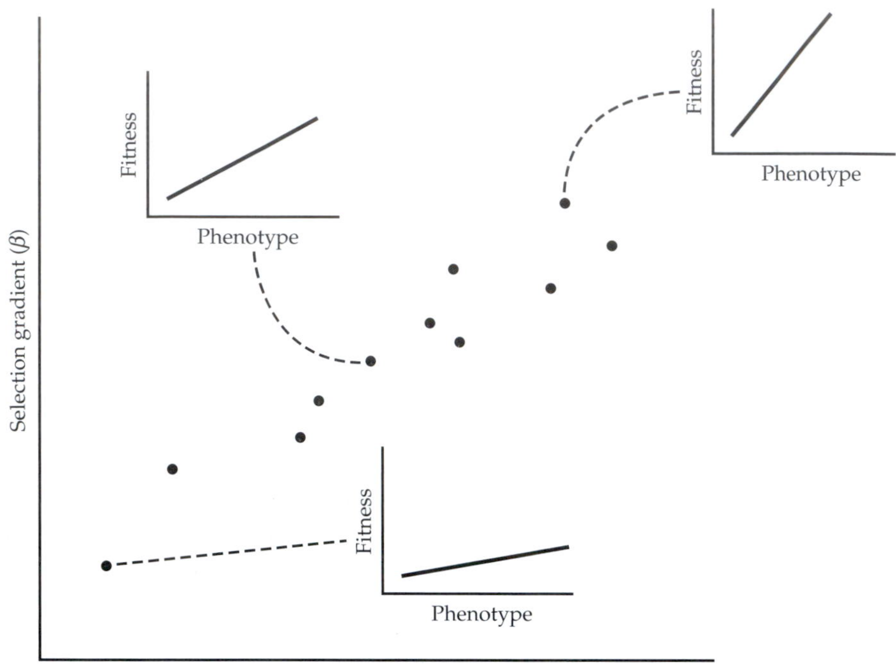

Figure 6.10 Hypothetical illustration of determining the selective agent by measuring selection in several populations that differ in an environmental variable (the putative selective agent). Each point in the graph represents an individual population. The position of each population reflects its value for the selection gradient (slope of the fitness function) on the *y*-axis and its value for the environmental variable on the *x*-axis. The positive relationship between these two variables is evidence that the environmental variable is a selective agent. The three smaller plots show the fitness functions used to estimate the selection gradient within three of these populations; note the steeper slopes for those placed higher along the *y*-axis.

selection; neither accounts for correlations among independent variables. In Figure 6.10, the positive relationship between the strength of selection and the environmental variable could actually be due to a different environmental variable that is correlated with the one plotted. As with targets of selection, one solution would be to perform multiple regression, in which one would regress the strength of selection in the different populations on multiple environmental variables simultaneously. To our knowledge this has not been done.

Note that these are observational studies; natural populations are observed, and there is no experimental manipulation. Therefore, these types of studies cannot provide definitive identifications of selective agents, for exactly the same reason that observational studies cannot definitively identify targets of

TABLE 6.3 *Summary of the different methods of determining selective agents and targets*

To determine	First step	Intermediate step	Definitive step
Selective agents	Multiplicative fitness components	Regress selection gradient on multiple environmental variables	Manipulate environment
Targets of selection	Selection gradients		Manipulate phenotype

selection: There could be unmeasured environmental factors correlated with the measured factor that are actually the selective agents. Once again, manipulative experiments provide complementary evidence, and such experiments are necessary to prove which environmental factor is the selective agent.

Dudley and Schmitt (1996) combined experimental manipulation of both the selective agent and the target of selection in a study of plant height as an adaptation to intraspecific competition. Working with a different species of jewelweed, *Impatiens capensis*, Dudley and Schmitt manipulated plant height by exposing the plants to different ratios of red to far red light; the differing light ratios are well known to produce differing levels of stem elongation. They then planted the resulting plants into natural populations at both low and high densities, thereby experimentally manipulating this environmental variable. The selection gradients for height were positive at high density, indicating that taller plants were better able to compete for light under crowded conditions, but negative at low density, indicating a cost to growing tall when light is not the main limiting resource. This is a good example of **density-dependent selection,** because the direction of selection depended on the density of the population. This is also excellent evidence for plant height as an adaptation for crowding because the experiments identified the selective agent and target through experimental manipulation. The different methods of identifying selective agents and targets are summarized in Table 6.3. The first step and intermediate columns in Table 6.3 are observational studies, and the definitive column is experimental.

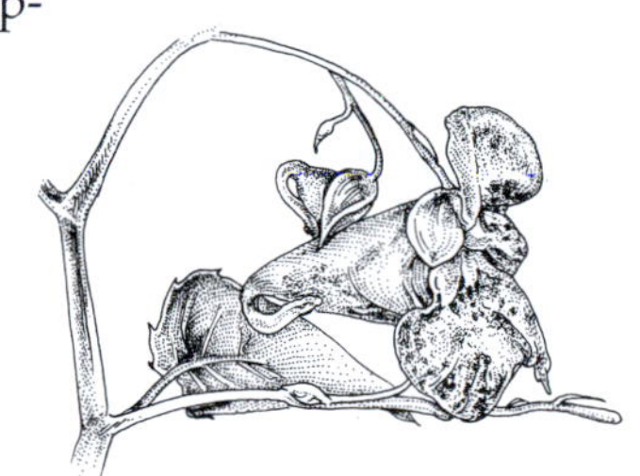

Impatiens capensis

Notes on the Study of Adaptation

Rigorously identifying selective agents and targets of selection, particularly in studies that combine observational and experimental approaches, provides excellent evidence that a trait is currently an adaptation. These studies show that individuals with certain values of the trait are better able to deal with a specific feature of their environment. In other words, certain values of the target of selection cause higher fitness because the organisms with those trait values can deal with the selective agent more effectively. It is hard to imagine better evidence for current adaptation. This evidence does not, however, prove that the trait evolved for this purpose originally, but rather that the trait serves the function of coping with this selective agent at the present time.

Often biologists believe that a trait must be heritable for it to be an adaptation, but the current amount of additive genetic variation for a trait has no bearing on whether the trait is currently adaptive. In fact, if there has been direct selection over time to create or improve an adaptation, this selection itself could very well deplete additive variance. To return to an example from Chapter 4, the possession of two eyes for a vertebrate is undoubtedly adaptive in most natural environments, but there is also little to no additive variance for this trait. Of course, for an adaptation to have evolved, there must have been additive variance for that trait at some point in the past. Similarly, the existence of additive variation in the present day is necessary for future evolutionary changes in the trait, and it is in the context of questions about future adaptation that measures of additive variation and heritability are most important. The final section in this chapter extends this idea to incorporate selection with additive variance and covariance to make quantitative predictions of evolution.

Synthesis: Predicting Short-Term Phenotypic Evolution

We have now examined a number of related topics in the evolution of phenotypic traits by natural selection. We discussed measuring the strength of selection using selection differentials and gradients, and showed that the former includes indirect selection within generations caused by phenotypic correlations. In the previous chapter we showed how the breeder's equation, $R = h^2S$, could be used to predict the evolutionary change across one generation; however, this equation is limited to one trait only and does not take genetic correlations into account. Also in the previous chapter we discussed correlated responses to artificial selection caused by genetic correlations, but in these experiments selection is typically applied to only one trait at a time—this is not how selection acts in nature. Now we are in a position to combine all these elements in one general equation that predicts evolution by natural selection for any number of traits:

$$\Delta\bar{z} = G\beta \qquad \textbf{6.5}$$

This is a multivariate version of the breeder's equation from the previous chapter, $R = h^2S$, and like the breeder's equation it is crucial for understanding

BOX 6.1 *Summary of Bivariate Relationships in Ecological Genetics*

In this chapter and the previous two we introduced four different types of bivariate plots, so a review of these may be useful (Table 6.4). The key to keeping them straight and to understanding what each one means is to carefully consider what each axis represents (key to understanding any graph) and what each point in the plot represents. In the first three types of plots in Table 6.4, the axes represent phenotypic traits. If each point represents means for a family, and the offspring mean is on the *y*-axis and parents are on the *x*-axis, then the plot represents additive genetic information. If the trait measured is the same in parents and offspring, then the slope of this relationship is the heritability; if the traits are different in parents and offspring, then the covariance of the relationship is ½ times the additive genetic covariance. If instead of family means, each

TABLE 6.4 *Review of four different types of bivariate plots*

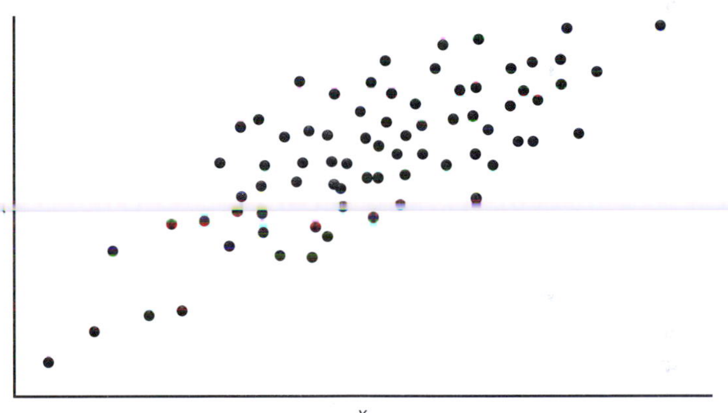

Relationship	Measures	*y*-variable	*x*-variable	Each point represents
Offspring–parent regression slope	Heritability	Offspring trait A	Parent trait A	Family
Offspring–parent cross covariance	One-half genetic covariance	Offspring trait A	Parent trait B	Family
Correlation between two traits	Phenotypic correlation	Trait A	Trait B	Individual
Slope and curvature of fitness function	Selection	Fitness	Phenotypic trait	Individual

Note: The scatterplot at the top is a representative example of one possible relationship.

Box 6.1 continued

point represents one individual in the population, then the plots convey phenotypic information. If the traits on the *x*- and *y*-axes are different, then the plot represents the phenotypic correlation. If a trait is on the *x*-axis and fitness is on the *y*-axis, then the plot is the fitness function.

phenotypic evolution by selection. In both cases, the term on the left side of the equation is the change in phenotypic mean across one generation ($\Delta\bar{z}$), and on the right side are terms representing additive genetic variance (**G**) and the strength of selection $\boldsymbol{\beta}$, respectively. Because Equation 6.5 is for multiple traits simultaneously, it also includes genetic covariances among traits. How is this accomplished? Equation 6.5 is a **matrix equation,** in which each term actually represents a set of related terms; this is shown in Equations 6.6 and 6.7. Matrix algebra is a systematic method to simplify mathematical expressions when a large number of terms are involved. Equations 6.5–6.7 represent exactly the same mathematical operation—Equation 6.5 is in a compact matrix form that is general for any number of traits, Equation 6.6 is in an expanded matrix form for three phenotypic traits as an example, and Equation 6.7 is a system of three standard (nonmatrix) equations for the three traits. The first and last terms ($\Delta\bar{z}$ and $\boldsymbol{\beta}$) of the matrix equation are called **column vectors;** they are nothing more than a list of variables (or the numbers they represent), three in this three-trait example. In all cases the subscripts refer to the trait in question: subscript 1 for trait 1, subscript 2 for trait 2, and so on. For example, $\Delta\bar{z}_1$ is the change in mean across one generation for trait 1, and β_2 is the selection gradient measuring the strength of selection on trait 2.

The **G** term is a **square matrix,** with an equal number of rows and columns, one row and one column for each trait. This is referred to as the **G-matrix** or the **additive genetic variance-covariance matrix.** Many studies have been devoted to estimating the **G**-matrix and understanding its role in evolution. The **diagonal** elements of the **G**-matrix (see Equation 6.6), in which both subscripts are the same, represent the additive variances of the traits. For example, $G_{11} = V_A$ for trait 1, $G_{22} = V_A$ for trait 2, and so on. The **off-diagonal** elements, those with different subscripts, are the additive genetic covariances between pairs of traits. Thus G_{23} is the additive genetic covariance between traits 2 and 3. Note that the **G** matrix is **symmetrical**, which means that each covariance appears twice, once above and once below the diagonal. Because the **G**-matrix consists of additive genetic variances and covariances, it is estimated using the techniques described in the last two chapters, usually offspring–parent regression or nested half-sibling analysis.

$$\begin{bmatrix} \Delta\bar{z}_1 \\ \Delta\bar{z}_2 \\ \Delta\bar{z}_3 \end{bmatrix} = \begin{bmatrix} G_{11} & G_{12} & G_{13} \\ G_{12} & G_{22} & G_{23} \\ G_{13} & G_{23} & G_{33} \end{bmatrix} \begin{bmatrix} \beta_1 \\ \beta_2 \\ \beta_3 \end{bmatrix} \quad \text{6.6}$$

$$\Delta\bar{z}_1 = G_{11}\beta_1 + G_{12}\beta_2 + G_{13}\beta_3 \quad \text{6.7a}$$

$$\Delta\bar{z}_2 = G_{12}\beta_1 + G_{22}\beta_2 + G_{23}\beta_3 \quad \text{6.7b}$$

$$\Delta\bar{z}_3 = G_{13}\beta_1 + G_{23}\beta_2 + G_{33}\beta_3 \quad \text{6.7c}$$

To solve for the predicted change across one generation $\Delta\bar{z}$ for each of the traits, each of the elements of the row of **G** adjacent to that $\Delta\bar{z}$ term is multiplied by each of the selection gradients in order. (Spreadsheet and other computer programs can be used to perform matrix operations like this.) The results of this process are the standard equations shown in Equation 6.7a–c. The shaded terms are the portions of evolutionary change that are due to direct selection on the trait. These portions equal the additive genetic variance of the trait times the selection gradient for that same trait; this represents adaptive evolution over one generation. The other terms are correlated responses to selection on other traits. For example, the second term in the equation for $\Delta\bar{z}_1$ is the selection gradient for trait 2 times the additive genetic covariance between trait 1 and trait 2. These equations are analogous to Equation 6.4, in which there was direct selection and also indirect selection caused by phenotypic correlations within generations. In Equation 6.5, there is a response to direct selection across generations, as well as correlated responses across generations due to genetic covariance. This equation ties much of quantitative genetics and phenotypic evolution together in a unified way, analogous to how *F*-statistics tied much of single-locus population genetics together.

A rare example of evolutionary predictions in a natural population comes from Campbell's (1996) study of floral evolution in scarlet gilia, *Ipomopsis aggregata* (Table 6.5). Campbell found three traits that were under direct selection—length and width of the floral corolla, and the proportion of time that the flowers had receptive stigmas (proportion pistillate). There was positive direct selection on all three traits and a corresponding prediction of an increase in the mean of all three in the next generation. Note, however, that the predicted changes for both corolla length and proportion pistillate are decreased by the negative covariance between these two traits. Thus, the adaptive evolution of these traits is slowed or constrained by the negative covariance.

TABLE 6.5 *Predicted evolutionary change in the mean of three floral traits ($\Delta\bar{z}$), calculated as the product of the genetic variance/covariance matrix (**G**) and the vector of selection gradients (β)*

	$\Delta\bar{z}$	G			β
		Corolla length	Corolla width	Proportion pistillate	
Corolla length	0.043	1.092	0.021	–0.039	0.05
Corolla width	0.035	0.021	0.025	0.004	1.22
Proportion pistillate	0.005	–0.039	0.004	0.002	0.96

Source: Data from Campbell 1996.

EXERCISE: Calculate $\Delta\bar{z}$ for each of the three traits in Table 6.5 using Equation 6.7a–c. Make sure your answers match the values given in Table 6.5.

Equation 6.5 provides a general framework for understanding genetic constraints on evolutionary change. We can define an **evolutionary constraint** as any factor that slows the rate of evolution of the most adaptive combination of traits. Genetic constraints are a special case, and in Chapter 3 we discussed how genetic drift and gene flow can constrain adaptive evolution. The simplest genetic constraint is lack of additive variance for the trait because without additive variance the trait mean will not change across generations. In terms of Equation 6.7, G_{ii} is close to zero, so the adaptive response to direct selection, $G_{ii}\beta_i$, is also close to zero regardless of the strength of selection. This means that even if there is strong direct selection, adaptive evolution is slowed or constrained by the lack of additive genetic variance. The low additive genetic variance for the proportion pistillate trait in Table 6.5 (0.002) is part of the reason for the small predicted change (0.005).

The other source of genetic constraint comes from the terms in the *G*-matrix representing the genetic covariances. For elytra length in fungus beetles and beak width in Darwin's finches, we have already seen how indirect selection within a generation causes the total selection to be in the opposite direction as direct selection (Table 6.1). In an analogous way, correlated responses to selection can cause nonadaptive or even maladaptive evolution. Nonadaptive evolution is a change in a trait that is not under direct selection ($\beta \approx 0$), and maladaptive evolution is a change in the opposite direction from the direction that would increase adaptation of a given trait. In other words, maladaptive evolution occurs when $\Delta\bar{z}$ and β have opposite signs.

TABLE 6.6 *Predicted changes across one generation for three body size traits in male flour beetles*

	$\Delta\bar{z}$			G ($\times 10^{-3}$)			β
	Total	Direct	CR	Weight	Length	Width	
Weight	0.01	0.05	–0.04	2.27	1.12	1.38	0.26
Length	0.03	0.00	0.03	1.12	0.41	0.45	0.00
Width	0.04	–0.03	0.07	1.38	0.45	0.46	–0.12

Note: Standardized values (in standard deviation units) are shown for $\Delta\bar{z}$ and selection (β) and unstandardized values are given for the G-matrix; however, all calculations were done on unstandardized values.
Source: Data from Conner and Via 1992.

Examples of constraints due to low additive genetic variance as well as constraints due to genetic correlations can be found in a study of selection on body size traits in a laboratory population of flour beetles (*Tribolium castaneum*; Conner and Via 1992). The **G**-matrix, selection gradient vector, and vector of predicted mean change across one generation are shown in Table 6.6. The total predicted change $\Delta\bar{z}$ is shown, and is broken down into the components due to direct selection and correlated responses (CR). There was strong directional selection for increased weight ($\beta = 0.26$), weaker selection for decreased width, and no selection on length. In spite of the strong selection for increased weight the predicted change in the mean was small, only 0.01 standard deviations. The first reason was low additive variance for weight—the heritability of weight was only 0.24, so that the predicted response to direct selection was only 0.05. The other reason for the low predicted change is the negative selection on width, which had a strong positive genetic covariance with weight, leading to a negative correlated response almost as strong as the direct response. Therefore, the adaptive evolution of increased weight was constrained by a lack of additive genetic variation as well as by a genetic covariance with another trait. Length and width show evidence for nonadaptive and maladaptive evolution, respectively, because both were predicted to increase due to their positive correlations with weight, in spite of the fact that the selection gradients indicate that length is at its optimum and width is above its optimum.

A graphical view of constraints caused by genetic correlations is given in Figure 6.11. The elliptical cluster of points represents a hypothetical negative genetic correlation between flower size and number, which could be caused by trade-offs in allocation of resources (Worley and Barrett 2000). The position of the cross in the middle is the current *bivariate mean*, in other words, the means for the two traits in the population. The letters A–C are placed at three possible evolutionary optima that represent values for both traits that maximize fitness. The three sets of arrows show evolutionary trajectories, which are the phenotypic paths that the population would traverse in evolving to

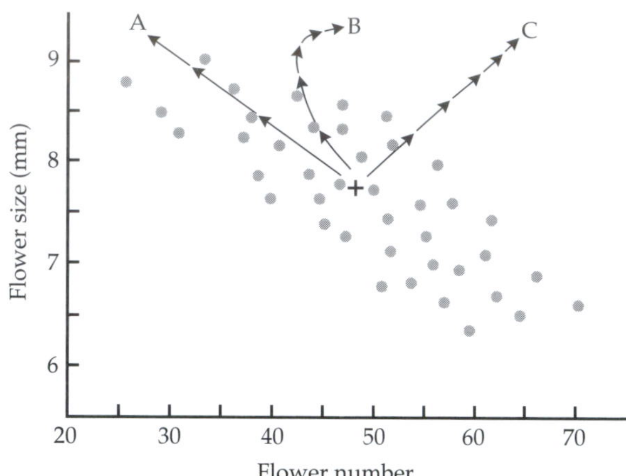

Figure 6.11 Hypothetical examples of evolutionary constraints caused by genetic correlation. The points represent breeding values; thus, the elliptical cluster of points represents a negative genetic correlation between the two traits. The cross represents the current mean for both traits in the population, and the letters A–C are three evolutionary optima. The three sets of arrows represent evolutionary trajectories to these optima; each arrow represents 50 generations.

these optima. Because each arrow represents a fixed amount of time (50 generations), long arrows denote rapid evolution and short arrows denote slow evolution. Evolution to the optimum marked by A (fewer, larger flowers) would be the fastest because it is along the major (longest) axis of the ellipse, meaning that it is in the direction of most genetic variance in bivariate space. Conversely, evolving to C (increased number and size of flowers) would be slowest because that is the direction of least variation in bivariate space. Thus, evolution to A is rapid but evolution to C is constrained.

Evolution to point B is a little more complex. The optimum at B is increased flower size with no change in flower number from the current mean shown by the cross. Since the population is far from the optimum for flower size, there is strong selection on flower size but no directional selection on flower number. Therefore, the population first evolves toward increased flower size by moving roughly along the major axis of the ellipse (as shown by the first arrow); this represents the direction of maximum genetic variance. This evolution actually draws the population away from the optimum flower number, thus creating maladaptation of flower number and consequently selection to reduce flower number back to its optimum. This situation is very similar to the predicted increase in length in Table 6.6, because in that example length was not under direct selection. As flower size approaches its optimum and flower number is drawn further from its opti-

TABLE 6.7 *Summary of constraints due to a genetic correlation between a pair of traits*

Sign of genetic correlation	Signs of selection gradients	
	Same sign	Opposite sign
Positive	Augmented	Constrained
Negative	Constrained	Augmented

mum (second arrow), the relative strength of selection on the two traits switches, so that at some point selection becomes stronger on flower number. This is why the trajectory eventually turns the corner (after the third arrow) and begins heading toward the optimum at B.

The isoclines, or lines of equal fitness (see Chapter 3), are not shown in Figure 6.11 for simplicity, but note that in all three cases the population trajectories climb an adaptive peak because population mean fitness is always increasing. This is true even in the trajectory towards point B; although the initial evolution of flower number is maladaptive, the increase in fitness due to increased flower size is greater, leading to increased mean fitness overall. Therefore, genetic constraints cause a slowing of the ascent of a population to an adaptive peak, but even with constraints, selection always increases population mean fitness. Also note that in all cases the arrows become shorter as the population approaches the optimum. There are two causes for this. First, as the population gets closer to the adaptive peak, selection becomes weaker because the population is better adapted. Second, the selection depletes genetic variation for the traits, which also slows progress toward the optimum.

To summarize constraints caused by genetic correlations (Table 6.7), if selection is acting in the same direction on two traits, there is a constraint only if the genetic correlation between the traits is negative. In contrast, the response to selection can be augmented if the genetic correlation is positive. Positive genetic correlations can cause a constraint if selection is acting in different directions on the two traits, in which case the selection gradients have opposite signs (e.g., weight and width in Table 6.6).

The Future of Ecological Genetics

In Chapters 4 and 5 we discussed two methods for studying the genetics of continuously variable phenotypic traits, statistical quantitative genetics and QTL mapping. In this chapter we have seen how methods of measuring natural selection on such traits can help us understand adaptation to the abiotic and biotic environment. We have seen how the selection gradients can be combined with statistical quantitative genetic estimates of the *G*-matrix to predict short-term evolutionary change and to identify evolutionary constraints. The results of QTL mapping cannot easily be used to predict evolutionary change, but they have the advantage over purely statistical methods of getting us closer to the actual genetic mechanisms underlying quantitative traits, such as the number of loci affecting the trait and the patterns of dominance, epistasis, or pleiotropy within and among loci. In most species it is difficult to use QTL mapping to study genetic variation within natural populations, the setting where evolution actually occurs. Therefore, statistical quantitative genetics and QTL mapping have complementary strengths and limitations.

Chapters 2 and 3 were devoted to population genetics, focusing on allele and genotypic frequencies at individual loci. Chapter 3 also discussed measuring the strength of natural selection on genotypes at single loci using s, the selection coefficient, whereas this chapter examined the use of S and β to measure the strength of selection on phenotypes. One goal for the future of ecological genetics is to find the loci affecting quantitative traits that are under selection, with QTL mapping often the first step. Then our goal is to be able to estimate selection coefficients on these individual loci to understand evolutionary changes in mean phenotype at the level of changes in allele frequencies at individual loci. This ties together two main definitions of evolution: the population genetic definition of evolution as a change in allele frequencies, and the quantitative genetic definition of evolution as a change in mean phenotype. The knowledge of how natural selection acting on phenotypic variation causes changes in allele frequencies at specific loci would give us a far better understanding of how organisms become adapted to their environments, and this is the primary goal of ecological genetics.

PROBLEMS

6.1 Johnston (1991) measured selection on several traits in cardinal flowers (*Lobelia cardinalis*). The standardized selection gradients for these traits are given in the top part of the accompanying table, and the phenotypic correlations among them are given in the bottom part of the table.

a. Calculate the selection differentials for these traits.

b. Which traits are likely targets of selection? Which trait is most strongly affected by indirect selection? How would having only the differentials and not the gradients mislead interpretations about adaptation?

c. For one of the targets of selection that you mentioned in your answer to part (*b*), briefly describe an experiment designed to prove that this trait is a cause of fitness differences.

Trait	β
Flower number	1.35***
Plant height	0.11
Median flower date	0.19*

Note: * $P < 0.05$; *** $P < 0.001$.

Phenotypic correlations	Flower number	Plant height
Plant height	0.71***	
Median flower date	–0.35***	–0.47***

Note: ***$P < 0.001$.

6.2 A hypothetical *G*-matrix for the traits in Problem 6.1 is shown below. Calculate $\Delta\bar{z}$ for the traits. Is there any evidence for constraints on the evolution of any of these traits? Describe these constraints, and be specific!

Trait	Flower number	Plant height	Flower date
Flower number	0.02	0.12	–0.10
Plant height	0.12	0.90	–0.42
Median flower date	–0.10	–0.42	0.84

6.3 The table below shows hypothetical selection gradients for leaf thickness in a plant that was grown experimentally at three different watering levels and three different light levels. Are water, light, both, or neither selective agents on leaf thickness in this experiment? Explain your reasoning.

	Light		
Water	Low	Medium	High
Low	0.55	0.54	0.52
Medium	0.28	0.31	0.29
High	0.15	0.14	0.16

6.4 In a study of a hypothetical butterfly, linear and quadratic selection gradients were estimated for three traits. These are shown below, with asterisks denoting statistical significance (gradient greater than zero) at $P < 0.001$. What can you infer about the shape of the fitness function for each? Make a sketch of what each fitness might look like based on these gradients. What other information would help further your understanding of the shape of the fitness functions?

Trait	β	γ
Wing length	0.50***	–0.62***
Proboscis length	0.01	0.49***
Wing spot size	–0.72***	–0.02

Note: *** $P < 0.001$.

SUGGESTED READINGS

Arnold, S. J. 1987. Genetic correlation and the evolution of physiology. Pp. 189–215 *in* M. E. Feder, A. F. Bennett, W. W. Burggren, and R. B. Huey, eds. *New Directions in Ecological Physiology*. Cambridge University Press, Cambridge, UK. A readable discussion of the evolutionary causes and effects of correlations among traits, with physiological examples.

Berenbaum, M. R., A. R. Zangerl, and J. K. Nitao. 1986. Constraints on chemical coevolution: Wild parsnips and the parsnip webworm. Evolution 40:1215–1228. One of the first studies to combine estimates of selection and genetic correlations to evaluate evolutionary constraint.

*Brodie, E. D., III. 1992. Correlational selection for color pattern and antipredator behavior in the garter snake *Thamnophis ordinoides*. Evolution 46:1284–1298. The clearest example of correlational selection to date.

Brodie, E. D., III, A. J. Moore, and F. J. Janzen. 1995. Visualizing and quantifying natural selection. Trends Ecol. Evol. 10:313–318. A succinct review of the regression-based methods for measuring selection as well as graphical techniques to better understand fitness functions and surfaces.

*Campbell, D. R. 1996. Evolution of floral traits in a hermaphroditic plant: Field measurements of heritabilities and genetic correlations. Evolution 50:1442–1453. A rare example of quantitative predictions of evolutionary change based on field estimates of selection gradients and the **G**-matrix.

*Conner, J. 1988. Field measurements of natural and sexual selection in the fungus beetle, *Bolitotherus cornutus*. Evolution 42:736–749. An example of the use of selection gradients and multiplicative fitness components to identify selective agents and targets of selection.

*Endler, J. A. 1986. *Natural Selection in the Wild*. Princeton University Press, NJ. An authoritative review of the concepts underlying, and the evidence for, natural selection, but it preceeded most of the many studies using the Chicago School methods.

Grafen, A. 1988. On the uses of data on lifetime reproductive success. Pp. 454–471 *in* T. H. Clutton-Brock, ed. *Reproductive Success*. The University of Chicago Press, Chicago. A critical review of the strengths and weaknesses of the Chicago School methods for studying adaptation.

*Grant, P. R., and B. R. Grant. 2002. Unpredictable evolution in a 30-year study of Darwin's finches. Science 296:707–711. A brief overview of some of the work of the Grants and their students on selection and evolution of Darwin's finches in the Galapagos Islands. This work represents the most comprehensive study of natural selection in the wild ever undertaken.

*Wade, M. J., and S. Kalisz. 1990. The causes of natural selection. Evolution 44:1947–1955.A readable discussion of methods for identifying selective agents.

**Indicates a reference that is a suggested reading in the field and is also cited in this chapter.*

SUGGESTED READINGS QUESTIONS

The following questions are based on papers from the primary literature that address key concepts covered in this chapter. For the full citation for each paper, please see Suggested Readings.

From Arnold, S. J. 1987. Genetic correlation and the evolution of physiology.

1. Why are genetic correlations important? How can they affect the evolution of traits? What is a correlated response to selection?
2. What are the two mechanisms that cause genetic correlations? Explain how each causes correlations.
3. Both parent–offspring regression and sibling-analysis can be used to measure genetic correlations as well as heritabilities. What additional information is needed to use these breeding designs to estimate correlations beyond the information needed for heritabilities?
4. How can artificial selection experiments be used to reveal the presence of genetic correlations?
5. Explain what each of the terms in Equation 9.1 represents.
6. In Figure 9.7, why does the evolutionary trajectory marked *b* differ from the one marked *a*?

From Berenbaum, M. R., A. R. Zangerl, and J. K. Nitao. 1986. Constraints on chemical coevolution: Wild parsnips and the parsnip webworm.

1. What breeding design (e.g., half-siblings, parent–offspring regression) did the authors use to estimate heritabilities and genetic correlations? What method to estimate selection?
2. For what traits were heritabilities, correlations, and selection measured? Are these appropriate traits to address the questions posed in the Introduction? In your opinion, are there other traits that should have been included?
3. What is the likely cause of the significant negative selection differential for flowering date?
4. How reliable are the estimates of genetic correlations given in Table 9? State your evidence.
5. Table 9 shows a negative genetic correlation between pBERs and SPHs. Could this cause a constraint, that is, a slowing of the evolution of the most adaptive combination of these traits (see also Table 5)?
6. On what did the authors base their conclusion that the evolutionary arms race between wild parsnip and the parsnip webworm has reached a temporary stalemate (page 1227)? Is this a reasonable conclusion to draw from the evidence presented in the paper?

From Brodie, E. D., III. 1992. Correlational selection for color pattern and antipredator behavior in the garter snake *Thamnophis ordinoides*.

1. Define correlational selection. Why is it important in evolution?
2. Explain the hypothesis of correlational selection on stripedness and reversal behavior. In other words, why would we expect a functional relationship between stripes and reversal behavior?
3. What was Brodie's measure of fitness? Is this a good measure of fitness for this study? Why or why not?
4. Do you think his choice of traits to study was reasonable? Was the rationale for each trait clear? Were there other traits that you think should have been included? If so, why?
5. Which traits were under significant directional selection? Which were under significant stabilizing/disruptive selection? Which pairs of traits were under significant correlational selection?

From Campbell, D. R. 1996. Evolution of floral traits in a hermaphroditic plant: Field measurements of heritabilities and genetic correlations.

1. Why is it important to measure heritabilities and genetic correlations in the field?
2. What breeding design (e.g., half-sibs, parent-offspring regression) did they use to estimate heritabilities and genetic correlations? What method to estimate selection?
3. For what traits were heritabilities, correlations, and selection measured? Are these appropriate traits to address the questions posed in the Introduction? Are there other traits that should have been included, in your opinion?
4. Which traits had significant heritabilities? Which pairs of traits were significantly genetically correlated?
5. Give two reasons why the predicted response to selection (based on the multivariate equation; Table 8) is so much less for proportion pistillate than it is for corolla length, when the selection gradient for the former was so much greater than for the latter.
6. Why is there a range of predicted responses for proportion pistillate even though the two estimates of the selection gradient for that trait are the same?

From Conner, J. 1988. Field measurements of natural and sexual selection in the fungus beetle, *Bolitotherus cornutus*.

1. What is direct and indirect selection? What are targets of selection and selective agents? What methods used in this paper help identify targets and agents of selection?
2. Explain how multiplicative fitness components (or episodes of selection) are defined so that they are independent of each other, and multiply to equal total fitness. What is the biological meaning of each of the five components used in this paper?

3. Focusing on the differentials and gradients for total fitness of the cohort of 67 males only (the bottom lines in Tables 3 and 5), what are the similarities and differences between the total selection on the three traits (selection differentials) and the direct selection on the traits (selection gradients)? What are the reasons for the differences between total and direct selection?
4. What are the main targets of selection identified by this study, and what are the likely selective agents causing this selection? Why is it important to identify targets and agents of selection?

From Grafen, A. 1988. On the uses of data on lifetime reproductive success.

1. Explain the distinction between adaptation and selection in progress. Is this distinction always clear, or are there areas where they merge? Are the methods of selection differentials and gradients developed by Lande and Arnold and used in Conner (1988) and Brodie (1992) useful for studying adaptation? Selection in progress?
2. What are the advantages and disadvantages of using natural and artificial variation to study adaptation and selection in progress? Does selection gradient analysis help prevent being misled by Grafen's three "malignant" causes of natural variation?
3. What are the three problems in the interpretation of selection gradients identified by Grafen? How serious are they likely to be, in your opinion? Were any of them problems for the selection gradient analyses presented by Conner (1988) and Brodie (1992)?
4. What are the problems in measuring fitness identified by Grafen? Again, how serious are they, and which of them apply to the Conner and Brodie papers?

From Wade, M. J., and S. Kalisz. 1990. The causes of natural selection.

1. From where does fitness arise, in the authors' opinion?
2. What should be experimentally manipulated to determine the target of selection? What should be manipulated to determine the selective agent? Recall the definitions of these terms.
3. Focusing on the top three panels in Figure 2, what causes the differences in slopes (selection gradients) across the three densities? What does this tell you about selective agents in this case?

CHAPTER REFERENCES

Andersson, M. 1982. Female choice selects for extreme tail length in a widowbird. Nature 299:818–820.

Arnold, S. J., and M. J. Wade. 1984a. On the measurement of natural and sexual selection: Applications. Evolution 38:720–734.

Arnold, S. J., and M. J. Wade. 1984b. On the measurement of natural and sexual selection: Theory. Evolution 38:709–719.

Bumpus, H. C. 1899. The elimination of the unfit as illustrated by the introduced sparrow, *Passer domesticus*. Biol. Lect. Woods Hole Mar. Biol. Sta. 6:209–226.

Conner, J., and S. Via. 1992. Natural selection on body size in *Tribolium*: Possible genetic constraints on adaptive evolution. Heredity 69:73–83.

Conner, J. K., R. Davis, and S. Rush. 1995. The effect of wild radish floral morphology on pollination efficiency by four taxa of pollinators. Oecologia 104:234–245.

Conner, J. K., S. Rush, and P. Jennetten. 1996. Measurements of natural selection on floral traits in wild radish (*Raphanus raphanistrum*). I. Selection through lifetime female fitness. Evolution 50:1127–1136.

Dawkins, R. 1982. *The Extended Phenotype*. Oxford University Press, New York.

deJong, G. 1994. The fitness of fitness concepts and the description of natural selection. Quarterly Review of Biology 69:3–29.

Dudley, S. A., and J. Schmitt. 1996. Testing the adaptive plasticity hypothesis: Density-dependent selection on manipulated stem length in *Impatiens capensis*. American Naturalist 147:445–465.

Grant, P. R. 1986. *Ecology and Evolution of Darwin's Finches*. Princeton University Press, Princeton, NJ.

Johnston, M. O. 1991. Natural selection on floral traits in two species of *Lobelia* with different pollinators. Evolution 45:1468–1479.

Kingsolver, J. G., H. E. Hoekstra, J. M. Hoekstra, D. Berrigan, S. N. Vignieri, C. E. Hill, A. Hoang, P. Gibert, and P. Beerli. 2001. The strength of phenotypic selection in natural populations. American Naturalist 157:245–261.

Lande, R. 1979. Quantitative genetic analysis of multivariate evolution, applied to brain:body size allometry. Evolution 33:402–416.

Lande, R., and S. J. Arnold. 1983. The measurement of selection on correlated characters. Evolution 37:1210–1226.

Price, T. D., P. R. Grant, H. L. Gibbs, and P. T. Boag. 1984. Recurrent patterns of natural selection in a population of Darwin's finches. Nature 309:787–789.

Schluter, D. 1988. Estimating the form of natural selection on a quantitative trait. Evolution 42:849–861.

Schluter, D. 2000. *The Ecology of Adaptive Radiation*. Oxford University Press, New York.

Schluter, D., T. D. Price, and P. R. Grant. 1985. Ecological character displacement in Darwin's finches. Science 227:1056–1059.

Smith, T. B. 1990. Natural selection on bill characters in the two bill morphs of the African finch *Pyrenestes ostrinus*. Evolution 44:832–842.

Stewart, S. C., and D. J. Schoen. 1987. Pattern of phenotypic viability and fecundity selection in a natural population of *Impatiens pallida*. Evolution 41:1290–1301.

Weis, A. E., and W. L. Gorman. 1990. Measuring selection on reaction norms: An exploration of the *Eurosta-Solidago* system. Evolution 44:820–831.

Wolf, J. B., and M. J. Wade. 2001. On the assignment of fitness to parents and offspring: Whose fitness is it and when does it matter? Journal of Evolutionary Biology 14:347–356.

Worley, A. C., and S. C. H. Barrett. 2000. Evolution of floral display in *Eichhornia paniculata* (Pontederiaceae): Direct and correlated responses to selection on flower size and number. Evolution 54:1533–1545.

7

Applied ecological genetics

In this chapter we discuss ways in which ecological genetics can be applied to problems facing human society. The first topic is conservation genetics, a large field devoted to using the principles of ecological genetics to aid in the preservation of species. Next we discuss the nascent field of invasive species evolution, which seeks to determine the role of rapid evolutionary change in the development of invasive species (that is, non-native species that become destructive to natural ecosystems). The third topic in the chapter is transgene escape, in which we discuss the processes by which foreign genes inserted into crops might escape and become established in natural populations. Finally, we apply the principles of ecological genetics to the problem of the evolution of resistance in pest species (such as insects, weeds, and disease-causing microbes) to the chemical agents (insecticides, herbicides, and antibiotics) that humans use to control them.

These topics are all tied together by the central importance of genetic variation and selection. Consequently, most of the key concepts and techniques covered in the first six chapters are relevant to these issues, so discussing these human problems provides an opportunity to review, integrate, and apply the material. There are also some direct ties between the four topics. For example, invasive species are one of the major threats to endangered species, and so they are important in conservation. Also, a major concern is whether the escape of transgenes for traits such as pathogen resistance or frost tolerance would create new invasive species. Finally, crops with anti-herbivore transgenes may cause rapid evolution of herbivore resistance to the transgene products.

Conservation Genetics

The field of conservation biology is concerned with producing scientific knowledge useful to preserving biodiversity. **Biodiversity** can be defined as genetic and phenotypic variation both within and among species, plus the variety of ecosystems created by these species. Most biologists believe that humans are causing a mass extinction similar in scope to the one that caused the extinction of much of the flora and fauna (including the dinosaurs) at the Cretaceous-Tertiary boundary. A major challenge for conservation biology is to integrate science into the practical, and often highly political, process of protecting habitats and species. In the United States, the major conservation law is the Endangered Species Act (ESA) of 1973, which provides protection to species placed on the endangered or threatened lists. Under the ESA, an **endangered species** is one that is likely to become extinct in all or a major portion of its range, while a **threatened species** is one that is likely to become endangered in the near future (Primack 2000). A great deal of effort in conservation biology has focused on conservation genetics, although some have argued that population ecology plays a larger role than genetics in species extinction (e.g., Caro and Laurenson 1994).

The questions addressed in conservation genetics can be grouped into three broad categories:

1. What is the unit or group to be conserved? Is it the population, subspecies, or species, and how are these units defined? Are two groups genetically distinct enough to each warrant protection?
2. How do genetic factors, especially random genetic drift, inbreeding depression, and mutation, directly affect current population viability and extinction risk?
3. How much genetic variation is available in endangered and threatened species for adaptation to future environmental change?

The first question deals with which populations are to be saved and the last two deal with how to save them. In other words, the major goal of conservation genetics is to maximize both representation and persistence of biodiversity (Moritz 2002). In addition, the twin goals of representation and persistence can interact, because one management technique is to move individuals among populations. The goals of these relocations are to bolster declining populations and to decrease the reduction in genetic variation and fitness caused by random genetic drift in small populations. This strategy is nothing more than human-directed gene flow, designed to convert among-subpopulation differentiation caused by drift to within-subpopulation variation (see Chapter 3). However, if the populations are unique units for conservation, then mixing them destroys their uniqueness, and therefore the units are no longer separately represented. Also, if there is differentiation caused by selection for adaptation to local conditions, then mixing these populations destroys the local adaptation and may therefore reduce persistence of the populations. These issues are central to conservation genetics and will appear repeatedly in the rest of this section.

Most research in conservation genetics to date has used single-locus population genetic techniques and selectively neutral genetic markers (see Chapters 2 and 3), but quantitative genetic techniques that focus on phenotypes (see Chapters 4 and 5) are also highly relevant to conservation genetics. Neutral markers have been used partly due to the history of the field and also because they are much less labor-intensive than quantitative genetic studies. In addition, quantitative genetic studies typically require controlled breeding and large sample sizes, which may not be possible with rare species. We will now consider each of three critical issues in turn; one recurring theme will be the relative utility of neutral markers versus quantitative genetic approaches for each question.

What is the unit to be conserved?

One use of molecular genetic markers in conservation is to identify populations of threatened or endangered organisms that need to be saved from extinction. These populations are often referred to as **stocks** in marine animals or **evolutionarily significant units (ESUs).** The identification of these units is fraught with difficulty but of great practical importance, because they determine which populations receive legal protection from the ESA and other laws, as well as qualify for financial resources for recovery efforts. ESUs have also been used as the units that should not be mixed in recovery plans. All species are ESUs, but in some cases drawing clear boundaries between species is difficult. Within species there are often populations or groups of populations that are named as different subspecies, which if they are distinct enough warrant protection under national and international law.

Defining ESUs has been controversial, but all definitions have included one or more of the following criteria:

1. Current geographic separation
2. Neutral genetic differentiation among ESUs caused by past restriction of gene flow
3. Locally adapted phenotypic traits caused by differences in selection

Much of the recent controversy has focused on the relative importance of criteria two and three. ESUs identified by these two criteria may or may not be congruent. Recall from Chapter 3 that a high degree of differentiation at neutral markers (differences in allele frequencies as measured by F_{ST}) is good evidence for a lack of gene flow, because with low gene flow random drift causes differentiation across all neutral or nearly neutral loci. On the other hand, only one to several migrants per generation are needed to prevent strong differentiation of neutral markers, so there can still be local adaptation due to differential selection in the absence of neutral differentiation. In Chapter 5 we said that common garden or reciprocal transplant experiments are necessary to test for genetic differentiation for phenotypic traits, and differences in selection gradients across habitats (see Chapter 6) are needed to determine whether this differentiation is locally adapted. This type of information is very useful before deciding whether two populations can be mixed as part of a recovery plan, because criterion 3 is very difficult to assess with neutral markers alone.

Criterion two has been used more than criterion three because it is easier and faster to score neutral markers than to do a reciprocal transplant, and transplant experiments may be problematic or impossible with very rare species. A serious problem with both criteria two and three is the question of how much differentiation is enough to warrant designation as an ESU. The amount of genetic differentiation among populations forms a continuum, and there is no objective way to convert this continuous variation into a dichotomy between ESU or not ESU (Crandall et al. 2000). Mostly for this reason, Moritz (1994) advocates using concepts from molecular phylogenetics (Graur and Li 2000) to define ESUs as historically isolated populations, but this view has difficulties as well (Crandall et al. 2000).

For example, the status of Kemp's ridley sea turtle (*Lepidochelys kempi*; Figure 7.1), whose numbers have declined drastically in the past several decades, was uncertain. Some researchers considered it a hybrid between the loggerhead turtle and another species, and others classified it as a subspecies of the more widespread olive ridley turtle. Thus, whether efforts to save the

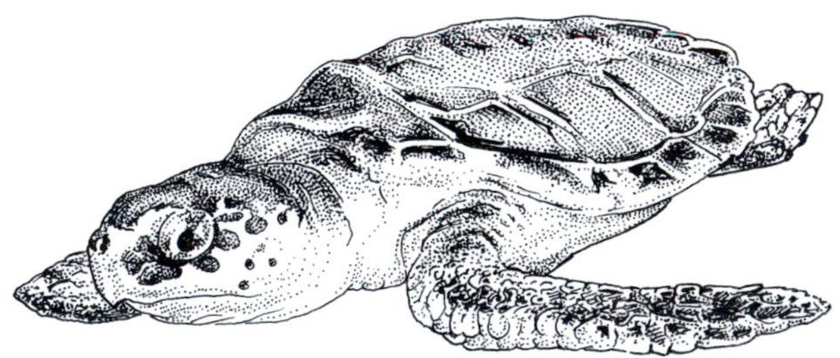

Figure 7.1 Kemp's ridley sea turtle (*Lepidochelys kempi*).

Kemp's ridley were justified depended on whether it was a separate ESU from these other species. Bowen et al. (1991) conducted an RFLP (see Chapter 2) analysis of the mitochondrial DNA (mtDNA) of these three species. They found all three were clearly genetically differentiated from each other, supporting efforts to preserve Kemp's ridley based on criteria one and two. However, since mtDNA is inherited primarily or exclusively from the mother, there may be extensive gene flow due to male movements that reduces differentiation at nuclear genes.

The opposite case is represented by the dusky seaside sparrow, a subspecies of seaside sparrow (*Ammodramus maritimus*) that went extinct in 1987. Using preserved DNA, Avise and Nelson (1989) showed that the dusky seaside sparrow was almost identical to other Atlantic coast seaside sparrows at a large number of mtDNA markers. Based on these data, the authors concluded that efforts to reintroduce dusky seaside sparrows from captive hybrids produced before 1987 had little purpose, again based on criteria one and two. Considering criterion three, it is possible that the darker plumage of the dusky sparrows was a local adaptation, but this is difficult to assess with extinct populations!

Some studies have evaluated all three criteria in defining ESUs. Cryan's buckmoth is found only in peatlands in the Great Lakes region of North America, and they feed only on the herb *Menyanthes trifoliata*. They are indistinguishable morphologically from related buckmoths, and they are not genetically differentiated from relatives at allozyme or mtDNA markers (Legge et al. 1996). However, Cryan's buckmoth is highly adapted to its host plant—they had 100% survivorship on *Menyanthes*, whereas two populations of close genetic relatives all died when reared on this plant. Therefore, the authors concluded that Cryan's buckmoth was an ESU based on criteria one and three. Gene flow seems to be sufficient to reduce differentiation at neutral markers but not sufficient to prevent local host adaptation.

How do genetic factors directly affect extinction risk?

The increase in homozygosity caused by true inbreeding and by random drift can cause the loss of fitness referred to as inbreeding depression (see Chapter 3). Inbreeding depression is common in nature, and crosses between inbred populations often restore fitness, i.e., produce heterosis (Keller and Waller 2002). The major concern in conservation is that habitat destruction or other human activities reduces effective population sizes of organisms so that heterozygosity decreases through random drift. For example, Heschel and Paige (1995) found lower fitness in smaller populations of scarlet gilia (*Ipomopsis aggregata*) compared to large populations, and that fitness increased when plants in the smaller populations were pollinated using

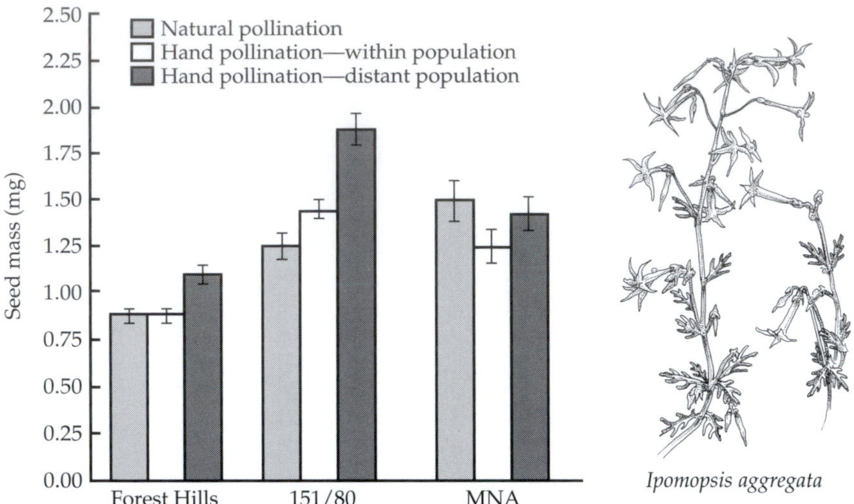

Figure 7.2 Mass of seeds resulting from different pollination treatments in scarlet gilia. Seeds produced from pollen brought in from distant populations were heavier than seeds produced from pollen from within the population (either natural or hand pollinated) in the two small populations (Forest Hills and 151/80). Pollen source did not affect seed mass in the large population (MNA). In no case did the mass of seeds produced by natural and hand pollination within populations differ significantly. (From Heschel and Paige 1995.)

pollen from distant populations (Figure 7.2). This "genetic rescue" effect (heterosis) did not occur when plants from larger populations were pollinated with distant pollen, because the larger populations were not suffering from inbreeding depression caused by random drift.

Richards (2000) found very similar results in white campion plants (*Silene alba*). Within isolated small subpopulations, sibling mating and outcrossing within patches resulted in similar fitness values (measured as germination success), while between-patch crosses increased fitness. In less isolated populations, sibling mating decreased fitness relative to plants produced by outcrossing either within or among subpopulations (Figure 7.3). These results can be interpreted using many of the concepts discussed in Chapters 2 and 3. All subpopulations lose variation due to random drift (including founder effects), and this variation is restored by gene flow in the less isolated populations but not in the more isolated populations. Therefore, true inbreeding does not cause additional inbreeding depression relative to outcrossing within isolated subpopulations, because the subpopulations already have increased homozygosity due to the fixation of deleterious recessive alleles by random drift. Crossing among isolated subpopulations restores heterozy-

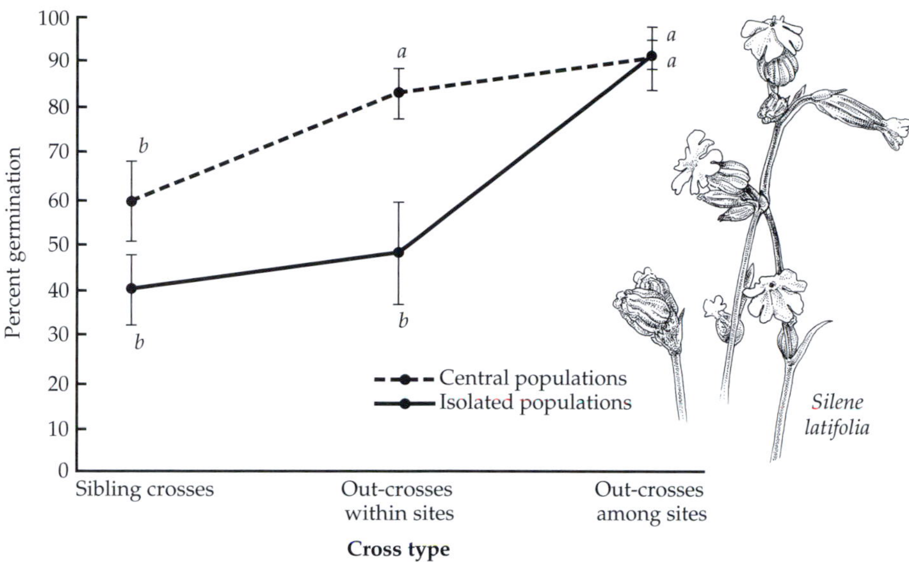

Figure 7.3 Percent germination of white campion seeds produced by three different crossing designs. Points marked with the same letter (*a* or *b*) are not statistically different from one another, but the points marked with *a* are significantly greater than the points marked with *b*. True inbreeding (sibling crosses) reduced fitness significantly in central but not isolated populations, whereas crosses among sites increased fitness significantly in isolated but not central populations. (After Richards 2000.)

gosity and thus creates heterosis. In the less isolated populations, natural migration replenishes the heterozygosity lost by random drift, so crossing among populations has little effect, but true inbreeding increases homozygosity and causes inbreeding depression.

Two parallel examples come from animals. Both adders (*Vipera berus*; Madsen et al. 1999) and prairie chickens (*Tympanuchus cupido*; Westemeier et al. 1998) have small populations that are isolated from larger populations of the same species, and in both species the isolated populations have lower levels of variation at neutral molecular markers and lower measures of fitness (low fertility, hatching success, and deformed offspring) consistent with inbreeding depression caused by random drift. The isolated populations in both species experienced steep declines in population numbers. All these negative effects were reversed by importing individuals from the larger, more genetically diverse, populations (Figure 7.4).

The message from these four examples seems clear: In small populations where fitness measures are lower than in larger populations, the introduction of individuals or pollen from larger natural populations can often ame-

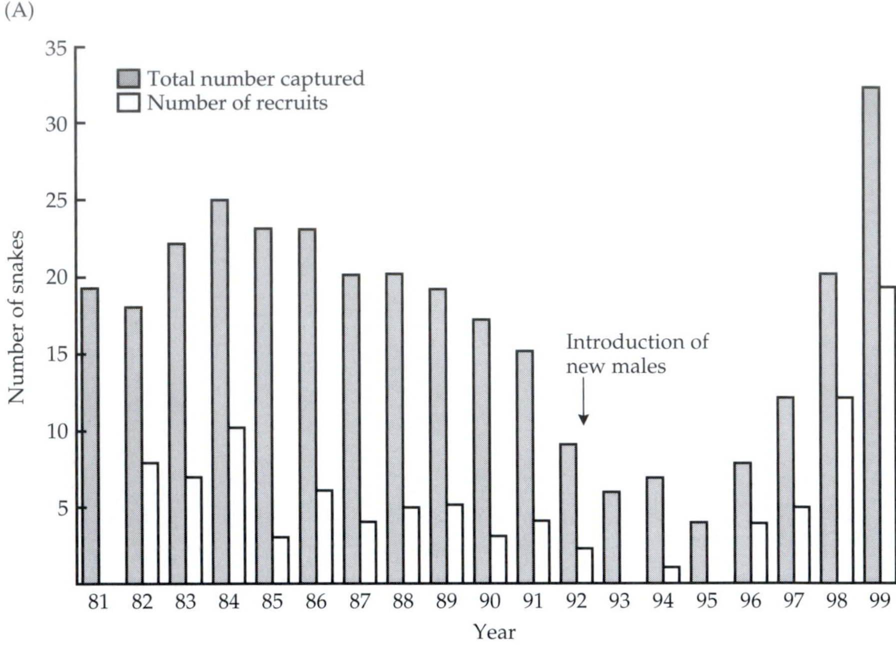

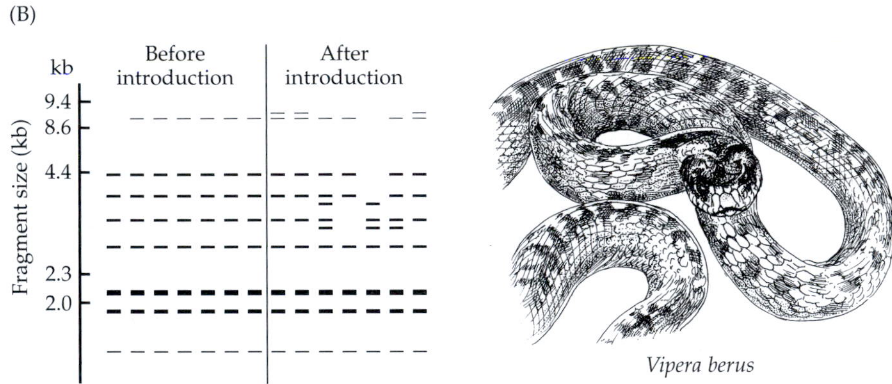

Figure 7.4 Genetic rescue in adders. (A) The number of snakes recorded in a small isolated population each year; the total number of snakes and the number of new snakes in the population (recruits) increased after the introduction of males from a large population. The time lag between the introduction and increase is due to the four years it takes for snakes to become reproductively mature. (B) Banding patterns from a Southern blot of major histocompatibility complex (MHC) class one genes showing the increase in marker variation after the introduction. Note that there are few differences in banding patterns among lanes (seven individual snakes) before the introduction. (After Madsen et al. 1999.)

liorate the effects of inbreeding depression. However, the source populations used must be chosen with care, because if the large and small populations are locally adapted to different conditions, it is possible that mixing them will lead to **outbreeding depression,** a decline in fitness from mating distantly related individuals (Lynch 1996). If the inbreeding depression is severe in the small populations, and only a few individuals are introduced from large natural populations in similar habitats, the benefits will likely outweigh the risks. Thus, the decision to mix two populations should be based on consideration of whether the populations represent genetically unique ESUs, as well as the relative likelihood of inbreeding and outbreeding depression. Note that inbreeding depression was much greater than any outbreeding depression in the four examples above.

These studies clearly show decreased fitness caused by small population size, and on average this decreased fitness will increase the risk of population extinction. The link between random drift and increased extinction risk was shown directly in a study of the annual plant *Clarkia pulchella* (Newman and Pilson 1997). Newman and Pilson planted 32 experimental populations of 12 plants into *Clarkia's* natural habitat in Montana. Half of the populations were constructed using unrelated individuals, while the other half of the populations consisted of three full-sibling families each. Therefore, all populations had the same actual population size ($N_a = 12$), but the populations made up of related individuals had a reduced N_e, as they were essentially founded by six individuals, the parents of the three full-sibling families. These lower N_e populations went extinct at a much higher rate over the three year study (Figure 7.5), likely due to inbreeding depression caused by random drift.

Based on the five studies discussed above and the concepts discussed in Chapter 3 (small population size, drift, and inbreeding depression), small population size and thus random drift can be expected to decrease population fitness and increase extinction risk on average. However, the studies above were all quite time consuming and involved manipulations that may not be possible in very rare endangered species (in fact, none of the five examples above involved an endangered species). For these reasons, conservation biologists, who are faced with a large number of endangered species, have usually measured variation in neutral molecular markers as a more convenient way of assessing the risk of extinction due to inbreeding depression. Random drift increases homozygosity by driving allele frequencies towards fixation, and the increased homozygosity tends to create inbreeding depression through expression of deleterious recessives and in some cases loss of heterozygote advantage (see Chapter 3). Therefore, conservation biologists often use H, the frequency of heterozygous individuals (see Chapter 2) averaged over several neutral markers, as an indicator of population fitness.

The problem with using neutral markers to assess fitness is that, because drift is a random process, it increases homozygosity across all loci *on average*.

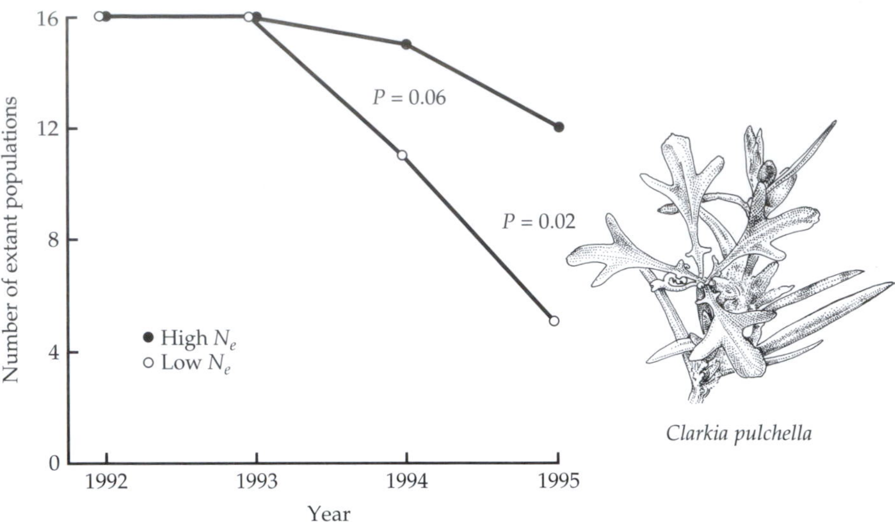

Figure 7.5 Survival curves for *Clarkia* populations established with the same actual size (N_a) but different effective sizes (N_e). Note that the small N_e populations went extinct at a significantly greater rate than the larger N_e populations. (After Newman and Pilson 1997.)

Homozygosity will be *decreased* by random drift at some fraction of individual loci. Because a reasonably sized set of neutral markers can only cover a very tiny fraction of any organism's genome, they might often not provide a good estimate of overall levels of inbreeding depression due to random drift. In other words, is there a relationship between estimates of neutral marker diversity (e.g., H) and population fitness/extinction risk? Based on current knowledge, the answer seems to be yes, but the relationship is fairly weak.

It has been shown that heterozygosity of neutral markers is often correlated with higher fitness (Mitton 1997). Schmidt and Jensen (2000) studied 13 subpopulations of the rare plant *Pedicularis palustris*. They found that subpopulation variation at 167 polymorphic AFLP loci was closely related to subpopulation mean fecundity and recruitment, and that this relationship was closer than the relationship between the AFLP variation and current census size. These results suggest that random drift has reduced both AFLP variation and fitness in some subpopulations, and that N_a is a poor predictor of N_e and thus the magnitude of random drift in these subpopulations. In Glanville fritillary butterflies (*Melitaea cinxia*) there was a significant negative relationship between the probability of extinction and average heterozygosity at seven allozyme and one microsatellite locus (Saccheri et al. 1998). Of 42 populations studied, the seven that went extinct had significantly lower H than the 35 surviving populations, even after correcting for differences in current actual population size and several other relevant ecological variables.

These individual examples seem convincing, but when many studies are analyzed using **meta-analyses,** the overall relationship between *H* and fitness is weak. (Meta-analysis is a statistical technique for jointly analyzing the results of many studies on the same topic.) Reed and Frankham (2003) found that the correlation between heterozygosity at molecular markers and population mean fitness was significantly greater than zero, but that the R^2 (see Chapter 4) of this relationship was only 0.2, meaning that only 20% of the variance in fitness can be explained by heterozygosity of neutral markers. Therefore, based on current evidence, it seems that measurements of *H* at molecular markers provide only a weak prediction of population fitness.

An additional genetic threat to small populations is new mutation. The whole-genome mutation rate (*U*) for fitness in *Drosophila* has been estimated to be about one per individual per generation, meaning that every individual has one new mutation somewhere in its genome that decreases fitness (Lynch et al. 1999). Selection will primarily determine allele frequencies at a locus if $s > 1/2N_e$, which means that as effective size of a population declines, mutations with larger deleterious effects (larger *s*) can be fixed by random drift rather than being removed by selection (see Chapter 3). This phenomenon could create a positive feedback loop, in which fixation of deleterious mutations causes a further decline in N_e, which further increases the fixation rate of deleterious mutations, and so on. This process has been called a **mutational meltdown** (Lynch 1996). The importance of mutational meltdowns in causing population extinctions in nature is not well known, but the threat from mutation accumulation seems to be weaker than other genetic factors in experimental populations of yeast and *Drosophila* (Gilligan et al. 1997).

Adaptation to environmental change in the future

A final question addressed by conservation genetics is whether a population will be able to adapt to future environmental change. The environmental change could be natural, but more worrisome are the much more rapid environmental changes caused by humans, such as habitat fragmentation and degradation, global warming, and introduction of invasive species. If key phenotypic traits in populations are unable to adapt to rapid change, the populations will likely go extinct. Recall from Chapters 5 and 6 that the rate of evolutionary change in mean phenotype (*R* or $\Delta\bar{z}$) depends both on the strength of selection (*S* or $\boldsymbol{\beta}$) and the amount of additive genetic variance and covariance (h^2 or $\mathbf{G}$). If environmental change is rapid, then this may cause rapid movement of adaptive peaks. When peaks move, populations may no longer be at an adaptive peak experiencing weak stabilizing selection, but rather on the slope below a peak experiencing strong directional selection (see Figure 6.5). Given the high likelihood of strong selection caused by environmental change, the main determinants of the ability of a population to respond to change are whether there is adequate genetic variance for rapid evolution, and how long this variance will be maintained in the face of continued strong selection.

Evidence from artificial selection studies (see Chapter 5) suggests that evolution in response to novel selection can be both rapid and sustained over at least several dozen generations, and continued selection can change the mean phenotype five or ten standard deviations (Lynch 1996). These studies are relevant to conservation genetics because the experimental populations tend to have small effective population sizes, mimicking the small subpopulation sizes of threatened and endangered species. However, because the artificial selection lines are typically founded with abundant genetic variation, these results are most relevant to subpopulations that have only recently declined in numbers. Therefore, these results provide some hope for endangered populations over medium time frames, but there may well be many populations that have lost much of their additive genetic variance through random drift. For adaptive evolution to continue over the long term (hundreds of generations of selection), or in populations that have lost genetic variation through random drift, new mutation or migration from differentiated populations is necessary to replenish the additive genetic variation.

How can we determine which populations harbor adequate additive genetic variation to respond to novel selection? The best test would be to use artificial selection, offspring–parent regression, or half-sibling analysis to directly measure additive variance for traits needed to adapt to the novel environment. Artificial selection is a more direct and powerful test, but only one trait at a time can be tested, in contrast to offspring–parent and sibling analysis (see Chapter 5). Unfortunately, these kinds of studies are time-consuming, and it is often difficult to determine which quantitative traits will be crucial in adapting to environmental change. In addition, the large sample sizes needed for these studies and the need for controlled breeding may be difficult or impossible to achieve with many rare and endangered species. For these and other reasons, conservation geneticists have often used neutral molecular markers to assess genetic variation for future adaptation.

Recall that, on average, random drift caused by small effective population size should reduce genetic variance at all loci (see Chapter 3), so reduced variance of neutral markers due to random drift should be reflected in reduced variance in quantitative traits as well. However, current evidence suggests that the correlation between heterozygosity (H) at neutral markers and heritability of quantitative traits is even weaker than the correlation between H and fitness previously discussed. In fact, a meta-analysis of 19 studies found a slightly negative (but nonsignificant) correlation between H

and heritability for a variety of traits (Reed and Frankham 2001). One possible reason for the lack of correspondence is the higher mutation rate for quantitative traits as compared to single marker loci. The higher mutation rate is expected because quantitative traits are affected by several to many gene loci (see Chapters 4 and 5), so that a mutation at any one of the loci will create variation for the trait.

A test of the ability of a population to adapt to future environmental change was conducted on the model (nonendangered) plant *Arabidopsis thaliana*. Ward et al. (2000) selected for increased seed production for five generations in *Arabidopsis thaliana* at two concentrations of CO_2, one low and one high. The low CO_2 concentration mimicked the low levels that occurred in the Pleistocene era, whereas the high concentration represented predicted levels by the end of this century due to human fossil fuel use. Ward et al. then performed a reciprocal transplant experiment with their selected lines, planting the final generation of each at both CO_2 concentrations, after first passing them all through a generation at intermediate CO_2 to minimize maternal effects (a critical consideration for all common garden and reciprocal transplant studies; see Chapter 5). They found that both lines produced the most seed at the concentration they had been selected in (Figure 7.6), but the difference was statistically significant only when the plants were tested at the lower concentration. The latter result indicates that *A. thaliana* can rapidly adapt to decreases in CO_2, but may have difficulty adapting to future increases.

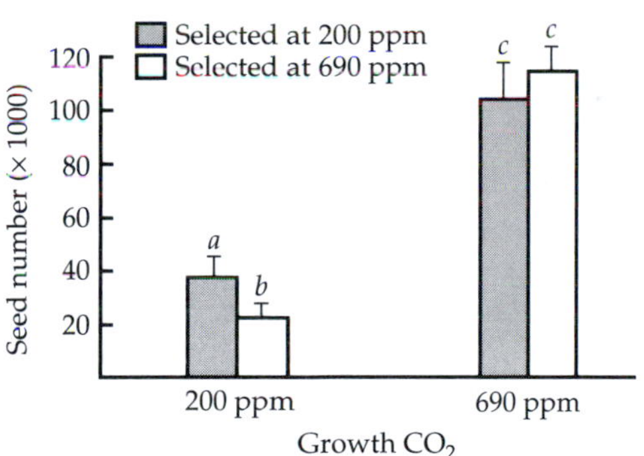

Figure 7.6 Adaptation to increased and decreased concentration of CO_2 in the plant *Arabidopsis thaliana*. Shown are fitness measures (number of seeds produced) by plants selected for five generations under low CO_2 (gray bars) and high CO_2 (white bars) when grown under low (200 ppm) and high (690 ppm) CO_2 concentrations. Plants selected at low CO_2 produced significantly more seeds than plants selected at high CO_2 under low CO_2 conditions (left pair of bars), while the difference at high CO_2 was not significant (right pair of bars). (After Ward et al. 2000.)

Thus far we have focused on within-subpopulation variation, which is appropriate because it is within subpopulations that adaptation to the environment takes place. However, another source of genetic variation for adaptive evolution is among-population genetic differentiation for quantitative traits. If genetic variation is lost due to random drift or continued selection, then migration from differentiated populations can restore the variation. The migration could be natural or controlled as part of a management plan. The best methods for determining genetic differentiation for quantitative traits (Q_{ST}) are common garden and reciprocal transplant experiments (see Chapter 5). However, like measurements of additive variance, these studies are time consuming and require large sample sizes and therefore may be difficult or impossible with rare species. Once again, conservation geneticists have turned to neutral genetic markers to study differentiation, typically by estimating F_{ST} (see Chapter 3).

But are F_{ST} and Q_{ST} correlated? That is, is differentiation of neutral markers related to differentiation for quantitative traits? If so, then neutral markers could help determine the overall degree of among-population quantitative variation. Unfortunately, the current evidence suggests that, like the relationships between H and fitness and between H and heritability discussed above, the correlation between F_{ST} and Q_{ST} is weak. In a meta-analysis of published data on 29 species, the R^2 of the relationship between F_{ST} and Q_{ST} was 0.14, so that only 14% of the variance in Q_{ST} is explained by F_{ST} (Latta and McKay 2002; McKay and Latta 2002). Other authors believe the data suggest a stronger relationship, with R^2 estimates as high as 0.35 (Crnokrak and Merilä 2002; Merilä and Crnokrak 2001), but even with the highest estimate about two thirds of the variance in Q_{ST} cannot be predicted by F_{ST} for neutral markers. As we saw in Chapter 5, when there is adaptive differentiation at quantitative traits, Q_{ST} is higher than F_{ST}; therefore, a lack of neutral marker differentiation may not mean that there is a corresponding lack of adaptive differentiation.

Conservation genetics in the future

Based on current information, it seems clear that conservation geneticists should rely less on neutral markers than they have in the past. Although measurements of population mean fitness, additive variance for quantitative traits within subpopulations, and genetic differentiation for quantitative traits among subpopulations are more difficult and time-consuming than population studies of neutral markers, the former provide greater insight into the current and future prospects of a population. One promising new method is to use molecular markers to estimate relatedness among individuals in an undisturbed natural population, and then use the relatedness measures, rather than controlled breeding, to estimate the additive genetic variance (Ritland 2000). Another possibility is to use molecular markers that have important effects on fitness or are linked to loci affecting fitness. One example of this is the major histocompatibility complex (MHC); variation at the MHC is critical to successful immune response in vertebrates (see pp. 207–210 in Frankham et al. 2002).

Evolution of Invasive Species

One of the major threats to the global environment is the spread of **invasive species,** which we define as non-native species that increase rapidly in numbers and have negative impacts on native species. Invasive species have been estimated to have a negative impact on about half of all threatened and endangered species in the U.S. (Wilcove et al. 1998). They also directly impact humans, with environmental and economic costs caused by invasive species in the United States alone estimated to be as high as 137 billion dollars per year (Pimentel et al. 2000). Most **introduced** (nonnative) species do not become invasive, and attempts to explain why a species becomes invasive have focused almost exclusively on ecological factors. For example, the most common explanation for invasiveness is that the species has few natural enemies (herbivores, pathogens) in the new habitat, but the evidence for this is equivocal (Keane and Crawley 2002). Another common idea is that invaders have high levels of phenotypic plasticity (see Chapter 5), and therefore can succeed in many habitats without adapting to them genetically (Parker et al. 2003). However, the observation that many invasive species do not become invasive for a least a few decades after they are introduced into a new region suggests that rapid evolutionary change could be important as well (Ellstrand and Schierenbeck 2000). An introduced species, although perhaps escaping some of its natural enemies, is still unlikely to be well adapted to other aspects of its new range such as climate, pollinators, nesting sites, food sources, and so on. Thus, an introduced species may need to evolve adaptations to these novel aspects of the nonnative habitat before it can take advantage of its lack of natural enemies and become invasive.

There are now a number of examples of evolutionary change in nature over very short time scales (Reznick et al. 1997; Thompson 1998). As always, the two determinants of the rate of phenotypic evolution are the amount of genetic variation and the strength of selection. When they are first introduced into a new region, introduced species may lose genetic variation through random genetic drift caused by founder effects (see Chapter 3). The loss will be minimized if the population sizes grow rapidly after introduction, shortening the period of the bottleneck, but rapid growth depends upon the nonnative species being already well-adapted to its new habitat.

Genetic variation can be augmented if the species is introduced multiple times, which may be a common precondition for species to become invasive (Ellstrand and Schierenbeck 2000). For example, some invasive plants were first introduced as garden ornamentals and continue to be imported for this use. Other invasives are common pests of grain crops shipped around the world, and still others are well-adapted to being moved undetected during shipping. If the populations founded by different introductions are connected by gene flow, then the multiple introductions will replenish genetic variation within populations lost through random drift (see Chapter 3). Another source of genetic variation, especially in plants, might come through hybridization between the introduced species and a closely related native species (Ellstrand and Schierenbeck 2000). Genes from native plants might help the invader adapt to its new habitat.

If there is adequate genetic variation, then evolution is likely to be rapid in a newly introduced population due to strong natural selection. Natural selection will be strong because the population will likely find itself farther from its adaptive peaks for many traits (see Chapter 6) for two reasons. First, the novel environment in the new location will mean the shape of the adaptive landscape is different from that in the native habitat. Second, drift due to founder effects may cause the mean phenotypes of the populations to change. Mean phenotype may also change due to plasticity in response to the new environment (see Chapter 5). The prediction of rapid evolution in introduced species is supported by the fact that most of the known examples of rapid evolution in nature involve some major change in the habitat (Thompson 1998). For example, in Chapter 3 we discussed the rapid evolution of melanism in moths in response to changes in air quality in Britain.

Common garden and reciprocal transplant studies (see Chapter 5) are good approaches to determining the role of rapid evolution in invasiveness. For instance, placing samples of populations from the native and nonnative habitats together in the nonnative habitat would test whether the non-native populations have higher fitness in their new habitat. If so, then this is consistent with rapid adaptation to the new habitat, but there are at least two alternatives. First, it could be that the populations sampled from the native range were not the source population for the invasive, and that the invasive species was derived from a population that was already pre-adapted to the nonnative habitat. This possibility could be minimized by studying a number of populations from across the native and nonnative ranges, but in a widespread species it will be difficult to rule out this alternative. In cases where the species was introduced deliberately, there may be records of where the source population came from. Finally, molecular genetic markers could be used to help identify the source populations of the invasives, although this may prove difficult in practice.

The second alternative hypothesis to rapid adaptation of the invasive species is that differences between native and nonnative populations are due to rapid evolution of the *native* populations after the introduction, rather than adaptation of the invasive. This seems unlikely because evolutionary changes are expected to be most rapid after a large habitat shift. Still, this alternative cannot be entirely ruled out, because natural or human alteration of the native population's habitat could have occurred. If viable samples of the original invaders are available, then in might be possible to rule out both alternative hypotheses. While uncommon, in some cases seeds, resting eggs, spores, or other persistent life stages of organisms might be present in the environment or preserved in museum collections. If so, then they could be compared to both the current native and current nonnative species in common garden studies. Even if not viable, DNA from museum specimens could be used in molecular genetic studies to identify source populations and determine the likelihood that these were samples from the populations that gave rise to the current invasives.

As an example, Blair and Wolfe (submitted) collected seeds from 20 native European and 20 introduced North American populations of white campion plants (*Silene alba*). These collections spanned much of the range of this plant on the two continents. Samples from all 40 populations were grown in two common gardens, one in the greenhouse and one in the field in Virginia. On average, the North American populations had higher fitness than their European counterparts (Figure 7.7). This finding suggests that rapid adaptation to North America has occurred, particularly because there is evidence for multiple introductions of *Silene* into North America, potentially increasing genetic variation in the nonnative populations. However, a few of the European populations had an average fitness value as high as the highest of the North American populations (see Figure 7.7), so the possibility of the successful nonnatives coming from these pre-adapted populations cannot be ruled out.

As described above, common garden or reciprocal transplant studies can help determine if adaptation to the new habitat has occurred because these experiments test for higher fitness in the introduced populations relative to the natives. If a number of phenotypic traits are measured during the experiments, this approach can also identify which traits have become genetically differentiated between the native and introduced populations. The differentiated traits may be those responsible for the rapid adaptation. To test whether rapid adaptation of a few key traits is responsible for the invasiveness of an introduced species, one can measure selection on these traits (see Chapter 6) in the native and introduced habitats. If there is differential selection in the two habitats that matches the differentiation, then these observations provide strong evidence for adaptive differentiation.

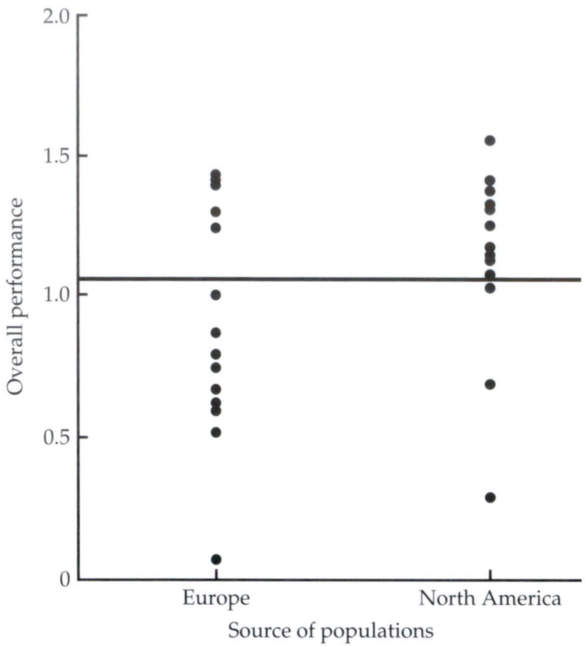

Figure 7.7 Fitness of *Silene alba* plants from Europe and North America grown in a common garden. Each point represents the mean fitness of plants from one source population, and the horizontal line shows the overall mean fitness for all populations. (From Blair and Wolfe, submitted.)

The role of evolutionary processes in invasiveness is only beginning to be investigated, with little information available on the most problematic species. This situation is likely to change in the near future as interest in this topic increases. One concern with conducting common garden or reciprocal transplant studies on destructive invaders is the possibility that the experiment itself will worsen the invasion by causing artificial migration (Parker et al. 2003). However, if conducted with proper care, the approaches of ecological genetics are likely to lead to important new insights into the problem of invasive species.

Transgene Escape

Recently, techniques for moving genes between species (**transgenics**) have been applied to the development of new crop varieties. The results of gene transfer are often referred to as **genetically engineered (GE)** or **genetically modified organisms (GM** or **GMOs).** In spite of considerable political resistance to transgenics, particularly in Europe, the area of fields planted with transgenic crops is rapidly increasing worldwide, from 11 million hectares in 1997 to 59 million hectares in 2002 (James 2002). Although for centuries plant and animal breeders have used hybridization between closely related species and artificial selection (see Chapter 5) to genetically alter domesticated species, the novel ability to move genes between very distantly related species could lead to new environmental threats. For example, genes that code for the production of insecticidal toxins in the bacterium *Bacillus thuringiensis* (*Bt*) have been transferred into several crop plants to protect them against insect herbivores. If these genes should become established in weedy species closely related to the crop, it could create more damaging and invasive weeds that are less controlled by insect herbivory. This scenario is similar to an introduced species becoming invasive due to escape from its herbivores (Marvier 2001).

There are two steps in the process of **transgene escape** into wild populations. First, there must be gene flow from the crop to a related wild species. Second, the transgene needs to become established in the environment; in most cases, establishment will only occur if the transgene confers a fitness advantage to the wild species. Based on current evidence, it is clear that gene flow from crops to related natural populations will occur in many cases (Hancock et al. 1996). However, how often transgenes will become established in the environment is less clear at present, mainly due to the small numbers of studies that have addressed this question, but most information suggests that transgenes will become established in some cases. The likelihood of establishment is a critical question, because once a transgene has become established in natural populations, it will be difficult or impossible to eradicate. Most of the evidence for escape of crop genes comes from nontransgenic crops.

There is extensive evidence for mating between nontransgenic crops and wild relatives in the same species or genus, and such mating leads to gene flow from the crops to natural populations. Evidence for crop to wild population gene flow has been reported in 12 of the world's 13 most important crops (Ellstrand et al. 1999). Studies of both radish and sunflower have shown mating between crop and wild plants separated by as much as a kilometer (Arias and Rieseberg 1994; Klinger et al. 1991). Crop-specific RAPD markers were found at frequencies over 0.3 in three populations of wild sunflower growing near sunflower crop fields (Linder et al. 1998). Transgenes for herbicide resistance were found in nontransgenic cultivars of canola (*Brassica napus*) almost 3 kilometers from the nearest transgenic canola fields (Rieger et al. 2002). Note that radish, sunflower, and to a lesser extent canola, are all highly outcrossing plants—gene flow distances could be much shorter for highly inbred crops (see Chapter 3).

Establishment of transgenes in wild populations

Given that gene flow from crops to wild relatives is likely, the next question is whether crop genes will become established in the wild populations. This depends first on the fitness of the hybrids between the crop and wild plants. Because crops have been artificially selected for traits that might be disadvantageous in the wild, and because sometimes the wild plants are a different species than the crop, we might expect that the hybrids would be less fit than the wild plants or even sterile. However, most studies suggest that crop-wild hybrids are not dramatically less fit than the wild species (Whitton et al. 1997). For example, F_1 hybrids between nontransgenic crops and weeds had equal or higher fitness than pure weeds in studies of radish and *Sorghum* (Arriola and Ellstrand 1997; Klinger and Ellstrand 1994), and lower but still substantial fitness in sunflower (Snow et al. 1998).

Therefore, it seems likely that fertile hybrids will be formed between crops and wild relatives quite often, but will the crop alleles persist in natural populations? If the crop continues to be planted near the wild populations, the crop alleles will likely persist in the wild simply through continued gene flow from the often very large numbers of crop plants. Even without continued gene flow, the limited evidence available suggests that crop alleles will persist, at least for a few generations. For example, Whitton et al. (1997) studied a population of wild sunflower growing adjacent to a field that was planted with cultivated sunflower for one year only. They found two RAPD markers that occurred in the crop but not in the wild plants prior to the time the crop flowered. In the next generation these crop markers were both present in the wild population. The frequencies of these markers in the wild plants remained fairly constant over the first five generations after the crop was present, with a possible increase in frequency of the crop marker alleles at generation five (Figure 7.8). Similar results were obtained in a study of crop radish (*Raphanus sativus*) experimentally hybridized with wild radish (*R. raphanistrum*). Equal numbers of hybrids and wild radish were planted in four experimental gardens, and crop alleles were still present at substantial frequencies in all populations three years later (Snow et al. 2001).

In the first study of persistence of a transgene in a wild plant, Mikkelsen et al (1996) grew oilseed rape (*Brassica napus*) with a transgene for herbicide resistance in field plots with a weedy relative, *B. rapa*. Fertile hybrids containing the transgene were spontaneously produced, and the transgene persisted through two generations of mating with pure weedy plants.

Fitness effects of transgenes

As we have seen, the evidence clearly indicates that crop genes will often escape and become established in wild relatives, at least for a few generations. However, long term persistence of a transgene is likely to be governed

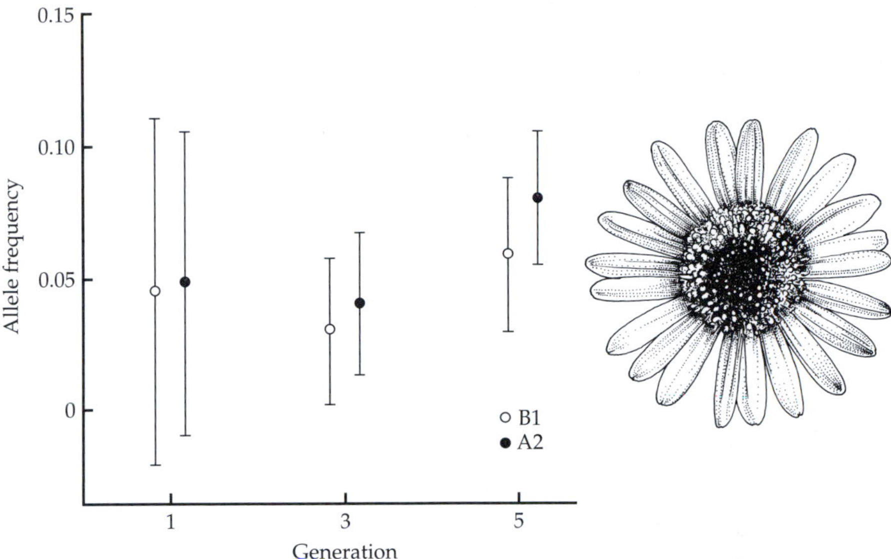

Figure 7.8 Frequencies of two crop-specific RAPD markers in a wild sunflower (*Helianthus annuus*) population at one, three, and five generations after a single planting of crop sunflowers in an adjacent field. (From Whitton et al. 1997.)

by the fitness effects of the transgene itself. If the transgene increases fitness of the wild plants, and the selection coefficient (see Chapter 3) is large enough, then the transgene is likely to persist and not be lost through random drift. If the transgene is deleterious in wild plants, selection will remove it from the natural population.

Evidence for fitness effects of transgenes in wild plants to date is limited and not consistent for different transgenes. The lack of consistency is not surprising, because selection on transgenes will depend on the specific phenotypic effect of the transgene as well as on the environment. A transgene in sunflowers (*Helianthus annuus*) conferring resistance to a mold did not significantly increase the fitness of wild sunflowers planted in California, Indiana, and North Dakota, even when plants were inoculated with the mold (Figure 7.9A; Burke and Rieseberg 2003). The transgene also did not significantly reduce the fitness of plants that were not inoculated (Figure 7.9B), so the transgene was apparently neutral in these sites, and its frequency and ultimate fate would be determined solely by random genetic drift (see Chapter 3).

In contrast, wild sunflowers with a *Bt* insecticidal transgene experienced reduced insect herbivory and increased seed production at field sites in

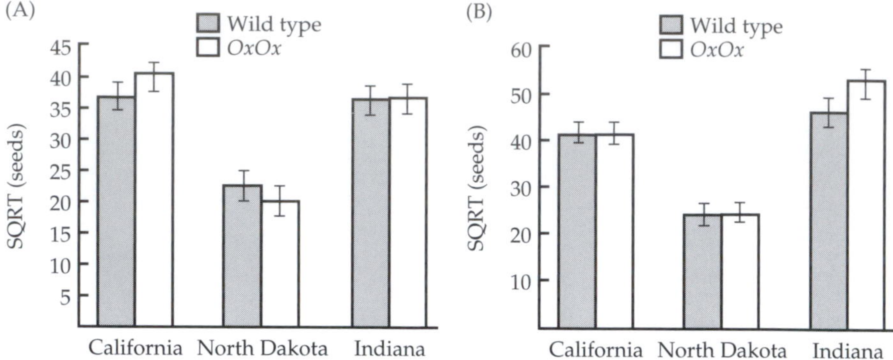

Figure 7.9 Fitnesses of hybrids between crop and wild sunflowers (*Helianthus annuus*), some with a transgene (*OxOx*) for resistance to white mold (*Sclerotinia sclerotiorum*) and some without (wild type). Fitnesses did not differ significantly between the two groups of plants, regardless of whether they were inoculated with white mold (A) or not (B). (From Burke and Rieseberg 2003.)

Nebraska and Colorado, although the increase in seed number was statistically significant only in Nebraska (Snow et al. 2003). This transgene seems highly likely to become established in wild sunflowers if sunflower cultivars containing it are widely planted. Because wild sunflower is sometimes a problematic weed, the *Bt* transgene may increase its weedy characteristics.

Strategies to reduce transgene escape risk

The simplest strategy to reduce the risk of transgene escape is to avoid planting transgenic crops in areas where wild relatives occur. Because gene flow between crops and wild relatives can occur over distances of at least a few kilometers, the distance to the nearest relative in the same genus should be much greater. In field tests of transgenic crops, border rows of nontransgenic plants of the same species have often been planted to "trap" pollen of the transgenic plants. However, in a study of cucumber, an outcrossing crop, border rows were effective only if there were many more trap plants than "donor" plants (Hokanson et al. 1997), so this strategy is not likely to be useful for large fields of outcrossing crops. The "isolation by distance" and border-row strategies could be more effective in highly inbred crops, but this strategy has not been thoroughly tested.

In addition to spatial isolation, it might be possible to isolate transgenic crops temporally by planting them so they do not flower at the same time as their wild relatives (Ellstrand and Hoffmann 1990). In species for which the edible part is not a fruit, but rather a vegetative part of the plant (e.g., radish, broccoli, lettuce, celery), it might be possible to harvest all the plants prior to flowering (Ellstrand and Hoffmann 1990).

Finally, it may be possible to make genetic changes in the crop to minimize transgene escape. Conventional breeding could be used to make crops more highly selfing or less compatible with wild relatives (Ellstrand and Hoffmann 1990). For vegetative crops, cultivars that are male sterile could be produced. A highly controversial technique under development, called either the "technology protection system" or "terminator technology," causes seed sterility in the offspring of crop plants (Daniell 2002). This could be an effective method of reducing transgene escape, but it prevents farmers from saving seeds to plant the next year's crop, which is a particular problem for subsistence farmers in developing nations. A potential ecological problem could occur if small populations of related plants adjacent to crop fields were overwhelmed with pollen containing the terminator genes; the resulting high seed sterility could drive the native population towards extinction. Other genetic strategies for containing transgenes are reviewed in Daniell (2002).

Evolution of Resistance to Pesticides and Antibiotics

Another major problem humans face is the evolution of pest species' resistance to the agents (usually chemical) used to control them. Most attention has been paid to three groups: microbial pathogens (i.e., the viruses, bacteria, fungi, and other organisms that make us sick); insects that feed on our crops and transmit diseases; and to a lesser extent weeds that compete with crops. Because the development of resistance represents nothing more than adaptive evolution of the pests in response to selection exerted by the chemical control measures (e.g., antibiotics, pesticides, herbicides), the concepts of ecological genetics are central to understanding and managing this problem. The problem of resistance is a major one. The U.S. Food and Drug Administration (FDA) estimates that antibiotic resistance costs 4–5 billion dollars per year in the U.S. alone (Amabile-Cuevas 2003), plus the incalculable cost in lost lives. Insecticide resistance has been estimated to cause yearly losses of 3–8 billion dollars worth of food crops in the U.S. (Palumbi 2001b). The number of resistant pest species in agriculture has increased steadily with time (Figure 7.10), and resistance to pesticides almost always appears within 25 years (and sometimes within one year) after a new pesticide is introduced (Palumbi 2001b).

For simplicity, in this section we will use the term "pest" to refer to all the organisms that have negative impacts on human health and welfare (e.g., pathogens, harmful insects, weeds) and the term "pesticide" to refer to the

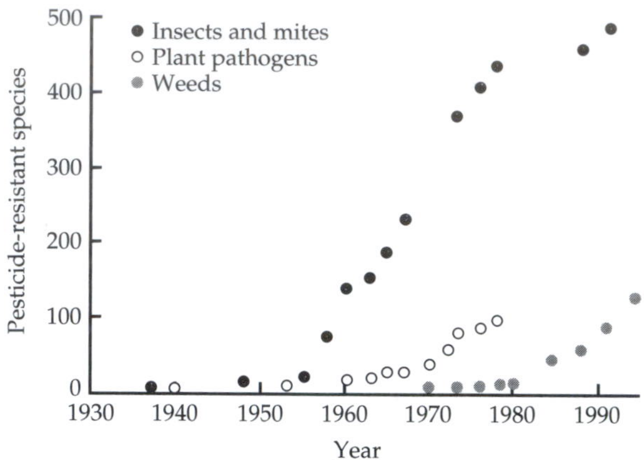

Figure 7.10 Increase over time in the number of crop pests that are resistant to pesticides. (After Gould 1991, and Heap 1997.)

chemical agents we use to control them (e.g., antibiotics, insecticides, herbicides). Note that in the literature, pesticide is often used to mean insecticide. The reason for our use of more general terms is that many of the principles are the same regardless of the particular pest or pesticide; however, there are differences, especially due to biological differences among pests, and some of these will be noted. As in the case of conservation genetics, most work on resistance has been based on single-locus population genetics, and we will use the common practice of using the symbol S to refer to an allele that makes the pest susceptible to the pesticide and R to refer to an alternative allele that confers resistance to the pesticide. There are some cases in which quantitative genetics is more useful than single-locus models, and these will be discussed where appropriate.

A major theme of this book is that the rate of adaptive evolution depends on the strength of selection and the amount of genetic variation for the trait. These two factors can, in turn, be affected by mutation, genetic drift, gene flow, and genetic correlations among traits. Therefore, we will discuss how each of these forces affects the evolution of resistance, and how a knowledge of these forces can help in the management of pest resistance.

Selection

The underlying cause of the evolution of resistance to pesticides is the selection exerted by the pesticide itself. By definition, pesticides kill (or at least reduce the fitness of) susceptible individuals, creating selection against susceptibility and for resistance. In general, the strength of the selection increases with the dose and potency of the pesticide. The main exception to this principle occurs when strong pesticide application is successful in eradicating the pest, or at least removing the *R* allele provided resistance is due to a single locus. Eradication is sometimes possible with great effort, but this outcome is very rare through the use of pesticides alone. Examples of anti-pest techniques that are sometimes used in concert with pesticides include vaccines, quarantines, natural enemies (predators or parasites) of the pest, hand picking of insects, and hand weeding or cultivation.

For example, in the U.S., antibiotic courses of treatment for minor infections are typically fairly long (10–14 days). Because this time period often continues past the time when the disease symptoms cease, it is common for patients to stop taking the antibiotic before the prescribed course is finished. Patients are urged to complete the treatment to minimize the evolution of resistance in the pathogens. This recommendation is based on two assumptions—first, that the full course of antibiotic will eradicate all the pathogens, or at least all those carrying *R* alleles; and second, that the patient's own immune system will not eradicate all the *R* individuals after the early cessation of treatment. In tuberculosis (which is *not* a minor infection) these assumptions appear to be true, because premature cessation of antibiotics leads to the evolution of resistant strains within the patient (Palumbi 2001b). Whether these assumptions hold in other diseases in not well known, because at present our knowledge of these and other key parameters in the evolution of antibiotic resistance remains rudimentary (Levin and Anderson 1999).

When resistance is due to a single locus, then the level of dominance of the *R* allele is crucial, because the *R* allele is initially at low frequency when it is introduced into the population through new mutation or migration. Recall from Chapters 2 and 3 that completely recessive beneficial alleles (the *R* allele is beneficial to the pest in the presence of pesticide) are almost invisible to selection because they are present almost entirely in heterozygotes (see Figure 2.10). Beneficial recessives are often lost through drift before they can increase in frequency enough to be present in an appreciable number of homozygotes. Resistance to insecticides seems to be at least partially recessive in most cases, leading entomologists to make use of this fact in managing resistance (Gould 1995). If the *R* allele is additive or dominant with respect to the *S* allele, then the beneficial effects occur in the heterozygotes, and the *R* allele can increase in frequency more quickly (see Figure 3.8). Herbicide resistance in weeds is typically either additive or partially dominant, complicating management efforts (Gould 1995). Because most microbial pathogens are bacteria and viruses, and therefore haploid, dominance is irrelevant. However, some pathogenic fungi and protists are diploid.

In Chapter 3 we saw that one way to accelerate the rate of increase of a beneficial recessive allele is through true inbreeding or random genetic drift, because both of these processes increase the frequency of homozygotes. Many weeds are at least partially selfing (Barrett 1988), and gene flow distances are usually limited in plants (e.g., Schaal 1980), which means that relatives are clustered and frequently mate with each other. This inbreeding could increase the rate of recessive resistance evolution in weeds. Insects cannot self-fertilize and tend to disperse widely, so inbreeding is less likely to be a factor for these pests. Note that the general pattern of recessive resistance in insects and additive or dominant resistance in weeds is the opposite of what would be predicted based on the mating systems of these pests; the reasons for this paradox are unknown.

Another key aspect of selection in understanding the evolution of resistance is the **cost of resistance,** defined as the fitness effects of the *R* allele (in both the homozygous and heterozygous state) in the absence of the pesticide. If resistance costs are present, then ceasing use of the pesticide for a time will cause the frequency of the *R* allele to decrease, and this strategy can be incorporated into resistance management. Based on current evidence, however, strong costs of resistance seem to be rare (Gould 1995). In most cases, insecticide resistance is only moderately costly in insects, and the costs are usually recessive (that is, present only in *RR* homozygotes). Therefore, the *R* allele could be maintained in heterozygotes for a long time in the absence of the insecticide. In weeds, the costs of resistance seems to be somewhat higher than for insects, so that this attribute might be useful in management (Gould 1995).

Costs of resistance to antibiotics are common in bacteria, but **compensatory mutations** or other genetic changes often occur in resistant strains, so that the resistant strains can develop equal or higher fitness than susceptible strains in the absence of antibiotics (Bouma and Lenski 1988). A similar compensatory genetic change eliminated the cost of resistance to the insecticide diazinon in the Australian sheep blowfly (McKenzie et al. 1982). Note that these are examples of epistasis for fitness (see Chapters 3 and 4), because the cost of the *R* allele is modified by compensatory changes elsewhere in the genome. In such cases, eliminating pesticide use will not by itself cause the frequency of *R* alleles to decrease. In spite of the laboratory studies showing an evolved reduction in the cost of antibiotic resistance in human pathogens, restricting antibiotic use has led to declines in levels of resistance (Bull and Wichman 2001; Levin and Anderson 1999). Regardless of the cost of resistance, reduced pesticide use will slow the rate of *increase* of resistance from initially low levels.

The recessive nature of the cost of resistance in insects raises the intriguing possibility of heterozygote advantage in an environment where the pesticide use is inconsistent. The *SS* individuals would have low fitness in the presence of pesticide, and the *RR* individuals would have low fitness in the absence of pesticide. Therefore, the *RS* individuals might have the highest fitness overall, as long as resistance is at least partly additive (i.e., heterozygotes have some resistance). This is the case for rats' resistance to warfarin discussed in Chapter 3 in which the cost of resistance is an increased need for vitamin K.

Genetic variance and covariance

As we have noted, besides selection, the other major determinant of the rate of evolution within populations is the amount and pattern of genetic variance and covariance for traits. The most important way that genetic variation impacts the evolution of resistance to pesticides is the extent of pesticide use. If pesticide use is widespread both in time and space, then a large population or metapopulation of pests is subject to selection. With a larger population size there is an increased probability of *R* alleles originating through mutation; in other words, the population can acquire more genetic variation for resistance. This issue is really at the interface between selection and genetic variance, because more widespread use of pesticides means that selection is being practiced on a larger population. However, the strength of selection as measured by the selection coefficient *s* on single-locus resistance or the selection gradient β on quantitative resistance is not determined by how widespread the pesticide use is, but rather on the dose and toxicity of the pesticide itself.

There is no doubt that more widespread use of pesticides leads to more rapid evolution of resistance. In Finland, the volume of antibiotics used was closely paralleled by the prevalence of resistance in bacteria responsible for ear infections in children (Figure 7.11). Perhaps the most serious concern

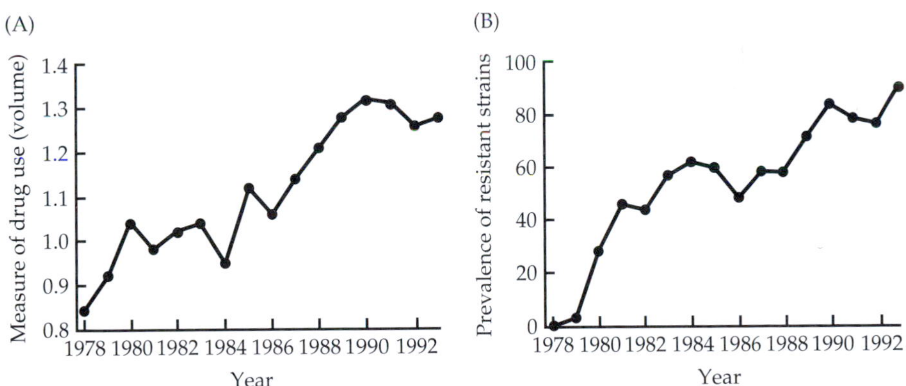

Figure 7.11 The relationship between antibiotic use and frequency of bacterial resistance, estimated from the incidence of ear infections in Finnish children. (A) Volume of antibiotic use. (B) Prevalence of bacterial strains that are antibiotic resistant. (After Levin and Anderson 1999.)

with widespread use is the addition of antibiotics to livestock feed. In the U.S., the FDA estimates that 10 times more antibiotics are used in agriculture than in treating humans, although new laws may begin to reduce this problem (Amabile-Cuevas 2003). Antibiotic resistant pathogens that have evolved in livestock can be transmitted to humans through direct contact with the livestock by farm workers, by eating the food products produced, or possibly through contaminated drinking water, where resistant strains have been found (Palumbi 2001b). In addition, the genes for antibiotic resistance can be transferred from their original host to human bacterial pathogens by plasmid transfer (discussed later in this chapter).

Antibiotics are often overprescribed by doctors, and sometimes even prescribed at the patient's request for viral ailments like the common cold, for which they are entirely ineffective (Palumbi 2001b). In other cases antibiotics are given prophylactically, that is, to prevent infection after surgery, or for mild infections that the patient's immune system could likely clear by itself. The overuse of antibiotics has proven difficult to control, partly because it is an example of Hardin's "tragedy of the commons" (Hardin 1968)—there is little cost of overuse to individuals because antibiotics tend to have few side effects and substantial benefit to individuals in the short term, but there is a much larger cost to society in the resulting increased evolution of resistance. The "tragedy of the commons" does not hold for some serious illnesses like tuberculosis and AIDS, in which resistance evolves within the infected individual causing treatment failure.

Resistance to pesticides is frequently due to single genes of major effect. This may be due to the fact that an extreme phenotype is needed to survive a high dose of pesticide (Macnair 1991). A single major gene mutation can produce an extreme phenotype instantly, particularly if it is dominant, but with polygenic traits it takes several to many generations to produce extreme phenotypes (see Chapter 5). Lower doses of pesticide could lead to the evolution of quantitative resistance through changes in many gene loci of smaller effect (Gould 1995). These predictions were confirmed in a laboratory study of diazinon resistance in sheep blowflies (McKenzie et al. 1992); at very high doses resistance due to a single major mutant evolved, while at lower doses polygenic resistance developed.

In the previous two chapters we learned that genetic correlations among traits cause correlated responses to selection; that is, selection on one trait causes an evolutionary change in correlated traits. Applying this concept to resistance, evolution of resistance to one pesticide often causes **cross-resistance,** or resistance to a second pesticide. Cross-resistance is a major problem, because the most common management response to the evolution of resistance is to change pesticides. Cross-resistance is caused by pleiotropy—for example, a gene that reduces transport of chemicals across the pest's cellular membranes is likely to provide resistance to more than one pesticide. This might be the mechanism behind the somewhat surprising observation that resistance to disinfectants in bacteria is often genetically correlated with antibiotic resistance (Amabile-Cuevas 2003). In some cases the genetic correlation between resistance to two control agents might be negative, producing an evolutionary constraint on the adaptation to both simultaneously (see Chapter 6; Jordan and Jannink 1997). When present, this negative correlation is very helpful in managing resistance.

Gene flow and population structure

In the previous section we discussed how more widespread use of pesticides selects upon increased genetic variation and thus leads to more rapid evolution of resistance. We now need to refine this idea by considering how resistance evolution is affected by structuring of the metapopulation into differentiated subpopulations with gene flow among the subpopulations (see Chapter 3). If there are spatial differences in the use of a particular pesticide, then this can give rise to local differentiation of loci affecting resistance. An excellent example of local differentiation comes from Taylor et al. (1995). They showed that resistance to pyrethroid insecticides in tobacco budworms (*Heliothis virescens*) varied geographically across the southern U.S., consistent with spatially variable selection due to differences in insecticide use. There was significant differentiation as measured by F_{ST} at the locus encoding the sodium membrane channel, *Hpy*, which is known to be involved in pyrethroid resistance, but little to no differentiation of 14 neutral markers. These results are consistent with variable selection creating differentiation at a locus under selection, in spite of high gene flow (as shown by the neutral markers).

This kind of variable selection by pesticides may be common, because farmers and doctors use pesticides and antibiotics differently from each other. With variable selection among subpopulations, simulation models suggest that intermediate levels of gene flow lead to the highest rate of resistance evolution in the entire metapopulation (Caprio and Tabashnik 1992). When gene flow is low, the partially recessive *R* alleles that reach high frequency in some subpopulations through selection and/or drift are not spread throughout the metapopulation. When migration rates are high, *S* alleles continue to flow into subpopulations with *R* alleles, keeping the latter at low frequency and thus mainly in heterozygotes, which have lower levels of resistance than *RR* homozygotes.

In general, gene flow is greater in insect pests than weeds, because the adults of most insect pests can fly whereas plants are sessile. The difference in gene flow is reflected in the greater average genetic differentiation of neutral loci (as measured by F_{ST}; see Chapter 3) in plants compared to insects

(Gould 1995). The lower gene flow in plants compared to insects slows the spread of *R* alleles in weeds. Therefore, nonuniform application creates fewer problems in weeds because farmers who practice good resistance management techniques are less likely to have their efforts ruined by *R* alleles coming in from farms with poorer resistance management. Insects with winged adult stages frequently move from farm to farm, spreading resistance alleles from poorly managed farms.

One can think of the pathogens infecting a single host individual as a subpopulation. In this model gene flow takes place through infections of new hosts (sometimes by vectors of a second species). Metapopulations of many pathogens are characterized by frequent extinction and recolonization events—extinctions when a host clears the pathogen or dies, and recolonizations due to new infections. Pathogens have many methods of transmission, which have been well-studied by epidemiologists. As mentioned above, antibiotic resistance genes in bacteria are often carried on **plasmids** (circular DNA that is often transferred between bacteria), which means that a unique form of interspecific gene flow involving only the resistance alleles can occur through **horizontal transfer,** the movement of genes between species.

An example that includes many of the concepts underlying resistance evolution is that of malaria, which sickens up to 500 million people and kills up to 2 million per year, mostly in Africa (Ranson et al. 2000). Malaria is caused by protozoans in the genus *Plasmodium*, primarily *P. falciparum*, that is transmitted to humans by mosquito vectors in the genus *Anopheles*, most notably the African species *A. gambiae*. *Plasmodium* has evolved resistance to a number of the drugs used to treat infections, including chloroquine, and the mosquitoes have evolved resistance to several insecticides used against it, including DDT and pyrethroids. Studies of the parasite have shown high levels of inbreeding in some subpopulations (F_{IS} = 0.92; Paul et al. 1995), which would facilitate the evolution of recessive resistance. Studies of another disease vector mosquito, *Culex pipiens*, which transmits the West Nile virus, have shown that esterase alleles that confer resistance to organophosphate insecticides (including DDT) have been spread worldwide through gene flow (Raymond et al. 1991). Gene loci causing resistance to DDT in *A. gambiae* have been localized using QTL mapping (Ranson et al. 2000).

Resistance management

One of the first techniques to manage resistance was **rotation** of pesticides, that is, systematically changing the pesticide used to control a pest. The idea is that resistance that develops to a given pesticide while it is being used will decline during the times other pesticides are in use. The amount of decline depends on a number of key factors. First, and foremost, rotation depends on the cost of resistance. As noted above, the costs seems to be moderate or can be compensated for in insects and microbes, reducing the effectiveness of rotation for these pests. Second, use of each pesticide has to be discontinued for a sufficient number of pest generations to allow the frequency of resistance alleles to decline significantly. Since resistance generally builds up faster than it declines, each pesticide can be employed much less than half the time. This consideration may limit the usefulness of rotation for weeds, in spite of the generally higher cost of resistance in weeds, because many weeds only have one generation per year. The third factor affecting the evolution of resistance during rotation is that, if there is cross resistance among some of the pesticides in the rotation, resistance will decline much more slowly. On the other hand, if there is a negative genetic correlation between resistance to two pesticides, rotation is extremely effective because selection imposed by one pesticide would decrease resistance of the other due to the correlated response (see Chapter 5). A final consideration in pesticide rotation is that the cycling must be synchronized over the entire metapopulation of pests. Otherwise, *R* alleles for each pesticide will be introduced by gene flow during the times when those pesticides are not in use, slowing the decline in resistance. Mathematical models suggest that rotation is not very effective in slowing the evolution of antibiotic resistance (Levin and Anderson 1999).

With the advent of crop plants genetically engineered to produce insecticides, the use of spatial variation in pesticide use to slow the evolution of resistance has become much more common. Farmers using *Bt* crops are often mandated to plant adjacent fields with non-*Bt* versions of the same crop as a **refuge** for *Bt*-susceptible insects. This strategy depends on resistance being recessive and rare. Given these two conditions, which commonly are true, most insects are susceptible, so most adult insects entering the mating pool will come from the refuges (few will survive to adulthood on the *Bt* crop). The rare *RR* individuals that are produced will mate mainly with *SS* individuals and will thus produce susceptible *RS* offspring. This strategy works best with larger refuges (less widespread selection for resistance; Figure 7.12) and greater recessivity of the *R* allele. The *R* allele can be made functionally more recessive by engineering plants with high levels of toxins, because the extreme toxicity tends to kill *RS* individuals even if resistance is only partially recessive. Because most *R* alleles are in heterozygotes when *R* is rare, killing these individuals is critical to slowing the evolution of resistance.

Another resistance management technique in common use is to employ a combination of two or more pesticides simultaneously. This is called a "tank mix" in agriculture (because the pesticides are mixed in the same sprayer tank), "combination therapy" in medicine, and "stacking" or "pyramiding" multiple toxins into a single GMO. The rationale is that pests that are resistant to one pesticide will be killed by the other pesticide, and vice versa. This strategy will not work if there is cross-resistance due to

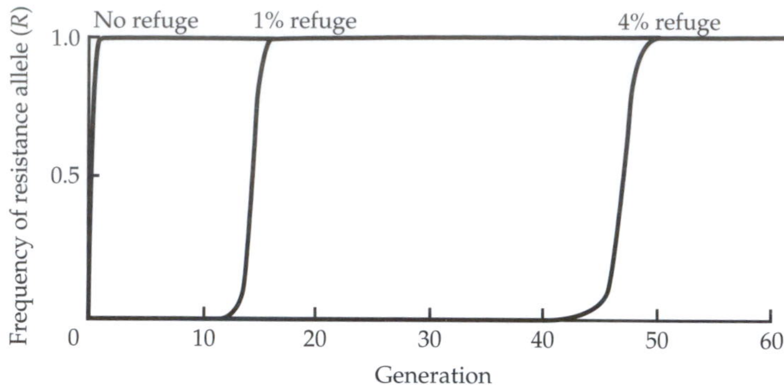

Figure 7.12 Theoretical rates of the evolution of *Bt* resistance with differing amounts of non-*Bt* crop refuges. The model assumes that the *R* allele is completely recessive, and that no *SS* and *RS* individuals survive on the *Bt* crop (i.e., *Bt* is completely effective). (After Bull and Wichman 2001.)

pleiotropy, unless it is negative cross-resistance, in which case combinations work very well. The mathematical models of antibiotic resistance mentioned earlier suggest that combination therapy is more effective than rotation in slowing resistance (Levin and Anderson 1999). In tuberculosis and AIDS, combination therapy is the only way to avoid treatment failure resulting from the evolution of resistance within the patient.

The Future of Applied Ecological Genetics

Until very recently, few people thought that the principles of ecological genetics were directly applicable to problems facing human society. This attitude is changing rapidly as the importance of short-term evolution becomes clearer in a number of areas. In some areas of applied ecological genetics, more research is needed to guide public policy. Examples include the role of rapid evolution in invasive species and the fitness effects of transgenes. In other areas, there is adequate scientific knowledge, but what is needed is the political or economic incentive to implement policy based on the research. Examples of this situation include the effects of unnecessary antibiotic use, especially in livestock, on the increase in antibiotic resistance, and the negative effects of small population sizes caused by habitat destruction and other human activities on the persistence of endangered populations. In these kinds of cases, it is important for scientists and others with scientific knowledge to make the knowledge more accessible to the policy makers and the general public, so that policies to reverse or prevent the negative impacts can be implemented.

SUGGESTED READINGS

Ellstrand, N. C. 2003. *Dangerous Liaisons?: When Cultivated Plants Mate With Their Wild Relatives*. Johns Hopkins University Press, Baltimore. A popular account of the threat of escape of transgenes and other crop genes to wild plants.

Fox, C. W., D. A. Roff, and D. J. Fairbairn. 2001. *Evolutionary ecology: Concepts and Case Studies*. Oxford University Press, New York, NY. An edited volume with chapters by many leaders in the field, which covers many of the topics in this chapter and the book as a whole.

*Frankham, R., J. D. Ballou, and D. A. Briscoe. 2002. *Introduction to Conservation Genetics*. Cambridge University Press, New York. An up-to-date, thorough, and authoritative review of conservation genetics.

*Newman, D., and D. Pilson. 1997. Increased probability of extinction due to decreased genetic effective population size: Experimental populations of *Clarkia pulchella*. Evolution 51:354-362. The first study to show that effective population size increases the risk of population extinction in the field.

Palumbi, S. R. 2001a. *The Evolution Explosion: How Humans Cause Rapid Evolutionary Change*. W.W. Norton & Company, New York. Written for a broad audience, this book describes many of the ways in which humans are causing evolution, including the evolution of resistance, and how this evolution affects humans in turn.

*Primack, R. B. 2000. *A Primer of Conservation Biology*, 2nd Edition. Sinauer Associates, Sunderland, MA. A very accessible introduction to the field of conservation biology.

*Snow, A. A., D. Pilson, L. H. Rieseberg, M. J. Paulson, N. Pleskac, M. R. Reagon, D. E. Wolf, and S. M. Selbo. 2003. A *Bt* transgene reduces herbivory and enhances fecundity in wild sunflowers. Ecological Applications 13:279-286. The first demonstration that a transgene can increase the fitness of wild relatives under open-field conditions.

Tabashnik, B. E., Y.-B. Liu, T. Malvar, D. G. Heckel, L. Masson, and J. Ferre. 1998. Insect resistance to *Bacillus thuringiensis*: Uniform or diverse? Philosophical Transactions of the Royal Society of London Series B-Biological Sciences 353:1751-1756. A review of evidence for population differentiation for diamondback moth resistance to *Bt* toxins.

**Indicates a reference that is a suggested reading in the field and is also cited in this chapter.*

SUGGESTED READINGS QUESTIONS

The following questions are based on papers from the primary literature that address key concepts covered in this chapter. For the full citation for each paper please see Suggested Readings.

From Newman, D., and D. Pilson. 1997. Increased probability of extinction due to decreased genetic effective population size: Experimental populations of *Clarkia pulchella*.

1. What are the three potential causes of small population extinction discussed? Which one is tested in this experiment?
2. What is the typographical error in the second sentence of the second paragraph?
3. Describe the basic experimental design. How was N_e manipulated independently of census N (= N_a, see Chapter 3)?
4. Why were pollinators excluded with netting, plants emasculated (anthers removed), and hand-pollinations performed?

5. Do you expect an increase in homozygosity *relative to Hardy-Weinberg expectations* in either treatment group? Why or why not?
6. Do the results support the hypothesis tested? Why or why not?

From Snow, A. A., D. Pilson, L. H. Rieseberg, M. J. Paulson, N. Pleskac, M. R. Reagon, D. E. Wolf, and S. M. Selbo. 2003. A *Bt* transgene reduces herbivory and enhances fecundity in wild sunflowers.

1. How did the investigators conduct open-field tests of transgenic plants while minimizing the risk of transgene escape? Could these techniques be useful for commercial sunflower fields? Why or why not?
2. What were the three treatment groups in the field, and what hypotheses do comparisons among them test? What hypotheses did the greenhouse experiment test?
3. Did the results support or refute the hypotheses from question 2? Support your answer with specific results from the paper.
4. What are some possible reasons for the nonsignificant effect of *Bt* on seed production in Colorado?
5. If the increase in fecundity of weedy *Bt* sunflowers is due to reduced herbivory caused by the Cryl1Ac toxin, are there evolutionary changes that might occur in wild populations that could ameliorate this problem?

From Tabashnik, B. E., Y.-B. Liu, T. Malvar, D. G. Heckel, L. Masson, and J. Ferre. 1998. Insect resistance to *Bacillus thuringiensis*: Uniform or diverse?

1. Why are *Bt* insecticides important in agriculture?
2. What are the physiological mechanism(s) of insect resistance to *Bt* toxins?
3. What are the limitations of *Bt* toxin switching (rotation) as a resistance management strategy?
4. Explain the three hypothesized scenarios for generating patterns of genetic differentiation in resistance that the research is testing.
5. On page 1753 the authors state that the two U.S. strains of moth (NO-QA and PEN) were resistant to the three CrylA toxins and cross-resistant to CrylF and CrylJ. On what basis did they conclude that resistance to the latter two toxins was cross-resistance rather than primary resistance?
6. How did they test for dominance of resistance? Summarize their results briefly. Why do you think they were surprised to find some dominant resistance?
7. Is there evidence for differences among the three populations in the patterns of cross-resistance to different *Bt* toxins? If so, describe this evidence.
8. What are the management implications for any population differentiation that they found?

CHAPTER REFERENCES

Amabile-Cuevas, C. F. 2003. New antibiotics and new resistance. American Scientist 91:138–149.

Arias, D. M., and L. H. Rieseberg. 1994. Gene flow between cultivated and wild sunflowers. Theoretical and Applied Genetics 89:655–660.

Arriola, P. E., and N. C. Ellstrand. 1997. Fitness of interspecific hybrids in the genus *Sorghum*: Persistence of crop genes in wild populations. Ecological Applications 7:512–518.

Avise, J. C., and W. S. Nelson. 1989. Molecular genetic relationships of the extinct dusky seaside sparrow. Science 243:646–648.

Barrett, S. C. H. 1988. Genetics and evolution of agricultural weeds. Pp. 57–75 *in* M. A. Altieri and M. Liebman, eds. *Weed Management in Agroecosystems: Ecological Approaches*. CRC Press, Inc., Boca Raton, FL.

Blair, A. C., and L. M. Wolfe (submitted). The evolution of a high performance invasive phenotype: An experimental study with a perennial plant.

Bouma, J. E., and R. E. Lenski. 1988. Evolution of a bacteria/plasmid association. Nature 335:351–352.

Bowen, B. W., A. B. Meylan, and J. C. Avise. 1991. Evolutionary distinctiveness of the endangered Kemp's ridley sea turtle. Nature 352:709–711.

Bull, J. J., and H. A. Wichman. 2001. Applied evolution. Annual Review of Ecology and Systematics 32:183–217.

Burke, J. M., and L. H. Rieseberg. 2003. Fitness effects of transgenic disease resistance in sunflowers. Science 300:1250.

Caprio, M. A., and B. E. Tabashnik. 1992. Gene flow accelerates local adaptation among finite populations: Simulating the evolution of insecticide resistance. Journal of Economic Entomology 85:611–620.

Caro, T. M., and M. K. Laurenson. 1994. Ecological and genetic factors in conservation: A cautionary tale. Science 263:485–486.

Crandall, K. A., O. R. P. Bininda-Emonds, G. M. Mace, and R. K. Wayne. 2000. Considering evolutionary processes in conservation biology. Trends in Ecology and Evolution 15:290–295.

Crnokrak, P., and J. Merilä. 2002. Genetic population divergence: Markers and traits. Trends in Ecology and Evolution 17:501.

Daniell, H. 2002. Molecular strategies for gene containment in transgenic crops. Nature Biotechnology 20:581–586.

Ellstrand, N. C., and C. A. Hoffmann. 1990. Hybridization as an avenue of escape for engineered genes. BioScience 40:438–442.

Ellstrand, N. C., H. C. Prentice, and J. F. Hancock. 1999. Gene flow and introgression from domesticated plants to their wild relatives. Annual Review of Ecology and Systematics 30:539–563.

Ellstrand, N. C., and K. A. Schierenbeck. 2000. Hybridization as a stimulus for the evolution of invasiveness in plants? Proceedings of the National Academy of Sciences 97:7043–7050.

Gilligan, D. M., L. M. Woodworth, M. E. Montgomery, D. A. Briscoe, and R. Frankham. 1997. Is mutation accumulation a threat to the survival of endangered populations? Conservation Biology 11:1235–1241.

Gould, F. 1991. The evolutionary potential of crop pests. American Scientist 79:496–507.

Gould, F. 1995. Comparisons between resistance management strategies for insects and weeds. Weed Technology 9:830–839.

Graur, D., and W.-H. Li. 2000. *Fundamentals of Molecular Evolution*. Sinauer Associates, Sunderland, MA.

Hancock, J. F., R. Grumet, and S. C. Hokanson. 1996. The opportunity for escape of engineered genes from transgenic crops. HortScience 31:1080–1085.

Hardin, G. 1968. The tragedy of the commons. Science 162:1243–1248.

Heap, I. M. 1997. The occurrence of herbicide-resistant weeds worldwide. Pesticide Science 51:235–243.

Heschel, M. S., and K. N. Paige. 1995. Inbreeding depression, environmental stress, and population size variation in scarlet gilia (*Ipomopsis aggregata*). Conservation Biology 9:126–133.

Hokanson, S. C., R. Grumet, and J. F. Hancock. 1997. Effect of border rows and trap/donor ratios on pollen-mediated gene movement. Ecological Applications 7:1075–1081.

James, C. 2002. Global status of commercialized transgenic crops: 2002. International Service for the Acquisition of Agri-biotech Applications, Ithaca, NY.

Jordan, N. R., and J. L. Jannink. 1997. Assessing the practical importance of weed evolution: A research agenda. Weed Research 37:237–246.

Keane, R. M., and M. J. Crawley. 2002. Exotic plant invasions and the enemy release hypothesis. Trends in Ecology and Evolution 17:164–-70.

Keller, L. F., and D. M. Waller. 2002. Inbreeding effects in wild populations. Trends in Ecology and Evolution 17:230–241.

Klinger, T., D. R. Elam, and N. C. Ellstrand. 1991. Radish as a model system for the study of engineered gene escape rates via crop-weed mating. Conservation Biology 5:531–535.

Klinger, T., and N. C. Ellstrand. 1994. Engineered genes in wild populations: Fitness of weed-crop hybrids of *Raphanus sativus*. Ecological Applications 4:117–120.

Latta, R. G., and J. K. McKay. 2002. Genetic population divergence: Markers and traits. Trends in Ecology and Evolution 17:501–502.

Legge, J. T., R. Roush, R. DeSalle, A. P. Vogler, and B. May. 1996. Genetic criteria for establishing evolutionarily significant units in Cryan's buckmoth. Conservation Biology 10:85–98.

Levin, B. R., and R. M. Anderson. 1999. The population biology of anti-infective chemotherapy and the evolution of drug resistance: More questions than answers. Pp. 125–137 *in* S. C. Stearns, ed. *Evolution in Health and Disease*. Oxford University Press, New York.

Linder, C. R., I. Taha, G. J. Seiler, A. A. Snow, and L. H. Rieseberg. 1998. Long-term introgression of crop genes into wild sunflower populations. Theoretical and Applied Genetics 96:339–347.

Lynch, M. 1996. A quantitative-genetic perspective on conservation issues. Pp. 471–501 *in* J. C. Avise and J. L. Hamrick, eds. *Conservation Genetics*. Chapman & Hall, New York.

Lynch, M., J. Blanchard, D. Houle, T. Kibota, S. Schultz, L. Vassilieva, and J. Willis. 1999. Perspective: Spontaneous deleterious mutation. Evolution 53:645–663.

Macnair, M. R. 1991. Why the evolution of resistance to anthropogenic toxins normally involves major gene changes: The limits to natural selection. Genetica 84:213–219.

Madsen, T., R. Shine, M. Olsson, and H. Wittzell. 1999. Restoration of an inbred adder population. Nature 402:34–35.

Marvier, M. 2001. Ecology of transgenic crops. American Scientist 89:160–167.

McKay, J. K., and R. G. Latta. 2002. Adaptive population divergence: Markers, QTL and traits. Trends in Ecology and Evolution 17:285–291.

McKenzie, J. A., A. G. Parker, and J. L. Yen. 1992. Polygenic and single gene responses to selection for resistance to diazinon in *Lucilia cuprina*. Genetics 130:613–620.

McKenzie, J. A., M. J. Whitten, and M. A. Adena. 1982. The effect of genetic background on the fitness of diazinon resistance genotypes of the Australian sheep blowfly, *Lucilia cuprina*. Heredity 49:1–9.

Merilä, J., and P. Crnokrak. 2001. Comparison of genetic differentiation at marker loci and quantitative traits. Journal of Evolutionary Biology 14:892–903.

Mikkelsen, T. R., B. Andersen, and R. B. Jorgensen. 1996. The risk of crop transgene spread. Nature 380:31.

Mitton, J. B. 1997. *Selection in Natural Populations*. Oxford University Press, New York.

Moritz, C. 1994. Defining 'Evolutionary Significant Units' for conservation. Trends in Ecology and Evolution 9:373–375.

Moritz, C. 2002. Strategies to protect biological diversity and the evolutionary processes that sustain it. Systematic Biology 51:238–254.

Newman, D., and D. Pilson. 1997. Increased probability of extinction due to decreased genetic effective population size: Experimental populations of *Clarkia pulchella*. Evolution 51:354–362.

Palumbi, S. R. 2001b. Humans as the world's greatest evolutionary force. Science 293:1786–1790.

Parker, I. M., J. Rodriguez, and M. E. Loik. 2003. An evolutionary approach to understanding the biology of invasions: Local adaptation and general-purpose genotypes in the weed *Verbascum thapsus*. Conservation Biology 17:59–72.

Paul, R. E. L., M. J. Packer, M. Walmsley, M. Lagog, L. C. Ranford-Cartwright, R. Paru, and K. P. Day. 1995. Mating patterns in malaria parasite populations of Papua New Guinea. Science 269:1709–1711.

Pimentel, D., L. Lach, R. Zuniga, and D. Morrison. 2000. Environmental and economic costs of nonindigenous species in the United States. BioScience 50:53–65.

Ranson, H., B. Jensen, X. Wang, L. Prapanthadara, J. Hemingway, and F. H. Collins. 2000. Genetic mapping of two loci affecting DDT resistance in the malaria vector *Anopheles gambiae*. Insect Molecular Biology 9:499–507.

Raymond, M., A. Callaghan, P. Fort, and N. Pasteur. 1991. Worldwide migration of amplified insecticide resistance genes in mosquitoes. Nature 350:151–153.

Reed, D. H., and R. Frankham. 2001. How closely correlated are molecular and quantitative measures of genetic variation? A meta-analysis. Evolution 55:1095–1103.

Reed, D. H., and R. Frankham. 2003. Correlation between fitness and genetic diversity. Conservation Biology 17:230–237.

Reznick, D. N., F. H. Shaw, F. H. Rodd, and R. G. Shaw. 1997. Evaluation of the rate of evolution in natural populations of guppies (*Poecilia reticulata*). Science 275:1934–1936.

Richards, C. M. 2000. Inbreeding depression and genetic rescue in a plant metapopulation. American Naturalist 155:383–394.

Rieger, M. A., M. Lamond, C. Preston, S. B. Powles, and R. T. Roush. 2002. Pollen-mediated movement of herbicide resistance between commercial canola fields. Science 296:2386–2388.

Ritland, K. 2000. Marker-inferred relatedness as a tool for detecting heritability in nature. Molecular Ecology 9:1195–1204.

Saccheri, I., M. Kuussaari, M. Kankare, P. Vikman, W. Fortelius, and I. Hankski. 1998. Inbreeding and extinction in a butterfly metapopulation. Nature 392:491–494.

Schaal, B. A. 1980. Measurement of gene flow in *Lupinus texensis*. Nature 284:450–451.

Schmidt, K., and K. Jensen. 2000. Genetic structure and AFLP variation of remnant populations in the rare plant *Pedicularis palustris* (Scrophulariaceae) and its relation to population size and reproductive components. American Journal of Botany 87:678–689.

Snow, A., P. Moran-Palma, L. Rieseberg, A. Wszelaki, and G. Seiler. 1998. Fecundity, phenology, and seed dormancy of F_1 wild-crop hybrids in sunflower (*Helianthus annuus*, Asteraceae). American Journal of Botany 85:794–801.

Snow, A. A., D. Pilson, L. H. Rieseberg, M. J. Paulson, N. Pleskac, M. R. Reagon, D. E. Wolf, and S. M. Selbo. 2003. A *Bt* transgene reduces herbivory and enhances fecundity in wild sunflowers. Ecological Applications 13:279–286.

Snow, A. A., K. L. Uthus, and T. M. Culley. 2001. Fitness of hybrids between weedy and cultivated radish: Implications for weed evolution. Ecological Applications 11:934–943.

Taylor, M. F. J., Y. Shen, and M. E. Kreitman. 1995. A population genetic test of selection at the molecular level. Science 270:1497–1499.

Thompson, J. N. 1998. Rapid evolution as an ecological process. Trends in Ecology and Evolution 13:329–332.

Ward, J. K., J. Antonovics, R. B. Thomas, and B. R. Strain. 2000. Is atmospheric CO_2 a selective agent on model C_3 annuals? Oecologia 123:330–341.

Westemeier, R. L., J. D. Brawn, S. A. Simpson, T. L. Esker, R. W. Jansen, J. W. Walk, E. L. Kershner, J. L. Bouzat, and K. N. Paige. 1998. Tracking the long-term decline and recovery of an isolated population. Science 282:1695–1698.

Whitton, J., D. E. Wolf, D. M. Arias, A. A. Snow, and L. H. Rieseberg. 1997. The persistence of cultivar alleles in wild populations of sunflowers five generations after hybridization. Theoretical and Applied Genetics 95:33–40.

Wilcove, D. S., D. Rothstein, J. Dubow, A. Phillips, and E. Losos. 1998. Quantifying threats to imperiled species in the United States. BioScience 48:607–615.

Appendix

For a sample of size n in which f_i = frequency of observations with value x_i:

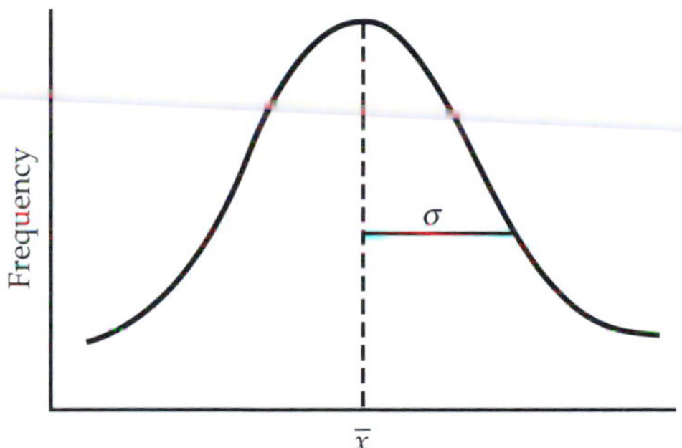

Mean = $\bar{x} = \dfrac{\sum_{i=1}^{n} x_i}{n}$

Variance = $V_x = s_x^2$ = Mean Square (MS) = $\dfrac{\sum_{i=1}^{n}(x_i - \bar{x})^2}{n-1} = \sum_{i=1}^{n} f_i (x_i - \bar{x})^2 =$

$\dfrac{SS}{n-1}$ where SS = "sum of squares" = sum of squared deviations from mean

Standard deviation $= \text{s.d.} = s_x = \sigma_x = \sqrt{V_x}$

Covariance of 2 variables =

$$\text{Cov}(x,y) = \sigma_{xy} = \frac{\sum_{i=1}^{n}(x_i - \bar{x})(y_i - \bar{y})}{n-1} = \sum_{i=1}^{n} f_i (x_i - \bar{x})(y_i - \bar{y})$$

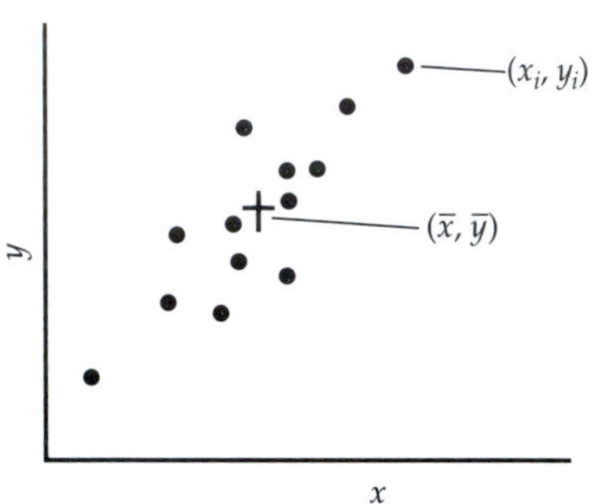

Each summed term is called the "cross product" and measures the extent to which the point deviates from the bivariate mean (the mean of x and y). Variance is the covariance of a variable with itself.

Correlation coefficient $= r_{xy} = \frac{\text{Cov}(x,y)}{s_x s_y} = \frac{\sigma_{xy}}{\sigma_x \sigma_y}$

Regression coefficient (slope) =

$$b = \frac{\text{Cov}(x, y)}{V_x} = \frac{\sigma_{xy}}{\sigma_x^2} = \frac{\sum_{i=1}^{n}(x_i - \bar{x})(y_i - \bar{y})}{\sum_{i=1}^{n}(x_i - \bar{x})^2} = r\frac{s_y}{s_x}$$

So if $s_x = s_y$, $b = r$

Standardized data (standard scores, *z*-scores) $= z_i = \frac{x_i - \bar{x}}{\sigma_x}$ so $\bar{z} = 0$ and $\sigma_z = 1$

Coefficient of variation $= \text{C.V.} = \frac{s \times 100}{\bar{x}}$

$\text{Var}(x + y) = \text{Var}(x) + \text{Var}(y) + 2\,\text{Cov}(x,y)$

For constants a and b, $\text{Var}(ax + b) = a^2\,\text{Var}(x)$

Glossary

Numbers in brackets refer to the chapters in which the term is introduced.

adaptation a phenotypic trait that has evolved to help an organism cope with an environmental challenge or to increase its mating success. [1]

adaptive landscape a three-dimensional depiction of population mean fitness as a function of genotype or phenotype, in which the horizontal axes are allele frequencies at two loci or two phenotypic traits, and the vertical axis is population mean fitness (synonym: adaptive topography). [3]

adaptive peak a population mean fitness maximum, defined by values of allele frequencies or phenotypic traits. [3,6]

adaptive topography a three-dimensional depiction of population mean fitness as a function of genotype or phenotype, in which the horizontal axes are allele frequencies at two loci or two phenotypic traits, and the vertical axis is population mean fitness (synonym: adaptive topography). [3]

additive genetic correlation (r_A) a measure of the degree to which two traits are affected by the same genes (pleiotropy) or pairs of genes (linkage disequilibrium). Selection on one trait produces an evolutionary change in all traits that have an additive genetic correlation with the selected trait. [5]

additive genetic variance (V_A) the magnitude of the phenotypic (and genotypic) variance that is due to additive effects of genes and that determines the degree to which the average phenotype of the parents is reflected in the average phenotype of their progeny. [4]

additive genetic variance-covariance matrix a square matrix with additive genetic variances for the traits on the diagonal and additive genetic covariances on the off-diagonal (synonym: *G*-matrix). [6]

additivity the type of gene action in which the alleles at a locus do not affect each other's expression or the expression of alleles at other loci; in other words, gene action with no dominance or epistasis. [3]

AFLP (amplified fragment length polymorphism) genetic markers detected by cleaving DNA with one or more restriction enzymes and then amplifying some of these fragments by PCR using primers with random nucleotide sequences. [2]

allele a particular form of a given gene; diploid organisms have two alleles at each locus, one from their father and one from their mother. [1]

allele frequency the proportion of a particular allele among all of the alleles at a locus present in a population or a sample from a population. [2]

allozymes different electrophoretic forms of the same enzyme coded for by different alleles at a given locus; commonly used as a genetic marker, especially between 1970 and 1990. The term allozyme is often used interchangeably with isozyme. [2]

analysis of variance (ANOVA) statistical technique for testing for differences among the means of several groups with respect to a continuous variable. [4]

analytical model a model in which the relationships among variables are defined using equations. [2]

artificial selection the process of selective breeding of organisms by humans to produce domesticated animals with more desirable traits; also used by evolutionary biologists to test for genetic variation and covariation. [5]

assortative mating non-random mating in which mate choice is based on a phenotypic trait. [2]

asymmetrical response a common result in artificial selection experiments in which there is a greater response to selection in one direction than there is in the opposite direction for the same trait. [5]

base a nitrogenous compound that differentiates the four nucleotides that make up DNA: A (adenine), G (guanine), T (thymine), or C (cytosine). [1]

base substitution a mutation that occurs when one nucleotide base is substituted for another in a DNA sequence. [3]

biodiversity genetic and phenotypic variation both within and among species, plus the variety of ecosystems created by these species. [7]

bottleneck a marked decrease in population size. [3]

breeding value the effect of an individual's genes on the value of a given trait in its offspring; sometimes called the additive genotype. It is equal to two times the deviation of the mean of the individual's offspring from the overall population mean. [4]

candidate gene approach a technique which attempts to determine if genes of known function affect complex phenotypic traits. [5]

causal variance components in a sibling analysis, the portions of phenotypic variance that are due to the underlying genetic and environmental sources of variance (e.g., additive genetic variance, dominance variance, environmental variance). [4]

centimorgan (cM) the usual unit for measuring distance on a genetic map. One cM is equivalent to a rate of recombination of 1% (i.e., $c = 0.01$). [5]

chromosomal rearrangements a class of mutations in which whole segments of chromosomes are involved, including inversions and translocations. [3]

chromosomes microscopic threadlike bodies each containing a single molecule of DNA complexed with histones and other proteins. [1]

codominant marker a genetic marker in which the heterozygotes can be distinguished from both homozygotes. [2]

codon a group of three adjacent nucleotides that specifies a corresponding amino acid subunit in a polypeptide chain. [1]

coefficient of relatedness (r_{IJ}) the proportion of genes identical by descent (IBD) among two individuals I and J. [2]

common garden experiment an experimental design in which individuals from multiple populations are raised together in the same environment in order to test for genetic differentiation in phenotypic traits. [5]

compensatory mutations a mutation that ameliorates the deleterious fitness effects of another mutation. [7]

correlated response to selection an evolutionary change in an unselected trait caused by an additive genetic correlation between the unselected trait and a trait under selection. [5]

correlational selection a type of selection in which two traits interact nonadditively to determine fitness, characterized by the finding that certain combinations of trait values have higher fitness than other combinations. [6]

cost of resistance the fitness effects of an allele that confers resistance (often denoted by R) to a pesticide or antibiotic in the absence of the pesticide or antibiotic. [7]

cross-fostering an experimental technique in which offspring are reared (fostered) by animals other than their genetic parents; cross-fostering is designed to reduce parental effects. [4]

cross resistance the condition in which resistance to one pesticide or antibiotic confers increased resistance to a second pesticide or antibiotic. [7]

crossing over during meiosis, the process in which portions of homologous chromosomes undergo physical exchange. [5]

dams the female parents in a quantitative genetic breeding experiment. [4]

deletion a mutation in which one or more base pairs is removed from a DNA sequence. [3]

deme a local interbreeding unit within a metapopulation (synonyms: local population, subpopulation). [2]

density dependent selection selection that differs according to population density. [6]

diploid organisms that have two copies of each chromosome (except for the sex chromosomes), and therefore two copies of each gene, one from each parent. The diploid chromosome number is denoted by $2n$. [1]

direct selection the type of selection in which there is a causal relationship between a phenotypic trait and fitness, which can result in adaptation. [6]

directional selection the form of selection characterized by a linear fitness function, with fitness increasing or decreasing in proportion to phenotypic value. [6]

discrete generations a life history, like that of an annual plant, in which the parental generation has died by the time the offspring generation reproduces. [2]

discrete polymorphism a phenotypic trait that exhibits only a few (usually two or three) distinct types or morphs (synonym: visible polymorphism). [2]

disruptive selection the form of selection in which fitness is lowest at some intermediate phenotype and higher at each phenotypic extreme. [6]

dominance any situation in which the heterozygous genotype for two alleles has a phenotype different from the mean of the corresponding homozygous genotypes; with complete dominance, the presence of one allele (the dominant allele) completely conceals the presence of an alternative allele (the recessive allele), and the phenotype of the heterozygous genotype is indistinguishable from the phenotype of the homozygous dominant. [3]

dominance variance (V_D) the magnitude of the phenotypic (and genotypic) variance that is due to dominance, that is, the interaction between alleles at the same locus. [4]

ecotypes locally adapted populations that are phenotypically and genetically differentiated for adaptive traits. [5]

effective population size (N_e) the size of an idealized population that would experience the same magnitude of genetic drift as the population of interest. [3]

electrophoresis a technique for separating macromolecules (proteins, RNA, DNA) on a gel using an electric field. [2]

endangered species a species that is likely to become extinct in all or a major portion of its range. [7]

environmental correlation (r_E) a measure of the degree to which two traits respond to variation in the same environmental factors. [5]

environmental deviation (E) the difference between the phenotypic and genotypic values caused by the environment. [4]

environmental variance (V_E) the portion of the phenotypic variance that is due to the environment, including all nongenetic sources of phenotypic variation. [4]

epistasis a form of gene action in which two or more loci interact nonadditively with each other to determine the phenotype; when epistasis is present, the phenotype associated with a particular genotype depends on which alleles are present at another locus. [3, 4]

epistatic selection the type of selection in which fitness depends upon nonadditive interactions between alleles at different loci. [5]

epistatic variance (V_I) the portion of the phenotypic (and genotypic) variance that is due to epistasis, that is, interactions among gene loci (synonym: interaction variance). [4]

equilibrium a state in which some value (in ecological genetics, often allele frequency or mean phenotype) remains unchanged. An equilibrium can be stable, in which case the value returns to the equilibrium after being slightly perturbed, or unstable, in which case the value moves to another equilibrium. [3]

evolutionarily significant units (ESUs) populations of threatened or endangered organisms that need to be saved from extinction. [7]

evolutionary constraint any biological factor that slows the rate of adaptive evolution. [3, 6]

evolutionary trajectories pathways that populations traverse across adaptive landscapes during evolution, tracing the ways that the joint allele frequencies or mean phenotypes might evolve. [3]

exons the portions of RNA transcripts that are retained in the messenger RNA, including those parts that code for proteins. [1]

fitness the ability of an organism to survive and reproduce, and thus have descendents in future generations. One good operational definition of fitness is lifetime number of offspring produced. [6]

fitness function the curve that describes the relationship between fitness and a phenotypic trait. [6]

fitness surface a three-dimensional representation of the relationship between two phenotypic traits and individual fitness. [6]

fixed a population in which all members are homozygous for the same allele at a given locus (antonyms: segregating, polymorphic). [2]

fixation index (F_{ST}) Wright's measure of population differentiation. [3]

founder event a type of bottleneck, defined as the creation of a new population by a small number of colonists. [3]

frame shift a mutation in a protein-coding region of a gene in which there is an insertion or deletion of some number of nucleotides that is not an exact multiple of three; the result is that all downstream codons are translated in the wrong phase. [3]

frequency dependent selection a type of natural selection in which the fitness of each genotype or phenotype depends on its frequency in the population. In positive frequency dependence, fitness increases as the genotype or phenotype becomes more common, and in negative frequency dependence, fitness increases as the genotype or phenotype becomes rarer. [3]

***G*-matrix** a square matrix with additive genetic variances for the traits on the diagonal and additive genetic covariances on the off-diagonal (synonym: additive genetic variance-covariance matrix). [6]

gamete haploid reproductive cell (sperm or egg). [1]

gametic phase disequilibrium a nonrandom relationship between the alleles present at two or more loci, which can cause a genetic correlation (synonym: linkage disequilibrium). [5]

gene a region of DNA (deoxyribonucleic acid) coding either for the messenger RNA encoding the amino acid sequence in a polypeptide chain or for a functional RNA molecule. [1]

gene action the manner in which genotype affects phenotype, including additivity, dominance, pleiotropy, and epistasis. [3]

gene expression the process of creating RNA transcripts and proteins from the genetic information contained in DNA. [1]

gene flow movement of genes between populations caused by migration and subsequent mating. [2, 3]

gene sequencing determining the complete sequence of a molecule or molecules of DNA, often through highly automated procedures. [2]

genetic code the set of 64 possible codon sequences and the amino acids (or translation-termination signals) that are specified by each codon. [1]

genetic differentiation differences between populations in allele frequencies at one or more loci, or in mean phenotype in a common environment (synonym: population differentiation). [2]

genetic drift fluctuations in allele frequency that occur by chance, particularly in small subpopulations, as a result of random sampling error in the choice of gametes that form the next generation (synonym: random genetic drift). [3]

genetic map a linear representation of the positions of loci (especially marker loci) along chromosomes, based on the frequency of recombination between the loci. [5]

genetic variation naturally occurring genetic differences among organisms in the same species. [2]

genetically engineered (GE) or genetically modified organism (GM or GMO) an animal or plant that has genes from a different species (i.e., transgenes). [7]

genome the totality of the DNA in a gamete or in each cell of an organism. [1]

genotype the particular alleles present in an organism for any specified number of gene loci. [1]

genotype-by-environment interaction (G × E) the phenomenon in which different genotypes respond differently to environmental variation; represents genetic variation for phenotypic plasticity (synonym: genotype-environment interaction (g-e)). [5]

genotypic frequencies the proportion of each of the various genotypes present in a population or sample of a population. [2]

genotypic value (*G*) the phenotype produced by a given genotype averaged across environments. [4]

genotypic variance (V_G) the magnitude of the phenotypic variance that is due to all genetic causes, corresponding to the sum of the additive, dominance, and epistatic variances. [4]

haploid a cell or organism with only one copy of each chromosome. The haploid chromosome number is denoted by *n*. [1]

Hardy-Weinberg equilibrium (HWE) the population condition characterized by the genotypic frequencies produced under random mating; for two alleles the genotype frequencies are given by p^2, $2pq$, and q^2. [2]

heritability (h^2) the proportion of the total phenotypic variance that is due to genetic causes; in other words, heritability measures the relative importance of genetic variance in determining phenotypic variance. Narrow-sense heritability is the additive genetic variance divided by the phenotypic variance (V_A/V_P), whereas broad-sense heritability is the genotypic variance divided by the phenotypic variance (V_G/V_P). [4]

heterosis phenomenon in which the F_1 generation has higher fitness than the parental strains or subpopulations that were crossed (mated) to produce them (synonym: hybrid vigor). [3]

heterozygosity the frequency of heterozygous genotypes, often symbolized as *H*. [2]

heterozygote advantage phenomenon in which the heterozygous genotype has a higher phenotypic value (especially for fitness) than either homozygous genotype (synonym: overdominance). [3]

heterozygous a diploid genotype or individual with two different alleles at a single locus. [1]

homozygosity the frequency of homozygous genotypes, often symbolized as *P* or *Q*. [2]

homozygous a diploid genotype or individual with two indistinguishable alleles at a given locus. [1]

horizontal transfer movement of genes between species. [7]

hybrid vigor phenomenon in which the F_1 generation has higher fitness than the parental strains or subpopulations that were crossed (mated) to produce them (synonym: heterosis). [3]

identical by descent (IBD) alleles that are descended by DNA replication from a single allele present in an ancestor. [2]

inbreeding mating between individuals in a population that are more closely related than expected by random chance. [2]

inbreeding depression the decrease in fitness that often accompanies inbreeding or random genetic drift. [2]

indirect selection a covariance between a trait and fitness within a generation that is caused by a phenotypic correlation between that trait and another trait that experiences direct selection. [6]

industrial melanism the increase in frequency of dark (melanic) pigmentation in insects as an adaptation to remain inconspicuous on surfaces darkened by soot from air pollution. [2]

insertion a mutation that occurs when one or more base pairs is added to a DNA sequence. [3]

interdemic selection the third phase of the shifting balance theory, in which subpopulations (demes) at higher adaptive peaks export migrants to subpopulations at lower adaptive peaks, causing the lower-fitness subpopulations to evolve toward the higher peak. [3]

introduced species a species that originated in a different region that becomes established in a new region, often due to deliberate or accidental release by humans (synonym: non-native species). [7]

introns intervening noncoding sequences that are present in many eukaryotic genes and that are removed from the primary transcript during RNA processing. [1]

invasive species nonnative species that increase rapidly in numbers and that have negative impacts on native species. [7]

inversions a type of chromosomal rearrangement in which a portion of a chromosome is flipped end-for-end and inserted back into the chromosome in the opposite orientation. [3]

island model the simplest model of gene flow, in which a proportion m of migrants are exchanged between discrete subpopulations in each generation. [3]

isozymes different forms of the same enzyme. Commonly used as genetic markers, especially between 1970 and 1990. The term isozyme is often used interchangeably with allozyme. [2]

linear regression a statistical technique of finding the best fitting straight line through a set of points representing joint values for two variables. [4]

linkage disequilibrium a nonrandom relationship between the alleles present at two or more loci, which can cause a genetic correlation (synonym: gametic phase disequilibrium). [5]

linkage groups chromosomes or portions of chromosomes in a genetic map, defined by groups of markers for which each adjacent pair has a recombination frequency less than 0.5. [5]

linked loci loci that are on the same chromosome and that show a recombination frequency less than 0.5. [1, 5]

local population a local interbreeding unit within a metapopulation (synonyms: deme, subpopulation). [2]

locus the position of a gene along a chromosome; often used to refer to the gene itself. [1]

log-odds ratio (LOD score) a log-transformed ratio of likelihoods, used for statistical tests, including testing for the presence of a QTL at a particular location on a genetic map. [5]

major gene a gene locus responsible for a large proportion of the phenotypic variation in a trait. [4]

map density the number of markers per centimorgan on a genetic map. [5]

map distance the distance between two markers on the same chromosome based on recombination frequency, usually measured in centimorgans (cM). [5]

mapping population an experimental population constructed by crossing, designed for the production of a genetic map. [5]

mark-recapture study a technique in which animals are captured, marked, and released back into nature. The frequency at which they are recaptured is used to estimate survival or migration rates. [3]

maternal effect nongenetic effects of mothers on traits present in their offspring. [4]

mean square (*MS*) the sums of squares divided by the degrees of freedom ($n - 1$). (synonym: variance). [4]

meiotic drive a process defined by a deviation from 1:1 Mendelian segregation among the functional gametes produced by a heterozygous genotype (synonym: segregation distortion). [2]

Mendelian segregation the production of equal numbers of gametes containing each allele from a heterozygous genotype. [2]

messenger RNA (mRNA) the product of transcription and RNA processing that is translated into a polypeptide chain. [1]

meta-analysis a statistical technique for jointly analyzing the results of many studies on the same topic. [7]

metapopulation a group of populations connected by some level of gene flow. [2]

metric trait a phenotypic character that is continuously distributed with more than just a few distinct types (synonym: quantitative trait). [4]

microsatellites genetic markers consisting of repeat units 2–9 nucleotides long. Also called simple sequence repeats (SSR), simple sequence repeat polymorphisms (SSRP), or short tandem repeats (STR). [2]

midparent the average phenotypic value of each pair of parents in an offspring-parent regression. [4]

migration the movement of individuals between populations; often used interchangeably with gene flow. [3]

minisatellites genetic markers consisting of tandem repeats of a repeating unit 10–64 nucleotides long. Also called VNTR (variable number of tandem repeats). [2]

minor gene a locus that determines a relatively small proportion of phenotypic variation in a trait. [4]

model a theoretical abstraction of the real world. [2]

mutation any permanent alteration of a DNA molecule. [3]

mutation-selection balance a process in which removal of variation by selection is balanced by the input of new variation into the population by mutation. [3]

mutational meltdown a process in which deleterious mutations are fixed due to random drift in small populations, which further decreases population size and thus increases the rate of fixation of deleterious mutations in a positive feedback loop. [7]

nested paternal half-sibling design a quantitative genetic design that is well-suited for estimating additive genetic variance, additive genetic correlation, and thus the *G* matrix. In this design, a few unique females (dams) are mated to each of a number of males (sires), and the traits of interest are measured on a few offspring from each dam. The data are analyzed with nested ANOVA. [4]

neutral a trait or locus that has a negligible effect on fitness. A trait or locus is nearly neutral or effectively neutral if the mean phenotype or allele frequencies are determined more by random genetic drift than by selection. [3]

non-additive genetic variance the sum of the dominance and epistatic variance. [4]

nucleotide one of the building blocks of DNA or RNA; each contains a nitrogenous base attached to a five-carbon sugar that in turn is attached to one or more phosphate groups. [1]

null allele an allele that does not produce a band after electrophoresis. Null alleles are common in many genetic markers, including allozymes, microsatellites, RAPDs, and AFLPs. [2]

observational variance components in a sibling analysis, the portions of the phenotypic variance that are due to the different factors (i.e., sires, dams) in the mating design. [4]

offspring–parent regression a quantitative genetic design that is well-suited for estimating additive genetic variance, additive genetic correlation, and thus the **G** matrix. In this design, the traits of interest are measured on a number of mothers, fathers, and their offspring. The slope of the regression line is an estimate of heritability. [4]

oligogenic trait a trait affected by a few loci. [4]

origin in electrophoresis, the location in a gel where the sample is placed. [2]

outbreeding depression a decline in fitness resulting from mating between distantly related individuals, especially those from different locally-adapted populations. [7]

outcrossing mating system of organisms that do not regularly undergo inbreeding. [2]

overdominance phenomenon in which the heterozygous genotype has a higher phenotypic value (especially for fitness) than either homozygous genotype (synonym: heterozygote advantage). [3]

panmictic a population in which any member of the species can potentially mate with any other member of the opposite sex; often used as a synonym for random mating. [2]

parental effect nongenetic effects of parents on traits present in their offspring. [4]

pesticide rotation systematic change in the pesticide used to control a pest. [7]

phenotype the outward appearance of an organism for a given characteristic. [1]

phenotypic adaptive landscape a three-dimensional representation of the relationship between two phenotypic traits and population mean fitness. [6]

phenotypic correlation (r_P) a standardized measure of the degree to which two traits covary among individuals in the population. [5]

phenotypic plasticity the condition in which the same genotype produces different phenotypes in different environments. [5]

phenotypic value (P) the measurement of a given quantitative trait for a given individual. [4]

phenotypic variance (V_P) a measure of the population variation among individuals for a given trait; the variance of phenotypic values. [4]

physical distance the actual distance between two loci in terms of the number of nucleotides in the DNA, usually measured in units of thousands of base pairs (kilobases, kb) or millions of base pairs (megabases, Mb). [5]

plasmids usually circular DNA molecules found in bacteria, many of which are capable of horizontal transmission. [7]

pleiotropy the phenomenon in which one locus affects more than one phenotypic trait, causing a genetic correlation. [5]

polygenic trait a trait affected by many gene loci. [4]

polymerase chain reaction (PCR) an automated method to synthesize many copies of a specific fragment of DNA by repeated rounds of DNA replication. [2]

polymorphic a population that has more than one relatively common allele present at a given locus (synonym: segregating; antonym: fixed). [2]

population a local interbreeding (panmictic) group that has reduced gene flow with other groups of the same species. [2]

population differentiation differences between populations in allele frequencies at one or more loci, or in mean phenotype, in a common environment (synonym: genetic differentiation). [3]

primer a short segment of DNA used to initiate DNA synthesis in the polymerase chain reaction. [2]

quantitative trait a phenotypic character that is continuously distributed with more than just a few distinct types (synonym: metric trait). [4]

quantitative trait locus (QTL) mapping a technique in which genetic markers are used to identify the chromosomal regions affecting quantitative traits. [4, 5]

random genetic drift fluctuations in allele frequency that occur by chance, particularly in small subpopulations, as a result of random sampling error in the choice of gametes that form the next generation (synonym: genetic drift). [3]

random mating the mating system in which the frequency of matings between two genotypes is equal to the product of the genotype frequencies. [2]

randomly amplified polymorphic DNA (RAPD) genetic markers based on amplifying DNA using primers with random nucleotide sequences. [2]

reaction norms graph showing the phenotypes produced by a set of genotypes or families in multiple environments. [5]

realized heritability the heritability estimated from artificial selection, equal to the slope of the line relating response to selection to the cumulative selection differential. [5]

reciprocal transplant experiment an experimental design in which individuals from multiple subpopulations are each reared in the habitats of all the subpopulations; this experiment is designed to test for genetic differentiation in phenotypic traits as well as phenotypic plasticity. [5]

recombination the mixing of maternal and paternal gene loci into new combinations in the gametes during meiosis. [1, 5]

recombination rate (*c* or *r*) the per-generation probability of recombination between two loci. [5]

refuge a stand of noninsecticidal crop plants adjacent to transgenic insecticidal plants, designed to slow the evolution of resistance to the insecticide by increasing the fitness of insecticide-susceptible genotypes. [7]

relative fitness (*w*) absolute fitness divided by the fitness of the most favored genotype or by the population mean fitness. [3, 6]

response to selection (*R*) the change in mean phenotype across one generation due to natural or artificial selection; sometimes used as a definition of phenotypic evolution. [5]

restriction fragment length polymorphisms (RFLPs) genetic markers in which differences in DNA fragment sizes are produced by genetic variation in the distance between restriction sites. [2]

restriction site the specific short DNA sequence recognized and cut by a restriction enzyme. [2]

RNA processing the process in which intervening sequences or introns are removed from the RNA transcript by splicing and the ends of the transcript are modified. [1]

segregating the state of a population that has more than one relatively common allele present at a given locus (synonym: polymorphic; antonym: fixed). [2]

segregation distortion a process defined by a deviation from 1:1 Mendelian segregation among the functional gametes produced by a heterozygous genotype (synonym: meiotic drive). [2]

selection a consistent relationship between fitness and the phenotype, or differences in fitness among genotypes. Natural selection can lead to adaptation. [3, 6]

selection differential (S) a measure of total phenotypic selection, both direct and indirect. With continuous fitness values, the selection differential can also be estimated as the difference in the mean of the selected group and the mean of the entire population before selection (especially in truncation selection). The selection differential can be estimated as the covariance of fitness and a trait, or as the slope of the relationship between relative fitness and standardized trait values. [5, 6]

selection gradient (β) a measure of direct selection on each trait after removing indirect selection from all other traits that are in the analysis. [6]

selection plateau a phenomenon in which a population ceases to respond to artificial selection (synonym: selection limit). [5]

selective agent an environmental cause of fitness differences among organisms with different phenotypes. [6]

sex ratio the relative proportions of males and females in a population. [3]

sexual selection selection caused by differences among phenotypes in mating success. [6]

short tandem repeats (STR) see microsatellites. [2]

simple sequence repeats (SSR) see microsatellites. [2]

simple sequence repeat polymorphisms (SSRP) see microsatellites. [2]

simulation model a model in which the relationships among the variables are programmed into a computer for analysis. [2]

sires the male parents in a quantitative genetic breeding experiment. [4]

Southern blot a method of nucleic acid hybridization used for visualizing differences in DNA fragment size, especially in RFLP, and minisatellite markers. [2]

spatial structure the condition in which the individuals in a species are not uniformly distributed in space. [2]

stabilizing selection the form of selection with an intermediate optimum for fitness, in which fitness is highest at some intermediate phenotype and lower at the phenotypic extremes. [6]

standard deviation the square root of a variance. [4]

statistically significant a result that is unlikely to be due to chance alone. [2]

subpopulation a local interbreeding unit within a metapopulation (synonyms: deme, local population). [2]

sum of squares (SS) the sum of squared deviations from the mean for a set of observations; it is the numerator of the variance. [4]

targets of selection phenotypic traits that selection acts directly upon. [6]

template a strand of DNA that is copied during DNA or RNA synthesis. [2]

threatened species a species that is likely to become endangered in the near future. [7]

tight linkage the state of two loci being close together on the same chromosome. [5]

transcription the process in which the sequence of nucleotides present in one DNA strand of a gene is used as a template to produce a molecule of RNA (ribonucleic acid) complementary in nucleotide sequence. [1]

transgene a gene that has been transferred from the genome of one species into that of another. [7]

transgene escape the process by which a transgene, especially one in a crop plant, can become incorporated into a natural population. [7]

transgenic techniques processes in which genes from one species are incorporated into the genome of a different species. [7]

translation the process of converting the information encoded in the messenger RNA into a polypeptide chain, which takes place on ribosomes in the cytoplasm. [1]

translocation a type of chromosomal rearrangement in which portions of two nonhomologous chromosomes are exchanged. [3]

transposable elements (transposons) mobile fragments of DNA that are an important cause of insertions and deletions in some organisms. [3]

truncation selection the usual form of artificial selection in which there is a discrete phenotypic value (the truncation point) above which the organisms have high fitness and below which they have zero fitness. [5]

unbalanced data an experimental outcome in which there are unequal numbers of observations within each group, especially in ANOVA. [4]

underdominance the condition in which the heterozygote has a lower phenotypic value (especially for fitness) than either homozygote. [4]

unlinked loci loci in different chromosomes or so far apart in one chromosome that their frequency of recombination is 0.5. [5]

variable number of tandem repeats (VNTR) regions of the genome that consist of several to many tandem copies of the same sequence, usually 10–64 nucleotides in length. Also called a minisatellite. [2]

variance the standard measure of quantitative variation equal to the average of the squared deviation of each observation from the mean; see Appendix. [3]

verbal model a logical model with the relationships between variables expressed in words rather than mathematical equations. [2]

visible polymorphism a phenotypic trait that exhibits only a few (usually two or three) distinct types or morphs (synonym: discrete polymorphism). [2]

wildtype allele any of the normal alleles of a gene present in a natural population. [3]

Index